Vanished Giants

Vanished Giants

THE LOST WORLD OF THE ICE AGE

Anthony J. Stuart

The University of Chicago Press CHICAGO & LONDON

The University of Chicago Press, Chicago 60637
The University of Chicago Press, Ltd., London

Published 2021
Paperback edition 2022
Printed in the United States of America

31 30 29 28 27 26 25 24 23 22 1 2 3 4 5

ISBN-13: 978-0-226-43284-7 (cloth)
ISBN-13: 978-0-226-82403-1 (paper)
ISBN-13: 978-0-226-43298-4 (e-book)
DOI: https://doi.org/10.7208/chicago/9780226432984.001.0001

Library of Congress Cataloging-in-Publication Data

Names: Stuart, Anthony J., author.
Title: Vanished giants : the lost world of the Ice Age / Anthony J. Stuart.
Description: Chicago ; London : The University of Chicago Press, 2021. | Includes bibliographical references and index.
Identifiers: LCCN 2020022828 | ISBN 9780226432847 (cloth) | ISBN 9780226432984 (ebook)
Subjects: LCSH: Extinction (Biology) | Paleontology—Pleistocene. | Glacial epoch. | Animals, Fossil. | Extinct animals.
Classification: LCC QE721.2.E97 S78 2021 | DDC 560/.1792—dc23
LC record available at https://lccn.loc.gov/2020022828

♾ This paper meets the requirements of ANSI/NISO Z39.48-1992 (Permanence of Paper).

CONTENTS

CHAPTER 1

Introduction

1.1 A LOST WORLD

What killed off the mammoths, woolly rhinos, giant ground sloths, sabertooth cats, and so many other spectacular large, often giant, animals that thrived on all continents (except Antarctica) during the "Ice Age" (Quaternary/Pleistocene periods)—some until only a few thousand, or a few hundred, years ago?[1]

If we could somehow travel back to those days, by means of a handy time machine, no doubt we would be amazed, delighted, and often terrified, by the wonderful, bizarre creatures of this "lost world." We are apt to think of these extinct creatures as almost unreal prehistoric monsters, perhaps rating alongside mythical dragons and mermaids as well as the decidedly real but far more ancient dinosaurs. However, this is looking at it entirely the wrong way around. The unusual times are now; these beasts should still be with us if something extraordinary and unprecedented had not happened. Their demise is a severe loss to the modern fauna, emphasizing the need to protect and cherish what we have left today.

With remarkable insight, the eminent biologist Alfred Russel Wallace (co-originator, with Charles Darwin, of the theory of evolution by natural selection) observed: "It is clear, therefore, that we are now in an altogether exceptional period of the earth's history. We live in a zoologically impoverished world, from which all the hugest, and fiercest, and strangest forms have recently disappeared; and it is, no doubt, a much better world for us now they have gone. Yet it is surely a marvellous fact, and one that has hardly been sufficiently dwelt upon, this sudden dying out of so many large mammalia, not in one place only but over half the land surface of the globe."[2]

Less widely quoted but similarly astute are Darwin's observations on extinctions in South America: "It is impossible to reflect on the changed state of the American continent without the deepest astonishment. Formerly it must have swarmed with great monsters: now we find mere pigmies, compared with the antecedent, allied races. The greater number, if not all, of these extinct quadrupeds lived at a late period, and were the contemporaries of most of the existing seashells. Since they lived, no very great change in the form of the land can have taken place. What, then, has exterminated so many species and whole genera? Did man, after his first inroad into South America, destroy, as has been suggested, the unwieldy Megatherium and the other Edentata? What shall we say of the extinction of the horse? Did those plains fail of pasture, which have since been overrun by thousands and hundreds of thousands of the descendants of the stock introduced by the Spaniards? Certainly, no fact in the long history of the world is so startling as the wide and repeated exterminations of its inhabitants."[3]

The term *megafauna* is conventionally used for terrestrial vertebrates (mainly large mammals, but also some birds and reptiles) with mean body weights of 45 kilograms or more, including giants weighing upward of several tonnes. In general, the larger the animal, the more it was at risk of extinction, as large size usually correlates with slower breeding and a smaller number of individuals in the population.[4] Earlier extinctions in the Pleistocene affected small as well as large species, and many megafaunal losses were replaced by the evolution or immigration of ecologically similar forms. In contrast, late Quaternary extinctions primarily affected terrestrial large mammals, together with a few large birds and reptiles, leaving marine biotas largely unscathed.[5] Sadly, with the exception of those in sub-Saharan Africa and southern Asia, the vast majority are now extinct. The striking differences between the faunas, (both living and extinct) of each zoogeographical region reflects the fact that they have been, to varying extents, separated from one another for tens of millions of years.

In the following pages, I discuss how megafaunal extinctions occurred against the background of the Quaternary/Pleistocene Ice Age, with its profound changes in climate, glaciations, vegetation, fauna, changing sea levels, and the emergence and flooding of land bridges. I also look at the spread of modern humans across the globe and how this has impacted the megafauna, and examine the pattern of global extinctions, both on continents and on islands. Particular emphasis is given to the dating evidence, especially radiocarbon dating, which is essential for establishing chronological patterns in time and space and seeking correlations on the one hand with climatic/vegetational changes and with the archaeological record on the other. All radiocarbon dates listed in this book have been calibrated—that is, converted to a close estimate of calendar years expressed as "thousands of years ago" (*kya*). (See the appendix for more on radiocarbon dating.) We can be increasingly confident in the dates when we have substantial sets of concordant results. As discussed in the appendix, outlying results should be checked against independent cross-dating at another laboratory. Ideally, all dates would be cross-checked; but radiocarbon dating is expensive, and for the present the cost of cross-checking every result would be prohibitive. Of course, it is virtually certain we will never find and sample the last individual of its species, so even the most accurate LADs (last appearance dates) can only be used to give estimates of when each went extinct.

Today, renewed interest in the topic of Quaternary extinctions is fueled by concerns for the future of global ecosystems and the continuing disappearance of many living species. Much credit for this interest must go to the late Paul Martin (1928–2010) for his enthusiastic advocacy of the subject over a period of more than forty years. The destructive effects of human activities on our planet today are all too evident. The big question is: were our ancestors thousands of years ago responsible for the extinction of woolly mammoths, ground sloths, sabertooth cats, and many other megafaunal species throughout the world? Moreover, was this event the beginning of another mass extinction—the so-called "Sixth Extinction"—in progress today and which shows every indication of accelerating into the future? What would the world and its wildlife be like if modern humans had never evolved—would megafaunal species that disappeared thousands of years ago have continued to exist to the present day? On the other hand, to what extent has climate change resulted in extinctions? Unraveling the complex relative contributions of humans and climate is a key and exciting issue for further investigation.

1.2 REWILDING

Rewilding can be described as conservation on a large scale, with the intention of restoring depleted ecosystems to a close approximation of their "original state," before they were depleted by human activities. It may involve expanding and reconnecting now-fragmented original areas of vegetation, and reintroducing keystone animal species. Rewilding appears eminently desirable in many cases where a species can be reintroduced to it former range, provided that the habitat remains suitable or can be recreated. On the other hand, *Pleistocene rewilding* is far more controversial. Assuming that for most of the Late Pleistocene megafauna of North America disappeared due to "overkill" by humans, Paul Martin proposed that the ecosystem could be recreated by substituting the extinct species with extant species that have similar ecological roles.[6] Taking up this theme, several authorities (including Martin) published a highly controversial editorial in *Nature* advocating the introduction of lions, cheetahs, Asian elephants, Bactrian camels, dromedaries (Arabian camels), and other species to protected areas in the Great Plains. Burros (donkeys) and mustangs (horses) had already been introduced, within the last few hundred years, by European colonists.[7] In 1988, Sergey Zimov initiated his ambitious "Pleistocene Park" project in northeastern Siberia. The intention is to introduce a range of mega-mammals whose activities will recreate the mammoth steppe ecosystem by transforming the present boggy tundra.[8] Species introduced so far include Yakutian horses, reindeer, musk ox, elk, moose, and Canadian wood bison. Again, the premise is that the original Siberian megafauna were wiped out by humans, leading to drastic changes in vegetation cover. Unsurprisingly, Pleistocene rewilding has attracted considerable criticism, principally because of the fear that attempting to modify ecosystems in this way could have drastic unforeseen consequences. Oliveira-Santos and Fernandez have warned against promoting "Frankenstein ecosystems," stating that "the biggest problem is not the possibility of failing to restore lost interactions, but rather the risk of getting new, unwanted interactions instead."[9] They advocate that rather than restoring a lost megafauna, conservationists should dedicate themselves to restoring existing species to their original habitats. Even more extreme is the idea of recreating ("de-extincting") woolly mammoths and other extinct megafauna, by cloning ancient DNA from preserved frozen tissue. Because even in the most favorable circumstances the DNA is preserved only in short fragments that are also often damaged, this is out of the question—at least, for the present.[10] Creating a creature that superficially resembles a woolly mammoth by modifying the DNA of an Asian elephant is a more realistic goal, although this also raises profound practical and moral issues.[11]

CHAPTER 2

Crises in the History of Life

As is often stated, extinction is the ultimate fate of all species, and the vast majority of the species that have ever existed have vanished from the Earth. In the 1960s, when I was an undergraduate geology student at the University of Manchester, I read a fascinating article by Norman Newell (1909–2005) in the magazine *Scientific American*, with the intriguing title "Crises in the History of Life."[1] From his analysis of the geological record, Newell recognized several episodes of "mass extinction" in the last 540 million years (the Phanerozoic Eon—Cambrian to present day). This was revolutionary stuff that, although little heeded at the time, was to be amply vindicated in the light of subsequent research as Newell's ideas were transmuted from the near heretical to almost universally accepted orthodoxy. Ironically, the idea of global deluges or other catastrophic events in the geological past, which held sway in the early nineteenth century (as notably advocated by the eminent French naturalist and paleontologist Georges Cuvier), had been almost entirely supplanted by "uniformitarianism." The latter theory was first proposed by the Scot James Hutton in the late eighteenth century and later enthusiastically taken up by a compatriot, Charles Lyell, in his famous *Principles of Geology*.[2] Lyell built on Hutton's theory that Earth had been shaped entirely by the same forces that we see today acting over an immensely long period of time; and his ideas—especially the great antiquity of Earth—had a major influence on Charles Darwin developing his theory of evolution. Newell's recognition that there had been a series of mass extinctions throughout geological time after all (though significantly different than the catastrophes envisaged by Cuvier) initiated a vast amount of subsequent research, which continues unabated to the present. Working with marine faunas, Raup and Sepkoski distinguished five mass extinction events superimposed on an overall trend of greatly increasing diversity.[3]

With minor amendments, these are currently recognized as:

1. Late Ordovician, ca. 445 mya (million years ago);
2. Late Devonian, ca. 367 mya;
3. End Permian, ca. 252 mya;
4. End Triassic, ca. 201.4 mya; and
5. Cretaceous–Paleogene (K–Pg or K–T), ca. 65.5 mya.

For the purposes of this book, I shall refer only to the three most recent, as these profoundly affected terrestrial as well as marine animals and therefore can be usefully compared with extinctions in the late Quaternary.

2.1 END PERMIAN EVENT (P–TR)

The prize for the biggest-ever mass extinction easily goes to the truly apocalyptic End Permian event, ca. 252 mya, when an estimated 80% to 90% of all species (marine and terrestrial) perished.[4] Most forests disappeared, as did all reefs until they were "reinvented" some 15 million years later in the Middle Triassic. The reef-building rugose and tabulate corals of the Permian disappeared forever, and when reefs eventually reappeared, they were built by new types of coral (scleractinians) that still build reefs today. Several other major marine groups disappeared, including giant sea scorpions (eurypterids) and the last of the trilobites. On land, forests almost disappeared, resulting in the "coal gap" of the Early and Middle Triassic, as nowhere on earth was there sufficient growth of forest to produce coal deposits. The loss of forests would have been accompanied by the loss of many dependent species of animals, from insects to vertebrates. The rich late Permian terrestrial faunas—from South Africa, the Perm region of Russia, and elsewhere—comprised many large herbivorous reptiles, including pareiasaurs, dicynodonts, and other therapsids (mammal-like reptiles), together with gorgonopsians (flesh-eating therapsids) and a range of large amphibians.[5] Many of the bizarre, fascinating and rather nightmarish reptiles from the southern Urals can be seen in the extensive collections of the Paleontological Institute Museum in Moscow, which also features Quaternary mammal fossils from many regions of Russia.

In the 1980s, as the impact hypothesis for the K–Pg mass extinction (see below) became widely accepted, it seemed likely that the End Permian event—and perhaps all other mass extinctions—would also prove to have resulted from extraterrestrial impact. However, in marked contrast with the End Cretaceous event, in spite of intensive searching no convincing evidence for such an impact has been discovered, whereas there is a strong case for linking the extinctions to the massive eruptions of flood basalts that occurred in Siberia (the Siberian Traps) around 282 mya. Basalt lavas poured from numerous fissures accompanied by vast outpourings of carbon dioxide, which resulted in drastic global warming. It has also been proposed that sea temperatures increased to the point of precipitating a sudden and violent release into the atmosphere of huge volumes of the even more potent greenhouse gas methane from frozen gas hydrates in the deep oceans.[6] In his book *When Life Nearly Died*, Mike Benton paints a profoundly depressing picture of a devastated postapocalyptic landscape in which volcanic gases mixed with water to produce acid rain, destroying the vegetation cover.[7] Trees and soils were swept away, the landscape denuded to bare rock, while most land animals perished as their food supplies and habitats disappeared. In addition, pulses of flash warming continued for 5 million years, delaying the recovery of life. Some "disaster taxa," such as *Lystrosaurus* (a pig-size herbivorous dicynodont), survived the extinction, but it took 10 to 15 million years for complex ecosystems to become re-established.

2.2 END TRIASSIC EVENT (TR-J)

The Triassic period saw the first dinosaurs, pterosaurs, and mammals on land, and the earliest ichthyosaurs and plesiosaurs in the sea. The mass extinction that occurred at the end of the period, ca. 201.4 mya, is comparatively poorly known, although it was a major event, estimated to have wiped out around 75% of species.[8] Terrestrial groups that disappeared include mammal-like reptiles, most labyrinthodont amphibians, and many dinosaurs. The End Triassic extinctions have been rather firmly linked to the massive outpouring of flood basalts and accompanying gases from the huge Central Atlantic Magmatic Province, which occurred as the supercontinent of Pangaea began to break up.[9]

2.3 END CRETACEOUS EVENT (K-PG OR K-T)

The best-known mass extinction, although certainly not the biggest, was the End Cretaceous event, now dated to ca. 65.5 mya. In this event non-avian dinosaurs (that is, dinosaurs other than birds), the last of the pterosaurs, and several other vertebrates, were wiped out on land, while mosasaurs, plesiosaurs, ammonites, belemnites, rudists (a highly successful group of reef-building bivalve mollusks), many species of single-celled foraminifera, and other groups disappeared from the oceans. Luis and Walter Alvarez (father and son), led the team that discovered enormously enhanced levels of the rare element iridium in clays at the K–Pg boundary at many localities throughout the world, and inferred that this could only have resulted from the impact of an extraterrestrial body, a 10-kilometer-wide asteroid (later amended to 4 to 6 kilometers) that had ejected vast quantities of dust and particles into the stratosphere.[10] Subsequently, the site of the impact—the "smoking gun"—has almost certainly been found in the 180-kilometer-diameter Chicxulub crater in the Yucatán Peninsula, Mexico (extending under the adjacent Atlantic Ocean), which is now buried under several hundred meters of younger sediments. Indications of impact include shocked quartz and glass spherules (microtektites that rained down from ejected melt). A recent drilling project revealed impact breccias and impact-melted rock in the crater fill. The impact scenario postulates that a plume of material was ejected high into the atmosphere, whence it spread around the globe, blocking out sunlight for several years and resulting in a sudden cooling—a so-called "impact winter." Large amounts of sulfur dioxide injected into the atmosphere from the vaporization of anhydrite deposits at the impact site is believed to have exacerbated the cooling. Other suggested effects include acid rain, global wildfires, sudden temperature changes, infrared radiation, tsunamis, and superhurricanes.[11] As photosynthesis is thought to have ceased over large areas, many plants would have died, as would many herbivores that consumed them, and carnivores that fed on the herbivores. Greatly increased levels of carbon dioxide resulting from firestorms and reduced plant cover are thought to have led to global warming, killing off many of the survivors.

This sensational scenario with its dramatic appeal has been eagerly seized upon by the media, the general public, and popular science writers, and is accepted by many—but not all—earth scientists. It is important to recognize that there exists an alternative hypothesis, wherein although the impact was a contributory factor, the K–Pg extinctions were primarily caused by vast eruptions of flood basalt lavas in India (the Deccan Traps), accompanied by vast outpourings of gases that would have profoundly affected the global climate.[12] Although Late Cretaceous terrestrial faunas are known from nearly all continents, so far the only known region where the terrestrial fossil record extends through the K–Pg boundary is in western North America (then separated from the eastern part of the continent by the Western Interior Seaway). The crucial fact that we don't have comparable evidence from other regions, where events might have been very different, is generally overlooked, especially in the more popular accounts. The dinosaur-bearing uppermost beds of the terrestrial Lance Formation, Wyoming; the Hell Creek Formation, Montana; and other related deposits represent the very youngest part of the Cretaceous period (late Maastrichtian stage, ca. 67 to 65.5 mya), immediately prior to the K–Pg boundary. Their non-avian dinosaur faunas—some twenty-three species in total—included *Tyrannosaurus*, *Triceratops*, *Ankylosaurus*, *Ornithomimus*, and *Pachycephalosaurus*.[13] So far there is no definite evidence from western North America that any non-avian dinosaurs survived beyond the K–Pg boundary, but perhaps such evidence will eventually be forthcoming from other parts of the world.

Following each of these mass extinctions as well as many lesser events, life bounced back, although recovery took many millions of years after the major events. On the positive side, each extinction episode created unique opportunities for the survivors to "inherit the Earth," evolving and radiating into many new forms—for example, the evolution of dinosaurs in the Triassic, following the massive End Permian extinction, and the rise of mammals in the Cenozoic following the End Cretaceous (K–Pg) mass extinction.

Soon after the traumatic K–Pg event, the surviving mammals were all small, reaching a maximum of only about 10 to 15 kilograms.[14] They were mostly omnivores; the role of large terrestrial flesh-eaters was occupied by flightless birds ("terror birds"), crocodiles, champsosaurus (crocodile-like reptiles), giant lizards, and snakes. However, taking full advantage of a world newly divested of non-avian dinosaurs, mammals rapidly diversified into larger forms and exploited a wide range of ecological opportunities. By ca. 55 mya (Eocene period), they reached a maximum of about 900 kilograms with the North American herbivore *Barylambda*, and at ca. 30 mya (early Oligocene period) approximately 17 tonnes, in the shape of the huge hornless rhino *Indricotherium* (aka *Paraceratherium*) from southwest Asia, probably the largest-ever land mammal. From about 23 mya to the present day (Neogene: Miocene, Pliocene, Quaternary), the largest terrestrial animals have all been elephants and their relatives (proboscideans).[15]

2.4 CAINOZOIC EXTINCTIONS

No mass extinctions are currently recognized from the Cainozoic (that is, Paleogene plus Neogene). Nevertheless, there were several lesser extinction events, generally on a regional rather than global scale.[16] In these cases, neither the impact of extraterrestrial objects (bolides) nor volcanism appear to have been involved, but instead are due to somewhat less dramatic causes, such as climate change and competition from new invading species. Major events in the Paleogene (Paleocene, Eocene, and Oligocene) include marked changes in climate—notably, the Paleocene–Eocene Thermal Maximum warming event and the rise of the Himalayas from the collision of the Indian and Asian tectonic plates, which took place in the Middle Eocene to Early Oligocene. About 30 mya (Oligocene), Antarctica finally parted company with the rest of the former supercontinent Gondwana, leading to progressive cooling as it became encircled by cold currents. This process eventually would have resulted in the climatically induced near-total extinction of its poorly known ancient terrestrial flora and fauna. In North America and Eurasia, several generalized "archaic" mammal groups—including uintatheres, dichobunids (both large herbivores), and the carnivorous mesonychids—were lost around the Middle–Late Eocene boundary, about 37.2 mya, and were replaced by more "modern" forms such as camels, rhinos, and canids, adapted to drier, less forested conditions.[17]

The largest of several impact craters of Eocene Age are the Chesapeake Bay Crater, eastern United States, dated to ca. 35.5 mya, and the Popigai Crater in central-north Siberia, dated to ca. 35.7 mya. Both have estimated widths of around 100 kilometers and thus record major impacts events. Nevertheless, no extinction events have been attributed to them—a significant fact that seriously questions the idea that most mass extinctions were caused by impacts.[18]

Another major event in the Neogene (Miocene, Pliocene, Quaternary), was the "Great American Biotic Interchange," whereby, after tens of millions of years of separation uplift of the Isthmus of Panama, about 2.6 mya (for the main pulse) allowed the faunas of North and South America to partly mix (see chapters 6 and 7).

2.5 LATE QUATERNARY EXTINCTIONS

In marked contrast to the earlier events, late Quaternary extinctions (excepting those in the last few hundred years) were almost entirely confined to terrestrial vertebrates—predominantly the larger species, or megafauna—leaving the marine realm almost unscathed. Because late Quaternary extinctions occurred relatively recently, the history of individual species can be resolved in far finer detail and with much greater accuracy (due to refined dating methods, especially radiocarbon) than for earlier events. Moreover, fossil data are not restricted to particular stratigraphic sections, as is the case for the earlier terrestrial events, but are available from most regions throughout the globe.

Today we are in the midst of an extinction event that can readily be seen as a continuation of a process beginning tens of thousands of years ago with not only the loss of much of the megafauna, but also very many other vertebrate, invertebrate, and plant species.[19] If this process continues, which seems only too likely, we are facing another mass extinction—due neither to volcanism nor asteroid impact, but in this case to human activities—the so-called "Sixth Extinction." If unchecked, the consequences for life on Earth, including our own species, will no doubt be catastrophic.

CHAPTER 3

The Ice Age and the Megafauna

3.1 THE QUATERNARY ICE AGE

Boulder clays—that is, deposits of jumbled boulders and pebbles in a silt and clay matrix—and erratics—rocks, sometimes weighing many tonnes, of types that are out of place in the landscape and which clearly have been transported from elsewhere—occur across large areas of Britain, northern Europe, the Alps, much of North America, and elsewhere. Until the first half of the nineteenth century, such deposits were known as "drift," according to the now-defunct theory that they originated from the debris melted out from drifting icebergs, at a time when vast areas of land were supposed to have been submerged beneath the sea. However, this theory did not explain the smoothed, polished, and scratched rock faces, nor the deeply sculpted landscape features such as U-shaped valleys that are commonly found in many upland regions. The breakthrough came when the Swiss paleontologist Louis Agassiz (previously renowned for his studies of fossil fish) published "Etudes Sur Les Glaciers": the first major publication to make the case that ice had played an important role in the shaping of the earth in recent geological time, and that the former presence of glaciers and ice sheets could account for all these features. However, he got rather carried away in envisaging an ice sheet that "extended beyond the shorelines of the Mediterranean and of the Atlantic Ocean, and even covered completely North America and Asiatic Russia."[1] He further believed that sudden global cooling had wiped out woolly mammoths and other megafauna. In reality, the true maximum coverage of ice was far less than Agassiz imagined, although it still occupied huge areas. At maximum extent, ice sheets several kilometers thick blanketed most of the northern half of North America and northwestern Eurasia, while the Greenland and Antarctic ice sheets and mountain glaciers in the Alps, Himalayas, Andes, and elsewhere were greatly expanded.[2] As vividly described by Charles Darwin: "The ruins of a house burnt by fire do not tell their tale more plainly than do the mountains of Scotland and Wales, with their scored flanks, polished surfaces, and perched boulders, of the icy streams with which their valleys were lately filled. So greatly has the climate of Europe changed, that in Northern Italy, gigantic moraines left by old glaciers, are now clothed by the vine and the maize. Throughout a large part of the United States, erratic boulders and scored rocks plainly reveal a former cold period."[3]

The evocative term *Ice Age* is misleading in the sense that it gives the false impression of a prolonged period of uninterrupted ice and snow, and indeed, Agassiz believed that there had been just a single colossal glacial episode. However, subsequent work revealed that in reality there had been multiple glaciations, with temperate (interglacial) episodes between. In the early twentieth century, the German geologists Albrecht Penck and Eduard Brückner distinguished four separate glacial advances in the Alpine region.[4] Their scheme, which was applied (often inappropriately) throughout the Northern Hemisphere and even beyond, held sway for more than half a century, until many lines of research began to reveal a much more complex picture of multiple cold and warm episodes.

A major advance in our understanding of the Quaternary Ice Ages resulted from the meticulous analyses by Nick Shackleton (1937–2006) at the University of Cambridge, of oxygen isotope values in microscopic shells (foraminifera) from deep sea cores, from which he demonstrated that fluctuations in isotope ratios reflected how much ice was locked up in ice sheets and glaciers (global ice volume) at any one time, and hence changes in global sea level (fig. 3.1).[5] In a groundbreaking paper, Hays and colleagues demonstrated that these fluctuations were driven by the variations in Earth's orbital geometry that had been predicted by Milutin Milankovitch (see below).[6] The deep-sea core record, dating back to well before the beginning of the Quaternary (at ca. 2.58 mya), has been divided into a series of marine isotope stages, the youngest designated MIS 1 (marine isotope stage 1). The Brunhes–Matuyama magnetic reversal (within MIS 19) dated at 780 kya is used to calibrate the entire core, assuming a uniform rate of sedimentation.

The Danish paleoclimatologist Willi Dansgaard (1922–2011) was the first to discover that the polar ice caps also preserved a long record of climate change. By analyzing oxygen isotope ratios from sequential annual layers of ice in an ice core from Greenland he produced an extraordinarily detailed records of air temperature changes over the past 100,000 years.[7] The record from the

NGRIP Greenland core, completed in 2003, extends back to 123 kya (Last Interglacial). An outstanding finding was that temperature changes could occur very rapidly—in some cases, a shift of more than 10°C in less than a decade. The Swiss physicist Hans Oeschger (1927–1998) pioneered research on changes in atmospheric composition through time by analyzing air bubbles trapped within the layers of ice from both Greenland and Antarctica. This revealed that atmospheric CO_2 concentrations in glacials were considerably less than in interglacials, including the Holocene. The European Project for Ice Coring in Antarctica ice core, known as EPICA, through the much thicker Antarctic Ice Sheet penetrated 3.3 kilometers of ice and reached back more than 800,000 years, beyond the Early/Middle Pleistocene boundary.[8]

On land, climatic changes are recorded, for example, in sediment cores from lakes, thick loess deposits (windblown silts), and speleothems (flowstones, including stalactites and stalagmites) from caves. Long sequences of fossil pollen from lakes record detailed changes in vegetation cover from which climatic changes can be inferred. Notable sites include La Grande Pile in northeastern France[9] and Monticchio in southern Italy,[10] both of which record vegetational changes from the penultimate glacial stage through the Last Interglacial and Last Glacial to the Holocene, and correspond well with the deep sea and Greenland curves, in turn related to Milankovitch cycles (see below). Variations in diatom content over the last 440,000 years from Lake Baikal in southeastern Siberia—far from oceanic influences—also tell a similar story to the deep-sea record.[11]

The Quaternary period experienced a long-term trend of global cooling combined with climatic fluctuation driven by variations in the energy received from the sun, due to periodic variations in Earth's orbit, known as Milankovitch cycles, after the brilliant and dedicated Serbian geophysicist and astronomer Milutin Milankovitch (1879–1958), who first recognized them.[12] The amplitude of the major fluctuations became more strongly marked from approximately 424 kya (fig. 3.1).[13] Here I am primarily concerned with the late Quaternary, from the Last Interglacial (ca. 130 to 110 kya); through the complex climatic changes during the Last Glacial (ca. 110 to 11.7 kya), predominantly cold, but including milder interstadials; to the temperate Holocene interglacial ("postglacial"), which began ca. 11.7 kya and continues to the present day. These climatic fluctuations drove profound shifts in the distributions of plants, animals, and humans, so that in Europe, for example, during the Last Glacial, arctic lemming, reindeer, arctic fox, and woolly mammoth lived as far south as the Alps and Pyrenees, whereas in the warmest part of the Last Interglacial hippo (*Hippopotamus amphibius*) was present in the north

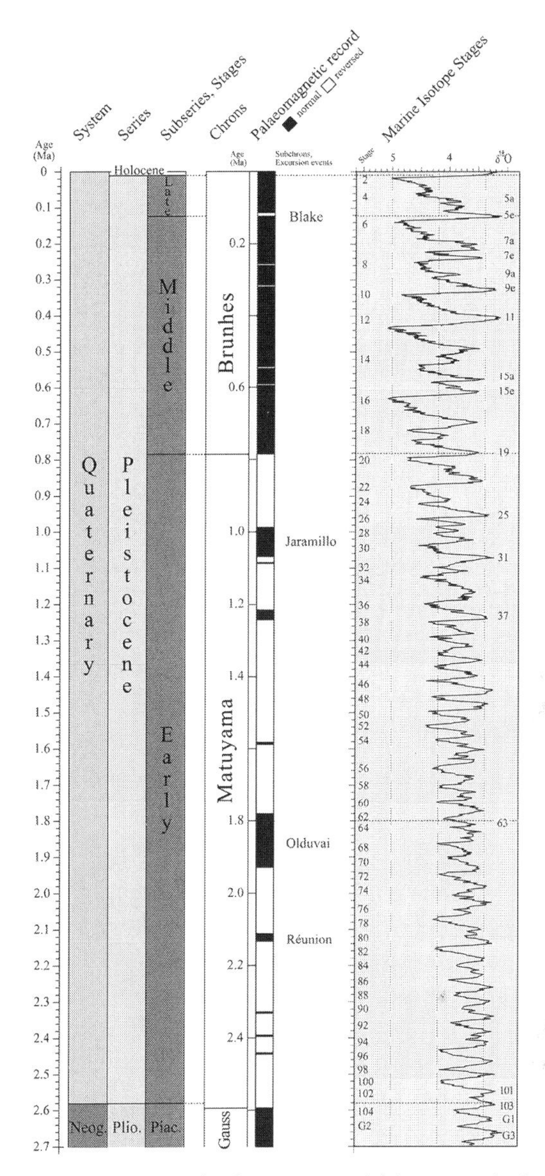

FIGURE 3.1. Stratigraphic scheme for the Quaternary/Pleistocene. The fluctuations in oxygen isotope ratios are a measure of how much water was locked up in ice sheets and glaciers (global ice volume) at any one time, and thus record changes in sea level. Sea levels were lowest in cold stages/substages and highest in warm stages/substages. Substage 5e corresponds to the Last Interglacial (Eemian: ca. 130 to 110 kya). This book focuses on the most recent 130,000 years (Late Quaternary/Late Pleistocene).

Key: ma: million years ago. Pleistocene: Early, Middle, and Late divisions shown. Paleomagnetic divisions (middle columns): black: normal polarity—as present day. White: reversed polarity. Neog.: Neogene, 23 to 2.58 mya. Plio.: Pliocene, 5.3 to 2.58 mya. Piac.: Piacenzian, 3.60 to 2.58 mya (International Commission on Stratigraphy). Marine isotope stages (right column): odd numbers (to the right): warm stages/substages. Even numbers (to the left): cold stages/substages.

of England.[14] Similarly, in North America south of the ice sheets, musk ox and caribou/reindeer reached to the southeast of the United States; while in the Last Interglacial, the ranges of more temperate species such as American mastodon, flat-headed peccary (*Platygonus compressus*), giant beaver (*Castoroides ohioensis*), and yesterday's camel (*Camelops hesternus*) extended as far north as Alaska and the Yukon (chapter 6).

At times within the cold phases, ice sheets covered much of North America and northwestern Eurasia, which together with expanded polar ice caps and mountain glaciers resulted in major drops in global sea level of up to 150 meters as vast quantities of water became tied up in the ice sheets.[15] This situation resulted in the exposure of several land bridges that are now submerged by sea. For example, New Guinea and Tasmania were connected to Australia, forming a single land mass (Sahul); the region of the Bering Strait (Beringia) connected northeastern Siberia (Chukotka) to Alaska; and Britain was joined to continental Europe across the southern North Sea. These land bridges allowed animals and humans to migrate, although often—due to climatic, vegetational, or topographic conditions—they acted as filters that prevented certain animals from crossing, while allowing the passage of others.

The rapid rise in temperature at the beginning of the Holocene, between ca. 11.7 kya and ca. 9.9 kya, was succeeded by a long period of relative stability, lasting up to the present day. However, there was a significant short-lived cooling episode: the 8.2 kya event, registered by isotopic changes in the Greenland ice cores, which is taken to mark the early-middle Holocene boundary.[16] The sudden cooling is attributed to an influx of meltwaters into the North Atlantic in connection with the decay of the North American Laurentide ice sheet. Although most prominent in the North Atlantic, its effects were global. The middle-late Holocene boundary is defined by an aridification event at 4.2 kya in mid and low latitudes. The term *Anthropocene* is likely to be adopted as the most recent division of the Holocene, in recognition of the profound global changes that have been and are resulting from human activities. There is currently much discussion on defining a formal global boundary.[17]

3.2 QUATERNARY FOSSIL VERTEBRATES

Finds of mysterious large bones and teeth have been reported since antiquity and were often interpreted as the remains of a vanished race of human giants or of mythical creatures. For example, the ancient Greek legend of the monstrous one-eyed Cyclops was plausibly inspired by finds of pygmy elephant skulls from one or more Mediterranean islands; the nasal opening, situated in the front of the skull, could readily be mistaken for a single eye socket. The fourteenth-century discovery of a woolly rhino skull from Klagenfurt, Austria, believed to be that of a dragon, is commemorated by the sixteenth-century sculpture, complete with wings, in the Neuer Platz in the city center. (Other examples are mentioned in chapters 5, 7, and 8.) By the eighteenth century, however, such finds came to be recognized as elephants, rhinos, and other more familiar beasts, and at first were assumed to be of existing species. The presence of such remains in Europe was seen by some as proof of the colossal biblical flood that had washed their corpses thousands of kilometers northward from the tropics.

The reality of extinction was slow to take hold, largely for religious reasons. For example, Thomas Jefferson (1743–1826; see chapter 6) believed in a "great chain of being," whereby once created a species would continue to exist, writing, "In fine, the bones exist: therefore the animal has existed." And: "For if one link in nature's chain might be lost, another and another might be lost, till this whole system of things should evanish by piece-meal."[18] However, by the turn of the eighteenth to nineteenth centuries, attitudes began to change, especially when the eminent Baron Georges Cuvier (1769–1832) demonstrated that woolly mammoth remains from Siberia represented an extinct species, distinct from modern elephants and adapted to cold climates.[19] He also described the huge ground sloth *Megatherium americanum* and other giant animals that were undoubtedly extinct. He argued that "there is little hope of one day finding those we have seen only as fossils"; if they had still been alive, such animals were just too big to be overlooked.[20] Ironically, it later transpired that there was no insurmountable obstacle after all to reconciling extinction with Christian belief; turning the argument on its head, it was seen as a measure of God's omnipotence to have created a world in which pieces could be removed without destroying the whole.

As described in the following chapters, Quaternary fossil vertebrates have been preserved in a variety of situations, including skeletal remains from caves; river and lake sediments; loess; asphalt ("tar pits"); and archaeological sites. Especially informative are the much rarer finds of frozen corpses from the permafrost of arctic and subarctic northern Eurasia and North America (chapters 5, 6), some of which have yielded preserved gut contents. Dung, skeletal remains, and desiccated mummified tissues from several megafaunal species have been recovered from dry caves in arid regions of North America, South America and elsewhere (chapters 6, 7). In addition to dating by radiocarbon and other methods, major advances in ancient DNA (aDNA) research and stable isotope analyses are also contributing significantly to our understanding of the genetic history and ancient diets of Quaternary megafaunal species. In combination with radiocarbon dating, aDNA is producing exciting results, including the estimation of changing past population sizes.[21] During the Last Glacial period in northern Eurasian and North America, several extinct megafaunal species, including cave lion and woolly mammoth, declined in genetic diversity. However, to what extent this contributed to their eventual extinction is less clear, as other species such as musk ox and saiga antelope also experienced similar bottlenecks but nevertheless have survived to the present day.[22]

CHAPTER 4

Cold Case: The Search for the Ice Age Killer

Half a century after Paul Martin reignited interest in this fascinating topic, the cause or causes of the late Quaternary extinctions remain highly controversial and continue to generate plenty of lively debate. The well-established contending hypotheses are "overkill" by human hunters, climatic/environmental change, or a combination of both. Both the overkill and climatic change hypotheses have their merits and demerits.[1] More recently proposed explanations such as "hyperdisease," extraterrestrial impact, and solar flare have serious shortcomings. Claims by some authors that the problem of cause has been solved, one way or another are definitely premature.[2] Megafaunal extinction was a complex phenomenon in which different factors dominated in different regions of the globe. My own views on the possible causes of megafaunal extinctions are discussed in the final chapter.

4.1 THE GLOBAL SPREAD OF *HOMO SAPIENS*

Most hominin evolution appears to have taken place in Africa, with *Homo erectus* the first species to expand into Eurasia about two million years ago.[3] The earliest modern humans (*Homo sapiens*) are recorded from Africa; a skull, mandible and other remains from Morocco have recently been dated to ca. 300 kya.[4] The oldest record of *H. sapiens* outside Africa is from Misliya in Israel ca. 180 kya—much earlier than had been thought previously.[5] Modern humans are known from Lida Ajer Cave, Sumatra, ca. 73 to 63 kya, northern Australia by about 65 kya, and Europe ca. 45 to 43 kya.[6] In Europe, they encountered Neanderthals (*Homo neanderthalensis*), who disappeared ca. 40 kya (possibly due to competition with *H. sapiens*).[7] Having colonized much of northern Asia, *Homo sapiens* reached northeastern Siberia, and crossed the Bering Land Bridge (or Bering Strait, if flooded at the time) to reach Alaska. Some have inferred that humans had entered North America by ca. 15 kya, possibly significantly earlier.[8] However, radiocarbon dating of cut-marked horse and other animal bones from Bluefish Caves, Yukon Territory, indicates that humans had arrived by ca. 24 kya.[9] This new evidence also supports the "Beringian standstill" hypothesis, which proposes that during the Last Glacial Maximum (LGM) a genetically isolated human population was present within the Bering Land Bridge for several thousand years.[10] Only when the Pacific coastal route became available through deglaciation, ca. 16 kya, were humans able to penetrate into North America south of the ice sheets and beyond. In migrating southward, taking advantage of marine resources, they presumably island-hopped, using rafts or boats, along the western coasts of North, Central, and South America, and had penetrated as far as the Southern Cone of South America by ca. 14.8 kya.[11]

Later, seafaring people spread westward from southeast Asia into the Indian Ocean, colonizing Madagascar by ca. 10.5 kya (chapter 9), while others voyaged eastward to the many islands of the Pacific, reaching New Zealand—the last unoccupied major land mass—about 610 years ago; best estimate: 1345 CE (chapter 10).

4.2 OVERKILL

It is generally accepted that the majority of the numerous extinctions of terrestrial vertebrates (large and small) on islands resulted from human colonization within the last few thousand to few hundred years.[12] Ecosystems that had evolved essentially in isolation were severely impacted by hunting, habitat destruction (especially burning of forests), and the introduction of alien species such as rats, dogs, pigs, and goats.

Martin enthusiastically promoted the "prehistoric overkill" hypothesis, according to which major extinctions of terrestrial vertebrates in the late Quaternary, both on continents and islands, resulted from the same process: the global spread of modern humans (*Homo sapiens*) who indulged in unsustainable levels of hunting.[13] In both overkill and environmental change hypotheses, the larger species are considered to have been especially vulnerable because they were slow-breeding and existed in relatively small populations. However, with his well-known but inappropriately named "Blitzkrieg" (German: "lightning war") hypothesis—a special case of overkill—Martin went much further and postulated that when modern humans first colonized North America, South America, Australia, New Zealand, and elsewhere in the late Quaternary, they encountered naive prey lacking behavioral and evolutionary adaptations to the novel, efficient, and aggressive human predator.[14] The rarity of "kill sites" (sites with direct association of projectile weapons and megafaunal remains) in North America was interpreted by Martin as evidence that the slaughter had occurred so rapidly as to leave little trace in the fossil record.[15] Martin envisaged that in "continents of colonization," human arrival was followed by catastrophic collapse of megafaunal populations and that humans initially underwent a population explosion fueled by superabundant food. On the other hand, more modest megafaunal losses in northern Eurasia, and only slight losses in Africa and Southern Asia, were attributed to the long coexistence of humans and megafauna in these regions and familiarity of megafaunal prey with human predators. One of several computer simulations of the hunting of large herbivores by humans in Last Glacial North America has been widely cited in support of the overkill hypothesis.[16]

An obvious chronological test of overkill, by means of radiocarbon or other dating methods, is that, for a given region, a marked increase in extinctions should follow extensive human colonization, as has been suggested for North America (see below). The fact that in many cases megafaunal taxa survived later on islands than on the continents and that extinctions occurred only after the first human arrivals is consistent with overkill. (This issue is explored further in chapter 13.) With variations, the basic "overkill" hypothesis continues to enjoy considerable support, especially with many researchers in North America and Australia, as we explore later. However, in most cases the "Blitzkrieg" hypothesis has not stood up well to further investigation.

Problems with overkill include:[17]

1. How could humans, with low population densities and relatively simple technologies, have exterminated so many varied large animals over their entire geographical ranges?
2. In northern Eurasia, most extinctions only occurred many millennia after the appearance of modern humans (chapter 5).
3. Late Quaternary extinctions, especially in North America, affected a number of smaller mammalian and avian species that are unlikely to have contributed significantly to the human diet, or to have constituted an important source of raw materials.
4. Many extinct carnivores were also very unlikely to have been extensively hunted.

5. For most of the subsequent Holocene period on the continents, few megafaunal species have disappeared, although they faced much higher human populations.
6. In northern Eurasia, species such as reindeer, red deer, and horse are the most abundant large mammals in archaeological sites and evidently extensively hunted; nevertheless, they survive to the present day.

Major objections to Blitzkrieg also include the supposed extreme rapidity of the process; the rarity of fossil and archaeological evidence that extinct megafaunal species were hunted (paradoxically cited as supporting the hypothesis); and the improbability that prey remained naive long enough to be exterminated.[18] However, an important exception should be made for the extinction of all species of moa in New Zealand. Moa had evolved for millions of years in the complete absence of ground-dwelling predators larger than tuataras (chapter 10). It is very likely that moa were naive prey—that is, lacked appropriate behavioral responses to the novel human predator—and consequently were rapidly wiped out.

It has been generally assumed that the agent of overkill was modern humans *Homo sapiens*. However, some earlier hominins (primate species ancestral to or closely related to modern humans) were also clearly capable of hunting large animals, giving rise to the very interesting question: could they also have had a significant impact on the megafauna? Previous ideas that Neanderthals (*Homo neanderthalensis*) were predominantly scavengers have been revised in the face of accumulating evidence that they were also active hunters. Outstanding evidence for hunting of large mammals as early as ca. 300 kya (MIS 9) has been provided by the renowned Paleolithic archaeological site of Schöningen in north-central Germany.[19] Here, nine wooden spears with sharpened tips, several exceeding 2 meters in length, were excavated from the deposits of an ancient lake shore. The spears, apparently intended as throwing weapons, were recovered in close association with the butchered remains of at least thirty-five individual horses, implying repeated hunting of horses at this favorable spot where animals would have been funneled between higher ground and the lake.[20] The hominins responsible are thought to have been either early Neanderthals or Heidelberg man (*Homo heidelbergensis*).[21] There is good evidence that at ca. 200 kya, Neanderthals successfully hunted woolly mammoths (*Mammuthus primigenius*) and woolly rhinos (*Ceolodonta antiquitatis*) on Jersey (Channel Islands, UK), then connected to the French coast.[22] There is also abundant evidence that at around 120 kya Neanderthals systematically hunted horses (*Equus ferus*) in Poland and aurochs (*Bos primigenius*) in France.[23] According to White and colleagues, "After many years of being characterised as predominantly, if not obligate, scavengers, Neanderthals have come to be seen as capable hunters, even top-level carnivores, possessing similar capabilities in the hunting realm as *Homo sapiens*."[24] Of course, this has important implications for the possible overkill of megafauna in Europe and southwest Asia before the arrival of modern humans.

4.3 ENVIRONMENTAL CHANGE HYPOTHESIS

The profound climatic changes and resulting disruption of biota that occurred globally around the Last Glacial–Holocene transition offer a plausible alternative cause for extinctions. Major changes in North America and northern Eurasia included the replacement of vast areas of open vegetation with grasses and herbs—the mammoth steppe—by forests, mainly boreal conifer forest and temperate deciduous forest.[25] In Alaska/Yukon and northern Eurasia, many species supposed to have been highly adapted to the mammoth steppe biome are thought to have become extinct because their habitat disappeared. Different kinds of vegetational changes also occurred on other continents, although generally these changes are less well known. However, a major problem with this hypothesis is that the geological record shows a complex pattern of climatic/environmental changes during most of the last 780 kya (Middle and Late Pleistocene), and featured repeated glacial–interglacial transitions that broadly resembled the most recent Last Glacial–Holocene transition (fig. 3.1), yet no such catastrophic extinction occurred previously. Accordingly, several researchers have proposed that the Last Glacial–Holocene cycle had unique characteristics resulting in unprecedented major extinctions—an interesting hypothesis that has yet to be tested by detailed comparisons with earlier cycles.[26] Another issue is, why were the animals that went extinct not able to find suitable habitat somewhere, as happened, for example, with survivors such as reindeer, horse, and saiga in northern Eurasia? In the case of those that went extinct, such as woolly rhinoceros, possibly the remaining areas were too small and fragmented; perhaps topographical or vegetational barriers prevented migration. The environmental change hypothesis predicts close temporal correlations between environmental changes and extinctions, and so is testable by comparison with climatic and vegetational proxies. These show convincing correlations for northern Eurasia (see chapter 5) but need more data for testing in other ecoregions.

On the basis of radiocarbon dates and ancient DNA megafaunal record from North America and northern Eurasia, Cooper and colleagues inferred that major faunal changes, including extinctions, corresponded with intersta-dial warming events.[27] However, currently available dates lack the necessary precision to test this hypothesis.

4.4 COMBINED HYPOTHESIS

Because neither overkill nor environmental change alone are able to explain all the observed facts, a number of authors have suggested a combination of the two hypotheses, in which extinctions resulted from human hunting of mega-faunal populations that were at the same time subject to habitat fragmentation and other stresses in response to climatic and vegetational changes.[28] Hunting pressure became critical only when populations were already reduced both in numbers and geographical range by environmental changes, and additionally the presence of humans in an area might have inhibited the natural migra-tion of megafaunal species responding to environmental changes. This sce-nario provides a plausible explanation as to how a widespread species such as woolly mammoth could disappear over its entire vast range. The combined hypothesis predicts a temporal and geographical pattern of range shrinkage for each megafaunal species—attributable to environmental changes—together with archaeological evidence that human populations were present in the refu-gial areas prior to extinction.

4.5 "HYPERDISEASE"

Ross MacPhee and Preston Marx postulated that extinctions could have been driven by a lethal pathogen introduced by humans (or their dogs) as they spread around the globe.[29] However, no compelling evidence—for example, preserved pathogens—has been found so far. Moreover, there is no known dis-ease at present that would preferentially select larger mammalian species or affect a range of orders.[30]

4.6 DEATH FROM THE SKY

It is now widely recognized that our solar system can be a dangerous place. In 1994 fragments of Comet Shoemaker-Levy 9 spectacularly collided with Jupiter. Extraterrestrial objects known as bolides—such as large meteorites, asteroids, and comets—have collided with Earth from time to time.[31] There is indisputable evidence of bolide impact during the late Quaternary, in the form of the modest-sized Barringer Meteor Crater (1,200 meters in diameter) in northern Arizona, dating from around 50,000 years ago. However, although no doubt most life in the near vicinity would have been wiped out by the impact, the effects of this event were strictly limited, and even in North America no extinction episode has been attributed to it. In 1908 a large meteorite exploded above the ground at Tunguska, in a remote area of central Siberia, flattening some 2,000 square kilometers of taiga forest. In 2013 the intensely bright trail of a bolide, estimated at 17 meters in diameter and weighing around 10,000 tonnes, was seen over the southern Urals; it exploded, causing extensive damage to buildings and extensive but largely minor injuries from the shock wave. A 650-kilogram chunk of this meteorite was recovered from near Chelyabinsk, Western Siberia. All these events brought home the message that Earth is subject to potentially lethal bombardment from space, and sensational articles frequently appear in the media reminding us of the vulnerability of our species to global disaster—even annihilation—by extraterrestrial objects striking Earth.

A sensational hypothesis for the cause of megafaunal extinctions was proposed by Firestone and colleagues. They described a range of characteristics (including magnetic grains with iridium, magnetic microspherules, charcoal, carbon spherules, and nanodiamonds) in sediment profiles across North America dating from around the onset of the Younger Dryas stadial (ca. 12.9 kya) and which were attributed to the impact of a comet or other extraterrestrial body.[32] This event is supposed to have caused widespread devastation, and to have served as a major contributor to megafaunal extinction in North America, as well as to the release of vast quantities of meltwater from the North American (Laurentide) ice sheet into the Atlantic Ocean, which in turn triggered the Younger Dryas cold episode. However, this idea has not been well received by most of the scientific community, as other researchers have been unable to corroborate their findings, either from the originally studied localities or elsewhere.[33] Since this hypothesis implies sudden extinction of North American megafauna at ca. 12.9 kya, it should be eminently testable by an extensive program of direct radiocarbon dating of megafaunal remains in North America.

Another dramatic idea proposes that ca. 12.8 kya, a massive increase in radiation from a solar flare caused global mass extinctions.[34] This hypothesis can be easily refuted because the extinctions were demonstrably not synchronous on a global scale, as would have been expected from such a catastrophic event. Moreover, if the radiation dose were high enough to wipe out mammoths and other megafauna, then many other terrestrial species and humans would also have been seriously affected, which clearly did not happen.[35]

CHAPTER 5

Northern Eurasia: Woolly Rhinos, Cave Bears, and Giant Deer

The importance of megafaunal extinctions in northern Eurasia (here defined as Europe, Russia, Kazakhstan, Mongolia, northern China, Korea, and Japan), has been downplayed by some authors, as fewer species disappeared here compared with, for example, North America. Nevertheless, losses here also had an enormous impact; since the Last Interglacial, about 120,000 years ago, approximately 41% of mammal species with body weights of 45 kilograms or more disappeared (fig. 5.1). Significantly, as elsewhere, the largest animals were the most affected, including three species of elephant, four rhinos, and a hippo—each at around 2 tonnes or more—together with eight other species of large mammal each weighing over 500 kilograms.

Moreover, northern Eurasia has by far the best coverage of radiocarbon dates for extinct (and surviving) megafaunal species, and in consequence their histories of range expansion and contraction in range and final extinction are better understood than for anywhere else.

During most of the Last Glacial, an ice sheet covered large areas of northwestern Eurasia; there were also extensive glaciers in the Alps and other mountain ranges.[1] Throughout this period, sea levels fluctuated but were consistently much lower than today—at the extreme Last Glacial Maximum (LGM), ca. 23 to 21 kya, by as much as 130 meters.[2] The depressed sea levels, for example, connected Britain to continental Europe via the dried-out bed of the southern North Sea, and at times connected Alaska to northeastern Siberia, allowing some animals to cross from one continent to the other. The climate was predominantly much colder and drier than today, although it featured complex and numerous fluctuations at different time scales, including on

FIGURE 5.1. Extinct and extant megafauna from Northern Eurasia (Palearctic Ecoregion). From the top, row 1, left to right: *Palaeoloxodon antiquus*; *Palaeoloxodon naumanni*; *Mammuthus primigenius*.[H] Row 2: *Stephanorhinus kirchbergensis*; *Stephanorhinus hemitoechus*; *Coelodonta antiquitatis*; *Elasmotherium sibiricum*. Row 3: *Equus hydruntinus*;[H] *Megaloceros giganteus*;[H] *Sinomegaceros yabei*; *Sinomegaceros pachyosteus*; *Camelus knoblochi*. Row 4: *Bison priscus*; *Ovibos moschatus*;[h] *Hippopotamus amphibius*;[*] *Crocuta crocuta*;[*] *Homo neanderthalensis*. Row 5: *Panthera spelaea*; *Ursus ingressus*; *Ursus spelaeus*; *Bos primigenius*.[H] Row 6: *Bos mutus*; *Capra ibex*; *Ovis ammon*; *Bison bonasus*; *Saiga tatarica*; *Sus scrofa*. Row 7: *Cervus canadensis*; *Cervus elaphus*; *Rangifer tarandus*; *Elaphurus davidianus*; *Alces alces*. Row 8: *Cervus nippon*; *Dama dama*; *Equus hemionus*; *Equus ferus*; *Camelus bactrianus*. Row 9: *Panthera leo*; *Panthera tigris*; *Panthera pardus*; *Ursus arctos*; *Ursus thibetanus*; *Ursus maritimus*.
Black: selected extinct species. Gray: selected living species. [*]Extirpated in Last Glacial. [h]Extirpated in Holocene. [H]Extinct in Holocene. The outlined *Homo sapiens* gives approximate scale.

the one hand milder (interstadial) phases and on the other phases of even more intense cold and expansion of ice sheets. Broadly speaking, much of the Last Glacial areas that are today covered by tundra, steppe, boreal, and temperate and mixed forest, at one time supported vast largely treeless areas of grasses, sedges, and forbs (the so-called "mammoth steppe"), while reduced areas of forest persisted in the south.[3]

Bones and teeth of Quaternary mammals, large and small, are found throughout most of northern Eurasia. Limestone caves are an especially rich source of remains, while many fossils occur at archaeological sites, or in river sediments where they may be recovered from naturally eroding riverbanks, in the course of construction work or quarrying for sand and gravel. Others have been trawled from the bottom of shallow seas, such as the southern North Sea and the Penghu Channel (between Taiwan and mainland China), dating from a time when these areas were dry land exposed by reduced sea level.[4] Such fossils are usually incomplete, having been subjected to a range of destructive processes, including chewing, trampling, and scattering by humans or other animals, dispersal and transport by flowing water, and exposure to the elements. Partial or complete skeletons occur much more rarely, mostly in lake clays and silts; while within the permafrost zone of northern Siberia, Alaska, and arctic Canada, frozen organic silts (known in Russian as *yedoma*) contain many beautifully preserved skeletal remains of woolly mammoth, bison, and other animals. Even more spectacular and scientifically informative are the frozen mummified carcasses of woolly mammoths, woolly rhino, bison, horse, and other animals that turn up occasionally as the permafrost melts. Mummification resulted when an animal buried in sediment and incorporated in the permafrost became desiccated as ice migrated out of the body, much as happens to meat stored too long in a freezer.

In northern Eurasia, the paleontologist is exceptionally fortunate in having not just the mortal remains of long-dead animals, but also paintings on the walls of caves (parietal art), and smaller portable sculptures and engravings of these animals.[5] It is a very moving experience to visit any of these painted caves in France or Spain and to realize that the people who visited these caves and created the art tens of thousands of years ago actually saw these animals alive and running around. This artwork is wonderfully evocative of the lost world of our Paleolithic ancestors. The animals depicted included such iconic extinct species as woolly mammoth, woolly rhino, cave lion, cave bear, bison, aurochs, and giant deer as well as the still-surviving reindeer, red deer, ibex, and horse. It is fortunate for us that these prehistoric artists produced this work, but exactly why they did is likely to remain a mystery.

Of all the known painted caves, Grotte Chauvet in southern France,[6] discovered in 1996 and designated a UNESCO World Heritage Site, is arguably the finest and probably also the oldest, although the latter claim is disputed by some specialists.[7] Learning from past mistakes (notably, mold-infested Lascaux), Chauvet has not been opened to the general public; instead, a full-size replica has been constructed as the next best thing.

5.1 HUMAN ARRIVAL

Hominins are known from Northern Eurasia as far back as around 2 million years.[8] Neanderthals (*Homo neanderthalensis*), Denisovans (see below), and perhaps other hominins who might have existed in the late Quaternary can be regarded as part of the extinct megafauna. The timing and causes of Neanderthal extinction have been much debated. The latest and most comprehensive study, using improved accelerator mass spectrometry (AMS) dating of key sites covering the Middle to Upper Paleolithic transition, indicates that they disappeared from different areas of Europe at different times but had entirely gone by about 41 to 39.3 kya.[9] The Denisovans, another extinct species of human related to Neanderthals, are known (so far) only from a finger bone and two molars that were found in a cave in the Altai Mountains in southern Siberia; with so little to go on, when they went extinct is so far unknown.[10]

The earliest known modern human (*Homo sapiens*), dated to ca. 300 kya, has been reported recently from Jebel Irhoud, Morocco.[11] The earliest evidence of *H. sapiens* from Europe is provided by Upper Paleolithic artifacts dating between 45 and 43 kya from Italy and the Balkans, indicating that there was an overlap of several millennia between Neanderthals and modern humans, which would have allowed ample time for cultural exchanges and indeed more personal interactions.[12] Recent studies found that a small amount of Neanderthal DNA is present in the genome of modern non-African humans, indicating that significant interbreeding occurred between the two species—probably in Asia between ca. 60 and 50 kya.[13] These recent findings indicate that the final disappearance of Neanderthals occurred before the extinction of cave bear, spotted hyena, and *Elasmotherium* (described below).

In this chapter I focus on the most iconic and best-studied extinct species from northern Eurasia, followed by a summary of extinction timings—as throughout this book, estimated from available radiocarbon dates made directly on megafaunal material—and a discussion of the temporal and geographical patterns that have emerged. In addition to megafaunal species that are globally extinct, I also include spotted hyena, hippopotamus, and musk ox, which have been extirpated from Eurasia but survive in other parts of the world.

5.2 GIANT DEER (AKA IRISH ELK): *MEGALOCEROS GIGANTEUS*

The first recorded discovery of giant deer was from County Meath, Ireland, in 1588. An Irish physician, Thomas Molyneux (1661–1733), gave us the first scientific account on the basis of skulls and antlers from Ireland.[14] However, believing that no created animal could have entirely vanished from the earth, he mistakenly equated his Irish beast with the North American moose, which understandably he had never seen. Ironically, he did compare his Irish material with elk from Sweden but—unaware that it in this case the Swedish animal *was* the same species as the North American moose—correctly realized that they were very different and thus remained happily untroubled by the reality of extinction. He wrote:

> That no real Species of Living Creatures is so utterly extinct, as to be lost entirely out of the World, since it was first Created, is the opinion of many Naturalists; and 'tis grounded on so good a Principle of Providence taking Care in general of all its Animal Productions, that it deserves our Assent. However great vicissitudes may be observed to attend the *Works* of Nature, as well as *Humane Affairs*; so that some entire *Species* of Animals, which have been formerly Common, nay even numerous in certain Countries; have in Process of time, been so certainly lost, as to become there-utterly *unknown*; tho' at the same time it cannot be denied that the *kind* has been carefully preserved in some other part of the World.

He was also struck by the fact of this "most large and stately Beast [there] remains among us not the least *Record* in Writing, or any manner of *Tradition* that makes so much a mention of its Name."[15]

The giant deer, *Megaloceros giganteus Blumenbach* (so-called "Irish elk") is one of the most well-known and evocative "vanished giants" of the Ice Age. Stags, some more than 2 meters at the shoulder, had enormous outspread palmate antlers spanning as much as 3.6 meters—easily the largest of any deer, living or extinct.[16] Nevertheless, although boasting the largest antlers, giant deer stags (at around 329 to 1228 kilograms[17]) weighed about the same as modern North American moose (*Alces alces*), which have much smaller antlers in relation to their body size. The astonishingly rapid annual growth of giant deer antlers would have required large quantities of calcium and phosphorus, which appears to have been achieved partly by drawing upon minerals, stored as bone, in the characteristically greatly thickened lower jaw and probably also elsewhere in the skeleton. When the antlers ceased growing in late summer food would still have been plentiful so that the depleted minerals could have been readily replaced.

As astutely observed by Stephen Jay Gould: "The 'Irish elk' like the Holy Roman Empire is misnamed in all of its attributes, being neither an elk nor exclusively Irish."[18] Ancient DNA work has shown that the giant deer is closely related to the much smaller living Eurasian fallow deer (*Dama dama*).[19] Although the range of the giant deer actually encompassed a large swathe of western Eurasia, its deeply rooted association with Ireland results from the extraordinary abundance of its remains on that island. The skulls, antlers, and bones of many hundreds of animals have been found; skulls bearing magnificent sets of antlers are to be seen not only in Irish and British museums and many country houses, but also in very many museums across the globe. There are also numerous mounted skeletons, most of them assembled from the remains of more than one individual. The reason for this abundance lies in the exceptional geological conditions for preservation in Ireland. Molyneux observed that skulls and antlers were frequently found by accident in similar "white marl" in several widely dispersed places. Large numbers were recovered in the nineteenth century by peasants who, when slowly hand-cutting the overlying peat for fuel, came across antlers or other fossils sticking out of the marl. The demand for "Irish elk" fossils was such that especially productive sites such as Ballybetagh near Dublin attracted commercial dealers who probed the bog with iron rods in order to locate bones or antlers.[20] Today sets of antlers are still very much sought after by museums and private collectors.

We now know that this marl was deposited in numerous shallow lakes across a large area of Ireland during the Late Glacial Interstadial period (from about 14.7 to 12.9 kya). These lake deposits were later buried by the growth of peat bogs during the Holocene (11.7 kya to present). Occasionally an individual giant deer may have drowned in a lake after breaking through thin ice, as happens fairly often with moose and Eurasian elk today, or simply died near or on the frozen surface, which, upon melting, dropped the disarticulated remains into the water.[21] There was no damage caused by flowing water, as might happen in a river deposit, and the lime-rich marl provided the perfect conditions for preserving the bones. The remains of stags rather than hinds predominate in both public and private collections, no doubt in part because their skulls (with attached antlers) were much more prized and more easily discovered than those of the antlerless females, which were probably mistaken for horses or cows and thus considered worthless. It is also possible, however, that males were more prone to becoming trapped, as they were encumbered by those huge antlers.

The huge antlers have long been a source of fascination, leading to now discredited ideas that over time some mysterious internal evolutionary drive (orthogenesis) caused them to undergo a progressive unstoppable increase in size, to the point where the handicap was so severe that it resulted in the extinction of the entire species. Nowadays, however, it is universally accepted that the size increase was an evolutionary process driven by sexual selection. Gould and others considered that the antlers functioned solely as a visual display, which served both to attract females and to intimidate other males, whereas Andrew Kitchener convincingly demonstrated that the antlers would also have been used in combat between stags. In 1985 Andrew Kitchener and Nigel Monaghan conducted a rather amusing but highly instructive practical experiment on the flat roof of an outbuilding of the National Museum of Ireland in Dublin. They were each "armed" with a genuine set of giant deer antlers and imagined themselves in the roles of rival stags wrestling one another for supremacy. When tilted steeply, the antlers locked together just above the second tine, as evidently was their function.[22]

Both the moderately low-crowned cheek teeth and studies of tooth wear of Irish specimens indicate that the giant deer was a mixed feeder, eating both browse and grass. The availability of high-quality plant food would have been essential to sustain the enormous annual antler growth in males. A major disadvantage of these huge antlers was that in autumn and winter, when fully grown, mature stags would have been excluded from even moderately dense tree cover, whereas moose, with its smaller, more compact antlers, is much better adapted to forest. A picture emerges of giant deer living in regions of predominantly open vegetation and escaping from predators by sustained running.

The rare depictions of *Megaloceros* in Paleolithic art include the remarkable wall paintings from Cougnac (Lot) and Chauvet (Ardèche), France, the latter radiocarbon-dated on the charcoal pigment to about 35 kya.[23] As is seen in many other animal depictions, the head (and antlers, where shown) are disproportionately small, the legs spindly and incomplete, in contrast to a plump body. At both sites the animals are shown with a prominent hump, and a dark stripe extending from hump to hock; one from Chauvet shows a dark band around the base of the neck.

Likely predators of giant deer include spotted hyena, cave lion, wolf, brown bear, and perhaps also Neanderthal and modern human. However, there is little evidence for hunting of giant deer by humans; it occurs only rarely in archaeological sites, although giant deer remains from Sosnovy Tushamsky (an early Holocene archaeological site in southern Siberia) appear to be food remains, which could represent either hunted or scavenged animals.[24] The large size of adult giant deer suggests that it would have been a formidable beast, able to run fast or, if cornered, to defend itself with its antlers or hooves (as in modern moose). No doubt, however, the young would have been much more vulnerable to predators.

During the Last Glacial, its maximum geographical extent was in the milder (interstadial) phases. Its distribution extended from Ireland and northern Spain in the west eastward to Lake Baikal in southern Siberia.[25] However, its range was restricted compared to many other extinct species, such as woolly mammoth, woolly rhino, and cave lion, and its absence from northern Siberia meant that it was never in a position to colonize North America. In eastern Asia, *M. giganteus* was replaced by (rather smaller) extinct species of giant deer: *Sinomegaceros yabei* (China and Japan), and *Sinomegaceros pachyosteus* (China).[26]

The reasons for the extinction of the charismatic *M. giganteus* have been much debated. Although giant deer remains are abundant in Ireland, across the rest of its range, material of this animal is rare in comparison with, for example, woolly mammoth or woolly rhinoceros. Adrian Lister and I had to work hard to assemble enough radiocarbon dates to trace an outline of its complicated history, including an indication of when and where it finally went extinct.[27]

The story of the demise of giant deer in Ireland is especially instructive. Most Irish records, including all those from the marl, date from the temperate Allerød interstadial (ca. 13.9 to 12.9 kya), when a rich herb vegetation occurred across most of the country. This notably included willow which was high in phosphorus and therefore important for the growing antlers. However, in Ireland giant deer did not survive the renewed cold episode known as the Younger Dryas (ca. 12.9 to 11.7 kya) which was accompanied by a severe deterioration in vegetation quality, including the disappearance of willow.[28] The youngest Irish date that we have, about 12.63 kya, falls within the early part of this cold phase. I had been about to write that the disappearance of giant deer in Ireland is an unequivocal instance of local extinction (extirpation) caused purely by climate changes as there is no evidence for humans in Ireland until around 10 kya (the Mesolithic site of Mount Sandel in Northern Ireland). However, in 2016 meager but credible evidence emerged to spoil this neat story. A humanly cut-marked brown bear patella (kneecap), which had long lain neglected in a cardboard box, was excavated a century ago from the delightfully named Alice and Gwendoline Cave, County Clare. It has now been independently radiocarbon-dated by two laboratories to around 12.5 kya, which puts it within the Younger Dryas cold phase,[29] and researchers are now actively looking for additional evidence of human occupation. Nevertheless, it would seem that very few people were in Ireland at this time, and at least for now climatic cooling remains the most likely reason for the demise of giant deer in Ireland.

Looking at the wider scene, giant deer disappeared not only from Ireland, but also from Britain, southern Scandinavia, and elsewhere in western Europe around the onset of the Younger Dryas cold phase (ca. 12.9 kya), and for many decades this was assumed to be when it went totally extinct. However, in 2004 a research program by the author, Pavel Kosintsev, Adrian Lister, and Tom Higham revealed that in the eastern part of its range it persisted for many more thousands of years, with latest known dates of about 7.67 kya and 7.84 from western Siberia (Kamyshlov and Redut localities) (figs. 5.2, 5.3).[30] Moreover, subsequent work has confirmed this picture with the discovery of many more specimens dating to the early Holocene, notably including another very young date—ca. 7.66 kya—from Maloarchangelsk in the west of European Russia.[31]

FIGURE 5.2. Giant deer *Megaloceros giganteus*, front view of skull and antlers, Holocene, from Kamyshlov, Western Siberia. (Courtesy of Leonid Petrov.)

FIGURE 5.3. Giant deer *Megaloceros giganteus* skeleton (same individual as fig. 5.2), Holocene, from Kamyshlov, Western Siberia. (Courtesy of Leonid Petrov.)

Nevertheless, we still don't know why giant deer failed to recolonize western Europe in the Holocene, or what caused its final disappearance. Since there are no marked environmental changes in the mid-Holocene to compare with those in the Last Glacial, the default might be to blame humans—and perhaps its demise was related to the appearance of Neolithic cultures in the region. However, the Sosnovy Tushamsky find might record scavenging by humans, so at present there is no unequivocal evidence that giant deer were hunted by people.

5.3 WOOLLY RHINOCEROS: *COELODONTA ANTIQUITATIS*

The woolly rhinoceros (*Coelodonta antiquitatis*) was an iconic member of the "mammoth steppe fauna" of the Last Glacial in northern Eurasia. There is a wealth of evidence for reconstructing its appearance and mode of life, not only from skeletal anatomy but also from many representations in Paleolithic art, frozen mummified remains (some with stomach contents), and stable isotope analyses.

With estimated weights of 1,038 to 2,958 kilograms (mean 1,905 kilograms), the woolly rhino was a formidable animal. It possessed an impressive battery of high-crowned (hypsodont) cheek teeth with thick enamel which were adapted to feeding on grasses and other low-growing vegetation with abrasive high silica content, exacerbated by incidentally ingested soil and grit.[32] The sloping back of the skull indicates a low-slung carriage of the head, much as in the similar-size living African white rhinoceros, also a grazer. The woolly rhino was evidently adapted to dry open landscapes and its short legs indicate that it would have found it very difficult to traverse deep snow.[33]

DNA studies indicate that of the five extant species of rhino, the Sumatran rhinoceros is its closest (although not very close) relative.[34] The probable origins of the woolly rhino are rather surprising. Its earliest known ancestor did not live in the north of Siberia, as might have been expected, but on the Tibetan Plateau, whereby 3.7 million years ago (Middle Pliocene) it had adapted to cold, open habitats at high altitudes. From there it spread to lowland areas during the Quaternary.[35]

Finds of woolly rhino mummies are much rarer than those of woolly mammoth, and until recently only partial woolly rhino carcasses had been recovered from the Siberian permafrost. However, in 2007 a reasonably complete frozen mummy of an adult female was found by gold miners on the lower Kolyma River in Yakutia, northeastern Siberia. The remains included the left side of the body, fore and hind legs (all with skin, but only sparse attached hair), and the skull with both horns and lower jaw (fig. 5.4).[36] A rib fragment gave a radiocarbon date of ca. 43.4 kya. Skin thickness varied from 5 millimeters to an impressively thick 15 millimeters on different areas of the body. The intestines, stomach, and their contents were also recovered. Pollen and spores from the stomach contents mainly consisted of grass and *Artemisia* (sagebrush)

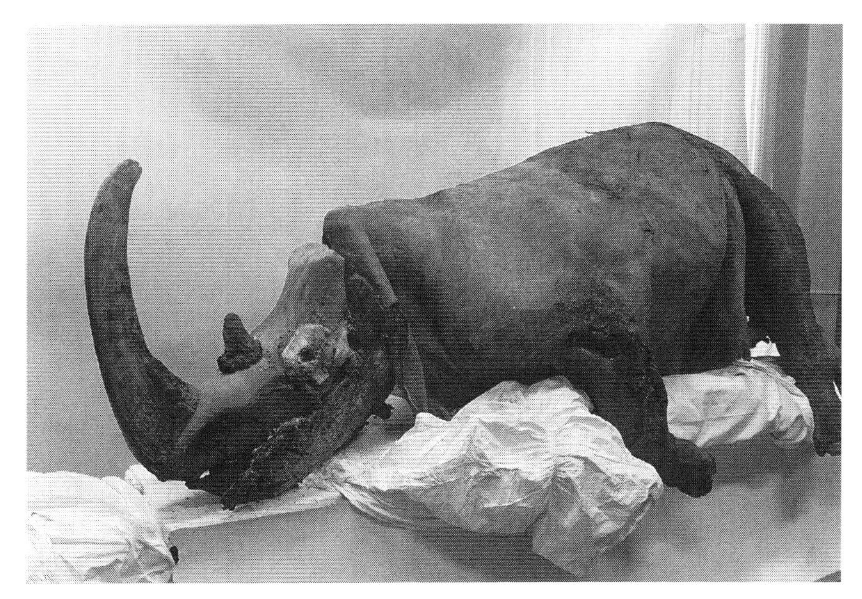

FIGURE 5.4. Woolly rhino *Coelodonta antiquitatis*, mummy, from Kolyma, northeast Siberia. (Courtesy of the Academy of Sciences of the Republic of Sakha [Yakutia].)

together with a variety of other forbs. Even more remarkable was the 2014 discovery, again in Yakutia, of a rhino calf—the first ever found—named Sasha after its finder, who spotted it eroding out of a riverbank. The back half of the animal is missing, but the front half is exquisitely preserved, with two tiny horns and a luxuriant hairy coat—amply justifying the name "woolly rhino."

However, these were not the first preserved carcasses to come to light. In 1929 two unique woolly rhino mummies were recovered from a mine in Starunia, southwestern Ukraine (then part of Poland). They had been preserved by a natural mixture of salt and hydrocarbons.[37] The mounted stuffed skin and skeleton of one of the rhinos are displayed separately in the Polish Academy of Sciences Museum of Natural History in Kraków, and a cast of the carcass as found can be seen in the Natural History Museum, London.

Several of the Siberian finds indicate that the woolly rhino had long, bristly guard hairs, especially on the neck and shoulders, and a dense, insulating underwool. The short fur on the limbs would have prevented large amounts of snow or ice from adhering to the legs and feet and hampering movement. Muzzles preserved in frozen carcasses and in the two mummies from Starunia clearly indicate that it possessed very wide lips, consistent with adaptation for nonselective grazing—again, as in the living white rhino. A modest number of well-preserved woolly rhino horns, composed of keratin as in modern rhinos, have been recovered from the permafrost of northern Siberia. These show that the large, long, and heavy skull bore two horns: a very large horn, up to 1.35 meters long in front, and a much smaller horn behind. These are also depicted in Paleolithic paintings and engravings. The front horn, which because of the low-slung position of the head would have been carried nearly horizontally, was laterally flattened, unlike the second horn or the horns of any of the five living species of rhinoceros which are rounded in cross section. Paired left-and-right wear facets strongly suggest that the front horn was used in a side-to-side motion, probably to clear thin snow cover to expose vegetation or to free plants frozen to the ground; by analogy with living rhinos, it is also very probable that the horn was used for both defense against predators and combat between males. The considerable weight of the head and horns was evidently supported by massive muscles and ligaments attached to the long neural spines on the anterior thoracic (chest) vertebrae.[38]

Although relatively uncommon in Paleolithic art, there are clear depictions of woolly rhino from a number of sites ranging from Aurignacian to Magdalenian, including for example, engraved sketches on portable pieces of slate from the open site of Gönnersdorf, Germany, and wall paintings and engravings from the French caves of Rouffignac, Font de Gaume, and Lascaux, and especially the many superb paintings from Chauvet Cave.[39] The hairy coat is indicated in many of the artworks but is lacking in some of those from Chauvet, which, however, show a broad vertical band across the animal's flank—possibly representing color patterning. In one Chauvet panel, a rhino is shown with multiple superimposed images, which could have given the impression of movement when viewed by flickering lamplight (as, of course, they would have been). Also from Chauvet, unique in Paleolithic art, is another outstanding and evocative charcoal drawing that shows two rhinos facing each other head on and locking horns as if in combat.[40]

More than 270 radiocarbon dates on woolly rhino remains have made it possible to follow changes in distribution leading to its final extinction (fig. 5.5).[41] Both the anatomical characters and the geographical pattern of dates indicate that woolly rhino was adapted to a dry climate, subject only to light snow cover, firm ground, and extensive low-growing vegetation. Before about 35 kya it was widespread across northern Eurasia, but from this time onward we see a general progressive reduction in its range—mainly in the form of a contraction toward the east, starting with its withdrawal from Britain at about 35 kya. The youngest dates that we have so far are about 14 kya from northeast Siberia. Its likely final extinction corresponds with the widespread replacement of grasses

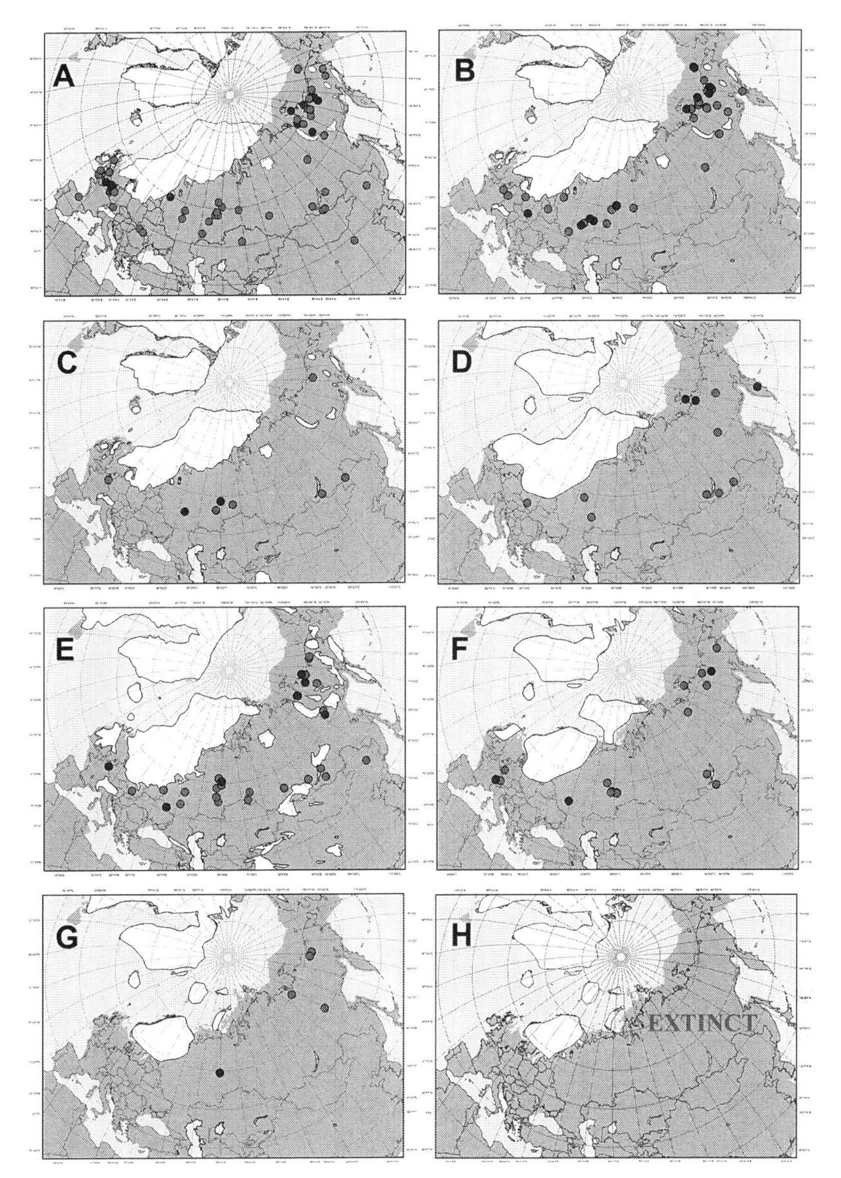

FIGURE 5.5. Time-sliced maps of radiocarbon-dated records for woolly rhino *Coelodonta antiquitatis* showing distribution changes prior to extinction. Gray: ultrafiltered dates. Black: non-ultrafiltered dates.

Note: Changing extent of ice cover (white) and of land vs. sea. A: 46–35 kya. B: 35.5–28.9 kya. C: 28.9–27.5 kya. D: 27.5–23.3 kya. E: 23.3–17.5 kya (LGM: GS-2b). F: 17.5–14.6 kya (LGM: GS-2a). G: L 14.6–13.9 kya (Late Glacial interstadial GI-1d, e). H: 13.9–12.8 kya (GI-1a,b). (Modified from Stuart and Lister 2012.)

and herbs (on which woolly rhino fed) by shrubs and trees with the onset of the warmer, moister Allerød interstadial phase. However, another important factor was probably increased snowfall, to which woolly rhino was evidently ill adapted. The relatively late survival of woolly rhinoceros in northeastern Siberia when it had disappeared further west may also relate to the later persistence of open vegetation in that region. An interesting question is why woolly rhinoceros failed to cross the Bering region and colonize North America; remarkably, no trace has been discovered in that continent. So far it hasn't been recorded from eastern Chukotka either, which suggests that perhaps for some unknown reason it didn't quite make it to the Siberian side of the Bering Land Bridge and was therefore unable to cross. Alternatively, we know that the land bridge clearly acted as a filter, so that American species such as the giant "short-faced" bear and scimitar cat failed to cross to Eurasia, whereas the musk ox and woolly mammoth were able to cross in the opposite direction. If woolly rhinoceros did manage to get this far, it may have been blocked from crossing because of wetter and boggy conditions on the land bridge. Dale Guthrie envisaged the Bering Land Bridge as a "mesic buckle" connecting the more arid regions to the west and east.[42]

So, the extinction of the woolly rhino can plausibly be attributed to climatic changes, whereas a human cause seems highly unlikely. There is almost no evidence to suggest that this large, formidable, and dangerous animal was hunted by humans, although possibly this happened occasionally and opportunistically.

5.4 "SIBERIAN UNICORN": *ELASMOTHERIUM SIBIRICUM*

The huge rhino *Elasmotherium sibiricum*, the last member of a long evolutionary line, was one of the oddest, most spectacular, and least familiar of the "vanished giants" (fig. 5.6). The name *Elasmotherium* refers to the distinctive structure of the cheek teeth (Greek, *elasmos*: "lamina"). Judging from the large bony protuberance on its skull, it bore a massive single horn quite unlike that of any Pleistocene or modern rhino. This solitary horn has led some to call it the "Siberian unicorn," although there was almost no resemblance to the slender, graceful creature of legend.

A superb mounted skeleton from Gaevskya Village in the Caucasus, and exhibited in the Stavropol Museum (fig. 5.7), measures approximately 2.5 meters at the shoulder and 4.5 meters long. From these measurements, the body weight has been estimated at 3.35 tonnes.

E. sibiricum was first described in 1808 by Gotthelf Fischer von Waldheim, the German director of the Moscow University Natural History Museum,

FIGURE 5.6. Artist's impression of the giant rhino ("Siberian unicorn") *Elasmotherium sibiricum* in life. (Painting by Kate Scott.)

FIGURE 5.7. Mounted skeleton of *Elasmotherium sibiricum*, shoulder height approximately 2.5 meters, from Gaevskya Village (Caucasus) in the Stavropol Museum. (Photo by Dr. Igor Doronin, courtesy of Zoological Museum, Saint Petersburg.)

on the basis of a left half-mandible with four cheek teeth donated to the museum in 1807 by Princess Ekaterina Dashkova, then president of the Russian Academy of Sciences.[43] The specimen had no locality information, but was presumed to come from "Siberia"—at that time, a label applied to anywhere east of the Volga River. Fischer von Waldheim recognized that he was dealing with an extinct animal entirely new to science, at a time when very few other extinct creatures had been described from anywhere. The subsequent history of this fossil is remarkable: in 1812, when Napoleon's invading armies occupied Moscow, nearly all the Dashkova collection was lost, presumably as a result of pillage or the fires that destroyed most of the city. Fortunately, the precious *Elasmotherium* jaw was saved by being evacuated to the city of Nizhny Novgorod. More than a century later it was returned to Moscow, by which time many more fossils of this extraordinary beast had been discovered.

The presence of a very large single horn in *Elasmotherium* is inferred from the prominent bony protuberance—much bigger than in any other rhino living or extinct—on the top of the skull. However, no example of *E. sibiricum* horn (which, as in other rhinos, would have consisted of keratin—the same substance as hair and nails) has been discovered so far. *E. sibiricum* was evidently a highly specialized grazer, indicated by the low carriage of the skull—lower even than in the woolly rhino and the living white rhino—as well as the high-crowned, rootless, permanently growing cheek teeth (unique among rhinos), with their distinctive intricately convoluted sheets of enamel. The tooth structure implies adaptation to hyper-grazing accompanied by the consumption of substantial quantities of dust or grit. Analyses of stable isotopes show high δ13C and δ15N values, typical for mammals inhabiting dry steppe or desert, since reduced precipitation tends to increase these values in the plants that they consume. Pollen, plant macrofossil, and faunal data suggest that *E. sibiricum* lived in open steppe habitats in the southern trans-Urals, while in other areas it inhabited forest-steppe landscapes with extensive grassy areas.[44] Its known distribution is limited to more or less within the area of modern steppe.[45]

Previously, *E. sibiricum* had been overlooked as a late Quaternary extinction for the simple reason that it was generally believed to have disappeared in the Middle Pleistocene, well beyond radiocarbon-dating range. However, a recent study using direct radiocarbon dates on *E. sibiricum* material demonstrated that it survived into the Last Glacial.[46] Eighteen samples were dated at the Groningen and Oxford AMS laboratories. As there were discrepancies between some of the results from the two labs (probably due to incomplete removal of consolidants), six samples were re-dated at Oxford, isolating the single amino acid hydroxyproline, which is unique to collagen, for greater reliability. The two youngest median calibrated radiocarbon dates are ca. 37.49 and 37.61 kya. Phase analysis (a statistical technique) gives a best estimate of 38.48 to 34.95 kya for the time of its extinction, indicating that, along with cave bears, spotted hyena, and Neanderthal, it had disappeared well before the onset of the Last Glacial Maximum. The extinction of *E. sibiricum* may have been related to its high degree of specialization, including extreme dietary adaptations.

Although the dates show that *E. sibiricum* was contemporary with modern humans, as yet no evidence for hunting or carcass utilization has been found, nor are there records from Paleolithic archaeological sites. There are only very sparse putative depictions in Paleolithic art, and I find them unconvincing. The best candidate—a simple black outline drawing from Rouffignac Cave in France—is most likely a stylized woolly rhino. Certainly, it has a single large horn, but this is shown well forward on the skull, unlike the central position in *Elasmotherium*. Another difficulty with this interpretation is that Rouffignac is more than 2,000 kilometers west of the recorded fossil range of *E. sibiricum*.

5.5 WOOLLY MAMMOTH: *MAMMUTHUS PRIMIGENIUS*

The woolly mammoth (*Mammuthus primigenius*) is universally the most widely recognized iconic symbol of the Ice Age. More has been written about it than any other "Ice Age giant," including numerous scientific papers, popular articles, and books.[47] We have a wealth of evidence—from skeletal remains (fig. 5.8), frozen mummies (sometimes with gut contents), Paleolithic art, ancient DNA, stable isotopes, and many more radiocarbon dates—than for any other extinct Ice Age mammal. Woolly mammoths were approximately the same size as modern African and Asian elephants, with an average shoulder height of about 3.2 meters and weighing around 3.8 tonnes. However, quite unlike their living relatives, they possessed long shaggy hair and strongly curved tusks.

For centuries, mammoth tusks have been recovered from the frozen ground of northern Siberia; the preservation is so good that much of this fossil ivory is suitable for intricate carving. In addition to large quantities of well-preserved bones, from time to time frozen mammoth mummies are discovered—first mentioned as long ago as 1692, but no doubt unrecorded finds go back very much further. The indigenous people of northern Siberia believed that giant tunneling creatures, who lived beneath the ground like huge moles, perished if they accidentally emerged at the surface into the light, thus providing a wonderfully neat explanation for the findings of mammoth carcasses. Such

FIGURE 5.8. Woolly mammoth *Mammuthus primigenius* skeleton, Taimyr Peninsula, north-central Siberia. (Courtesy of Zoological Museum, Saint Petersburg.)

discoveries were regarded with dread as inevitable bringers of misfortune, although in many cases this superstition didn't stop the finders removing and selling the valuable tusks.

In 1806 the "Adams mammoth," the world's first near-complete mammoth skeleton (with attached remnants of flesh and skin and more than 16 kilograms of hair), was recovered from the mouth of the Lena River in northeastern Siberia. In 1808 it was reassembled and mounted in the Zoological Museum in Saint Petersburg, where it can still be seen. The dried flesh on the head includes a remarkably small ear, as might have been expected for a species adapted to minimizing heat loss in such a cold environment.

Nearly a century later, in August 1900, came the exciting discovery, by a hunter, of the carcass of an adult male on the right bank of the Beresovka River, a tributary of the Kolyma, in a remote area of northeastern Siberia. In May 1901, an epic expedition mounted by the Imperial Academy of Sciences and led by Otto Herz and Eugen Pfizenmayer set out from Saint Petersburg to the location of the find, a four-month journey. Pfizenmayer left us his vivid impressions upon finally reaching the site:[48]

> Then around a bend in the path, a towering skull appeared, and we stood at the grave of the diluvial monster. . . . We stood speechless in front of this evidence of the prehistoric world, which had been preserved almost intact in its grave of ice throughout the ages. For long we could not tear ourselves away from this primeval creature, so hung about with legend, the mere sight of which fills the simple children of the woods and tundras with superstitious dread.

The expedition members worked for over a month under very arduous conditions to retrieve the precious find. Fires were lit to thaw the carcass so it could be dug out and cut into manageable sections, which were then allowed to refreeze and transported in twenty-seven cases on ten sleighs by reindeer and horses to Irkutsk on the Trans-Siberian Railway. From there they were taken by freezer car to Saint Petersburg, where they arrived February 18, 1902. Tsar Nicholas II and his wife, Alexandra, went to see the newly reassembled carcass (although the tsarina was apparently less than impressed with the evil-smelling remains). Subsequently, the stuffed original skin and hair, with modeled head and tusks, was mounted in the Zoological Museum in a sitting posture—as it was found (fig. 5.9). The intact skeleton was mounted separately. They are still treasured exhibits today.

However, the idea that woolly mammoths are preserved in huge blocks of ice (clear ice, at that) belongs to the realm of cartoons. In reality their mummified remains occur in frozen silts, clays, and sands. The myth that mammoths were frozen almost instantaneously helps fuel another potent belief that DNA is so perfectly preserved in the frozen carcass that it could be used to resurrect a living mammoth. Indeed, in some mammoth mummies, the tissues are exquisitely preserved, allowing detailed study of the organs. Nevertheless, although large amounts of DNA can be recovered from these remains, thus allowing determination of aspects of the animal's appearance and physiology, the DNA

FIGURE 5.9. Woolly mammoth mummy from Beresovka, northeast Siberia, discovered in 1900. (Courtesy of Zoological Museum, Saint Petersburg.)

is degraded into short fragments so that any suggestion of cloning a mammoth (or other Ice Age animal) is unrealistic—at least, for the foreseeable future.[49]

One of the very few benefits of global warming is that it is probably responsible for increased numbers of finds of mummified animals as more permafrost melts. In 1977 the well-preserved carcass of a mammoth calf was found in Magadan in northeastern Siberia. Named Dima, it went on a world tour, and in 1979 I was thrilled to see it as part of the Soviet Exhibition at Earl's Court, Olympia, London. It now resides in the Saint Petersburg Zoological Museum.

However, in 2006, on the banks of the Yuribei River in the Yamal Peninsula of Arctic northwest Siberia, a reindeer herder named Yuri Khudi and his sons came across by far the best-preserved carcass of a woolly mammoth (or indeed of any extinct animal) yet found (fig. 5.10). It was a tiny female calf (1.2 meters long), which he named Lyuba (Russian: "love"), after his wife. A count of the daily growth rings of a first molar showed that she was only a month old when she died, some 45,000 years ago.[50] The presence of silt in her trunk airways, throat, and lungs showed that she had died from asphyxiation after being trapped in sediment. Lyuba has proved to be an extraordinarily rich source of information on mammoths, as the external anatomy (amazingly including intact eyelashes) and internal organs are so beautifully preserved. The external genitalia, which are very like those of a modern elephant, identified her as female. The presence of two nipples between the front legs—the first time seen in a mammoth—also closely resembles modern female elephants. She has very small ears and tail to conserve heat, but the function of the curious flanges on either side of the trunk is unclear.

The trunk tip has two protrusions, as does a modern Asian elephant, but mammoths uniquely have a long front "finger" and a short back "thumb"—adapted for picking low-growing plants. Lyuba's stomach and intestines contained traces of her last meal: milk residue and fragments of plant matter—the latter probably from ingesting her mother's dung, as baby elephants do today.

FIGURE 5.10. Mummy of Lyuba, the baby female woolly mammoth discovered in 2006 on the banks of the Yuribei River, Yamal Peninsula, northwest Siberia. (Courtesy of Zoological Institute, Saint Petersburg.)

Adult woolly mammoths had coarse, shaggy outer hair up to 1 meter long, covering a layer of shorter finer hair and insulating inner wool close to the skin. Lyuba had around 1 to 4 centimeters of fat beneath the skin, providing further insulation. As in many mammoth mummies after death, most of the fur had slipped off Lyuba's body, which retained just remnants on the feet. The typical reddish-brown or orange coloration of the hair in these mummies almost certainly results from chemical alteration during burial, as is commonly seen in (much more recent) human mummies. Ancient DNA studies suggest the hair was generally dark but variable in shade.[51] Several adult carcasses also include gut contents. Most are dominated by grasses and sedges, but food recovered from one mammoth also included large quantities of moss, while another had eaten significant amounts of willow.

As might have been expected for such an impressive animal, woolly mammoths are featured prominently in Paleolithic art, both portable and mural. Visitors to the extensive Rouffignac Cave (Dordogne, France), can view engravings and drawings of no fewer than 158 mammoths, while other fine examples of mammoth cave art include Chauvet, Cussac, Grande Grotte Arcy-sur-Cure, and Pech-Merle.[52]

Perhaps unsurprisingly, direct evidence for hunting of woolly mammoths by humans (in the form of spear points embedded in bones) is very rare; only a handful of examples are known, mostly from Siberia. At the Yana site, in arctic northeastern Siberia, which has yielded hundreds of mammoth bones, one piece of scapula (shoulder blade) was found with a stone fragment embedded in it, while a second contained pieces of stone and ivory—evidently part of a composite spearhead. Both date to about 32 kya. At Lugovskoye, in southwest Siberia, a mammoth vertebra from the chest region, dated to 16.5 kya, had also been pierced by a stone spearhead. In the above cases, the hunters appear to have targeted the vital heart/lung region. Finds from Nikita Lake (northeast Siberia), dated to about 13.9 kya, include a mammoth rib with an embedded stone tool fragment.[53]

Evidence of mammoth hunting in Eastern Europe was found at Kostenki 14 (Russia), dated ca. 40 kya, where a pointed ivory artifact was found to have penetrated to the interior cancellous bone of a mammoth rib. At the Kraków Spadzista site in Poland, dated to between 29.9 and 28.8 kya, thousands of Late Gravettian stone tools were recovered in association with the remains of more than a hundred mammoths. Many mammoth bones are cut-marked—evidence for the stripping of meat from carcasses—while others had been broken by hammerstones, presumably to access the marrow. "More than 50% of the site's flint shouldered points and backed blades bear diagnostic traces of hafting and impact damage from use as spear tips," and a pointed fragment of a flint implement found embedded in a mammoth rib, had penetrated to a depth of 7 millimeters.[54]

Abundant mammoth remains at many other archaeological sites also suggest hunting. However, the famous huts—each constructed from dozens of mammoth bones and skulls—such as Mezhyrich (Ukraine) and Kostenki (River Don, Russia)—do not represent mass slaughter of mammoths, as might have been supposed, but instead comprise remains of different ages gathered from the treeless landscape.[55]

Most woolly mammoth remains are bones, teeth, and tusks, which can also yield a vast amount of scientific information. Ancient DNA studies suggest that woolly mammoth populations, widespread in the previous cold period, became drastically reduced during the warm Last Interglacial around 120,000 years ago as a result of the massive expansion of forests. However, they survived to re-expand during the Last Glacial period.

The availability of hundreds of radiocarbon dates—very many more than for any other extinct animal—together with aDNA allows us to trace the complex changes in geographical range and population size that occurred from about 44,000 years ago to its final disappearance.[56] From 44 to 24 kya, the woolly mammoth roamed over a vast area, from Britain and Spain in the west, across Eastern Europe to Siberia, China, and Japan, and via the Bering region into Alaska and the Yukon and as far south as the Great Lakes. Within this period, some temporary range shifts occurred; one such event, around 35 to 30 kya, may be linked to a decline in genetic diversity (from aDNA evidence) that possibly reduced its ability to survive further changes.[57]

Around 21.5 kya, within the Last Glacial Maximum, the woolly mammoth disappeared from most of Europe for some 2,000 years, apparently in response to deterioration in vegetational productivity, although, remarkably, woolly rhinos show no such response. After this episode mammoths re-colonized most of Europe from Siberia, although aDNA evidence suggests a huge drop in mammoth numbers, even though its range remained extensive. The warming that began ca. 14.7 kya did not produce an observable response in terms of range shifts, as the vegetation remained largely open because trees took a long time to migrate from their glacial refuges in southern Europe and elsewhere. Then, at about 13.8 kya, mammoth disappeared entirely and rather rapidly from Europe and most of northern Asia—a dramatic event that corresponds with the replacement of open vegetation ("mammoth steppe") by birch and pine woodland at the onset of the Allerød interstadial phase. Subsequently, in response to the renewed cold of the Younger Dryas (12.9 to 11.7 kya) and modest re-expansion of open vegetation, mammoth populations re-occupied limited areas of northwest Siberia and northeastern Europe. However, the overall shrinkage and fragmentation of populations continued with the Holocene warming from 11.7 kya. The latest mainland populations are recorded from the fringes of the Arctic Ocean in north-central Siberia: the Taimyr Peninsula ca. 11.1 kya, and the New Siberian Islands (then connected to the mainland) ca. 10.7 kya.[58]

Surprisingly, this is by no means the end of the story. In 1993 Sergey Vartanyan and colleagues shook the scientific world when they published their sensational finding that woolly mammoths had survived thousands of years longer on Wrangel Island, in the Arctic Ocean off northeast Siberia.[59] Here, as shown by a large series of radiocarbon dates, an isolated population of small mammoths continued to around 4.02 kya, contemporary with early Egyptian and Mesopotamian civilizations. There is good evidence that open mammoth steppe vegetation persisted much longer on Wrangel than elsewhere. Perhaps the eventual demise of the woolly mammoth resulted from the arrival of people, although there is a gap of some 400 years between the last known mammoths and the first evidence of humans. Moreover, at the Chertov Ovrag archaeological site—the earliest site on the island—people were hunting marine mammals, not mammoths.[60] As described in chapter 6, a population of woolly mammoth also survived very late on another Arctic island: Saint Paul Island in the Bering Sea off Alaska. It probably disappeared within about a century of 5.6 kya.

Thus, both the radiocarbon and aDNA evidence clearly show that the extinction of the woolly mammoth was far from a sudden event, but instead reveal complex range changes over many millennia, eventually fragmenting into smaller and smaller populations and culminating in final extinction.

5.6 CAVE BEARS: *URSUS SPELAEUS* GROUP

For centuries, large bones found in various caves in Central Europe were thought to be the remains of dragons or unicorns and were collected and ground down for use as a medicine ("*unicornu verum*") to treat a variety of ailments. However, by the eighteenth century, several authorities realized that most of the remains were of some kind of bear; the little-known polar bear seemed the most likely candidate. Then, in 1794, a young physician, Johann Christian Rosenmüller, made a bold and inspired breakthrough when he described a single well-preserved skull from the Zoolithenhöhle in Bavaria as belonging to a previously unknown *extinct* species of bear, which he named *Ursus spelaeus* (*Höhlenbär*, or "cave bear").[61]

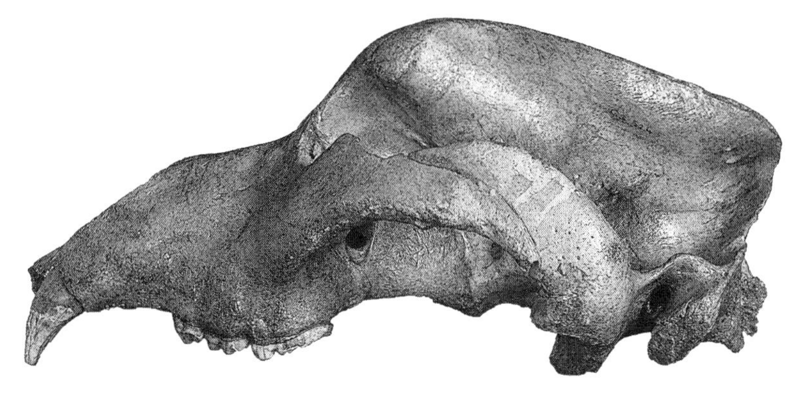

FIGURE 5.11. Cave bear *Ursus ingressus* skull, Gamssulzen Cave, Austria. Note the domed forehead characteristic of all cave bears. (Courtesy of Gernot Rabeder.)

In the Last Glacial period, cave bears were confined to upland areas in western and central Europe, with an apparently isolated population in the Urals. They were absent from northern Europe and also Britain, where only the living brown bear (*Ursus arctos*) occurred. Recent research on mitochondrial DNA by Mathias Stiller and colleagues has shown that cave bears were genetically diverse, very likely the result of their fragmented geographical distribution and complex migration history.[62] Several taxonomic groups of cave bears have been distinguished on both morphological and genetic grounds. *Ursus ingressus* tends to be larger and has more complex molar patterns than *U. spelaeus*. *Ursus rossicus* is a small form, and phylogenetic analyses suggest that it forms a sister group to *U. ingressus*. *Ursus spelaeus* is known only from western Europe, whereas *Ursus ingressus* had a more easterly distribution, including the population in the Urals. There was a narrow band of overlap between the two in central Europe. *Ursus kudarensis*, distinguished both by mitochondrial DNA and archaic dental features (like the Middle Pleistocene *Ursus deningeri*), was isolated in the Caucasus region.

Ursus spelaeus and its relatives were large, heavily built bears (fig. 5.11) with estimated maximum weights for males at around 1,000 kilograms (1 tonne), compared with 800 kilograms for the largest living brown and polar bears. Only the extinct American giant "short-faced" bears (*Arctodus*, *Arctotherium*) came close in size. Tooth-wear studies and isotope analyses from bones and teeth indicate that most cave bear populations were predominantly vegetarian, unlike modern brown bears which also eat substantial amounts of animal food.[63] However, isotope work indicates that cave bears from Carpathian sites (Romania) enjoyed a broader diet. As with living brown bears, the cave bear menu may have included berries, nuts, fungi, leaves, shoots, forbs, and new growth of grasses. However, if cave bears were predominantly vegetarian, then the plants eaten must have been highly nutritious and abundant for the animals to have attained such a large size. The problem of seasonal unavailability of high-quality plant food was solved by winter hibernation.

In 1999 Adrian Lister and I began research on megafaunal extinctions in northern Eurasia, which entailed removing a small piece of megafaunal bone or other tissue from selected specimens for radiocarbon dating. When I requested samples of less common species, helpful German colleagues pointed to piles of cave bear bones in their institutions, saying, "But we have plenty of these." Alas, in those early days our brief did not include cave bears, although we made up for this omission later. However, this abundance of remains highlights a striking feature of cave bear ecology: many generations of adult bears hibernated during winter in the caves, and probably fairly often an individual died from various causes such as old age, insufficient fat reserves to last through the hibernation period, or problems in giving birth. As might be expected, numerous remains of very young individuals also occur. Over many centuries the sum of these fatalities was the accumulation of the skeletal remains of thousands of individuals in the larger bear caves of central Europe (fig. 5.12), to the extent that at one time many caves were mined for fertilizer.

An innovative study by Gloria Fortes and colleagues using ancient DNA of Ice Age bears from caves in northern Spain suggests some very interesting aspects of their behavior.[64] In the case of cave bears, each cave contains unique but closely related haplotypes (a group of genes inherited from one parent) that show consistent genetic differences from bears in other caves, even those nearby. Significantly, this pattern is not seen in the brown bear. The implication is that individual cave bears, both male and female, returned each year to hibernate in the same cave in which they were born.

As with many other extinct species, our Paleolithic ancestors "thoughtfully" recorded a few images of cave bear for posterity in the form of a handful of beautifully executed wall paintings at Chauvet.[65] The characteristic domed forehead is clearly shown, and so far these are the only prehistoric depictions from anywhere that can be unambiguously attributed to cave bear. Numerous cave bear bones and skulls lie scattered on the cave floor, many encrusted with calcite flowstone. One skull was deliberately placed (presumably by humans) in a prominent position on a fallen limestone block. There are also many claw marks on the cave walls, some made before the paintings, others after, showing that humans and bears both used the cave within the same time period, although probably not simultaneously. There are also paw prints in the mud

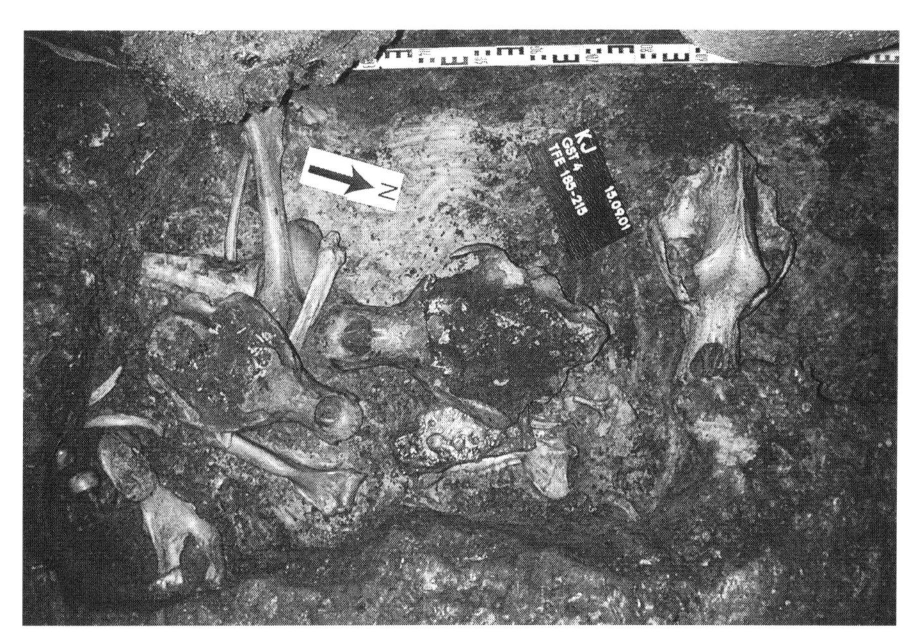

FIGURE 5.12. Skulls and limb bones of cave bear *Ursus ingressus in situ*, Križna Jama Cave, Slovenia. (Courtesy of Gernot Rabeder.)

on the cave floor and hollows that had been scooped out and evidently used for hibernation. An extensive program of radiocarbon dating shows that cave bears visited Chauvet Cave over a very long period, from about 42 to 33 kya, but there is no evidence for conflict or other interaction between bears and humans.

It was formerly widely believed that the cave bear survived well beyond the Last Glacial Maximum (broadly ca. 27.2 to 14.7 kya) to the Late Glacial (ca. 14.7 to 11.7 kya) or even into the Holocene (after 11.7 kya). However, research by Martina Pacher and the author, using radiocarbon dates made directly on cave bear remains, indicated that they disappeared very much earlier, well before the onset of Greenland Stadial GS-2; and subsequently this picture has been substantially confirmed by other researchers.[66] Recently published dates of ca. 25.27 and 26.1 kya on *Ursus ingressus* (identified by aDNA) from Stajnia Cave, Poland, extend the known temporal range of cave bears to within the early part of Greenland Stadial GS-3 and support the inference that their demise was related to climatic cooling and, crucially, the resulting deterioration in quality and quantity of their plant food.[67] The youngest dates from the Urals suggest that they had all gone from this region around 16,000 years earlier, but why this occurred we don't yet understand.

The possible role of humans in the extinction of the cave bear is uncertain. Numerous cut marks on cave bear bones from the cave sites Casamène (French Jura) and Hohle Fels (Swabian Alb, southern Germany) clearly show that their carcasses were exploited by humans, and cave bear bones are present in a few open-air archaeological sites in the Czech Republic and elsewhere. But direct evidence that they were killed by people is harder to come by. The only putative example known so far is from Hohle Fels, where a thoracic vertebra (either cave bear or brown bear) was found with a small triangular stone fragment embedded in it—probably broken off a much larger spear point.[68] Some researchers have proposed that cave bears would have been subject to competition by Neanderthals and especially modern humans for caves (a limited resource), and that this would have been an important contributor to their ultimate extinction. So, can this help explain why brown bears survived whereas cave bears went extinct? The argument goes that cave bears would have been especially vulnerable when hibernating, whereas brown bears would have been less so, as they were not reliant on caves. However, it seems to me that there are problems with this hypothesis. Firstly, it is highly unlikely that cave bears hibernated exclusively in caves. In dispersing to or from the Urals they would have encountered extensive regions with few or no caves, so that the only option would have been to dig a den in the earth under a rock or tree roots, as does the living brown bear. Moreover, the threat from humans cannot have been critical; we know that cave bears outlived Neanderthals by more than 13,000 years and disappeared at least 18,000 years after the arrival of modern humans in Europe.

5.7 CAVE LION: *PANTHERA SPELAEA*

Of the many large beasts that our ancestors in Europe and Siberia had to contend with during the Last Glacial, the extinct cave lion (*Panthera spelaea*) must have been one of the most feared and dangerous. Larger and more powerfully built than the modern African lion, during the Last Glacial its vast geographic range extended from Spain and Britain in the west across Europe to the far northeast of Siberia (Chukotka) and also crossed the Bering land bridge into northwest North America (Alaska and the Yukon).[69] The name "cave lion" is largely misleading, as this species only used caves occasionally and opportunistically.

Recent research based mainly on aDNA analyses has demonstrated that the extinct cave lion was a distinct species from the modern lion. Matted yellowish fur found in association with a cave lion skeleton in permafrost on the

Malyi Anyui River (Chukotka, northeast Siberia) showed that cave lions differed from modern lions by having had stronger and more robust hair and a very thick, dense, downy undercoat, as might have been expected in an animal living in cold climates.[70] The larger size of the cave lion suggests that it could have tackled larger prey more frequently than does modern lion. In northern Eurasia, the principal prey species would probably have included horse, reindeer, giant deer, red deer, musk ox, steppe bison, and perhaps occasionally young woolly rhino and young mammoth. (Dramatic evidence that cave lions had preyed on steppe bison in Alaska is discussed in chapter 6.) Competitors of the cave lion are likely to have included spotted hyena (*Crocuta crocuta*), wolf (*Canis lupus*) and brown bear (*Ursus arctos*). Where the ranges of cave lion and hyena overlapped, no doubt they would have disputed kills in much the same way as occurs in Africa today.

There are a number of depictions of cave lion in Paleolithic art.[71] Some can be attributed to the Aurignacian (ca. 43 to 26 kya), such as the beautifully carved head from Vogelherd Cave and the enigmatic lion-headed human figure carved out of mammoth tusk from Hohlenstein Stadel, both in southern Germany. The superb multiple images of lion from Chauvet Cave have also been attributed to the Aurignacian. Others, attributed to the Magdalenian (ca. 17 to 12 kya), include the engraved sketches from Gönnersdorf, Germany, as well as cave paintings from Lascaux, Les Trois Frères, and Les Combarelles; engraved stone slabs from La Marche; and the small portable sculptures from La Vache Cave—all in France. Interestingly, all the representations indicate that male cave lions as well as female lacked manes, in marked contrast to the large and impressive manes seen in modern male African and Asiatic lions. This difference is thought to reflect smaller pride size in the Ice Age populations.

A compilation of 111 radiocarbon dates indicates that cave lion went extinct across northern Eurasia about 14 to 14.5 kya (and in Alaska/Yukon about a thousand years later).[72] The timing of its extinction suggests that it coincided with the climatic warming that began around 14.7 kya accompanied by a spread of shrubs and trees and reduction in open habitats.

5.8 SPOTTED HYENA: *CROCUTA CROCUTA*

During the Ice Age the spotted hyena (*Crocuta crocuta*), now confined to Africa south of the Sahara, was also present over a large area of northern Eurasia, extending from Iberia and Ireland to the Russian Far East and China, but avoiding the colder climates of Scandinavia and northern Siberia.[73] These Ice Age hyaenas were larger than those in Africa today and presumably could have

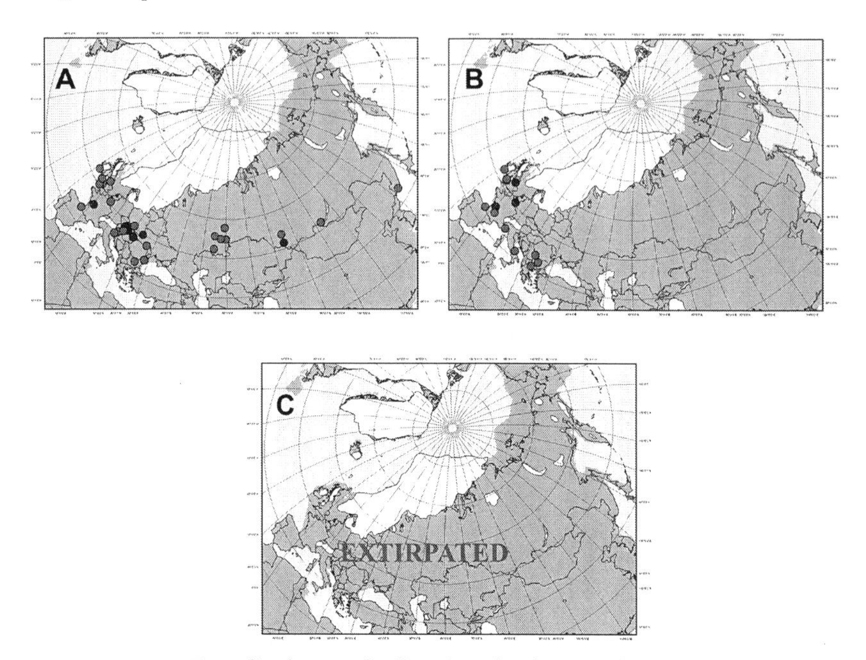

FIGURE 5.13. Time-sliced maps of radiocarbon-dated records for spotted hyena *Crocuta crocuta* showing distribution changes prior to extirpation in northern Eurasia. A: > 40 kya. B: 40 to 29 kya (youngest available date, ca 31 kya). C: 28.9 to 27.5 kya. Gray: ultrafiltered dates. Black: non-ultrafiltered dates. (Modified from Stuart and Lister 2014.)

taken larger prey. The living animals are active hunters of zebra, antelope, and other animals, as well as scavengers of almost anything of animal origin. The powerful jaws and massive teeth have evolved for specialized bone crushing, thus allowing the maximum nutrition to be extracted from a carcass. Eurasian spotted hyena probably hunted mainly horse, reindeer, and red deer, and occasionally giant deer, bison aurochs, and baby mammoth.

There are only about one hundred available radiocarbon dates for spotted hyaena, which is far fewer than for most other megafaunal species from the region.[74] These suggest that hyena was extirpated around 40 kya from Central Europe and Russia, but considerably later (ca. 31 kya) from northwest and southern Europe, so that it was probably restricted to the Mediterranean and the Atlantic seaboard after 40 kya (fig. 5.13). The fact that it persisted longest in areas with less extreme climate strongly suggests that cold intolerance was a major factor in its disappearance when faced with cooling climate—probably combined with reduced prey abundance, in turn caused by reduced vegetational productivity.

5.9 NEANDERTHAL: *HOMO NEANDERTHALENSIS*

The timing of and causes of *Homo neanderthalensis* extinction have been much debated. The latest and most comprehensive study, using improved AMS dating (see the appendix) of key sites covering the Middle to Upper Paleolithic transition, indicates that Neanderthals disappeared from different areas at different times but had gone from all of Europe by ca. 41 to 39.3 kya.[75] The related Denisovans are known only from very sparse material found in a single cave in southern Siberia, and their extinction chronology is unknown.[76] The earliest anatomically modern humans in Europe may be represented by the Uluzzian technocomplex in Italy and the Balkans, dating between 45 and 43 kya, indicating that there was an overlap of several millennia between Neanderthals and modern humans, which would have allowed ample time for cultural exchanges between a mosaic of populations. Recent studies, which demonstrate that a small amount of Neanderthal DNA is present in the genome of modern non-African humans, indicate that some interbreeding occurred between the two species,[77] probably between 60 and 50 kya in Asia. These recent findings suggest that the final disappearance of Neanderthals might relate to the onset of Greenland Stadial 9 (GS 9, ca. 40 kya) but clearly occurred well before the extinction of *Crocuta crocuta* and *Ursus spelaeus* at around the onset of Greenland Stadial 3 (GS 3, ca. 28 kya). It seems likely that Neanderthals gradually succumbed to competition from modern humans, possibly in combination with climatic deterioration.

5.10 HIPPOPOTAMUS: *HIPPOPOTAMUS AMPHIBIUS*

Hippopotamus (*Hippopotamus amphibius*)—now confined to sub-Saharan Africa (and North Africa in historic times)—occurred widely in southern Britain during the Last Interglacial period (ca. 130 to 117 kya), reaching as far north as County Durham in northern England.[78] At this time Britain enjoyed mild winters and average summer temperatures 2°–3°C warmer than now. These British hippos were enormous; with an approximate average weight of 2,600 kilograms and a maximum of 3,200 kilograms,[79] they would have been about 50% heavier than hippos living in Africa today. An impressive composite skeleton recovered from river gravel at the village of Barrington is displayed in the Sedgwick Museum in the University City of Cambridge (fig. 5.14).

During the Last Interglacial, *H. amphibius* ranged from Africa to Mediterranean Europe and Iberia northward to Britain. Intolerant of cold, it probably disappeared from Europe at the beginning of the Last Glacial, although at present we have no direct dating evidence.

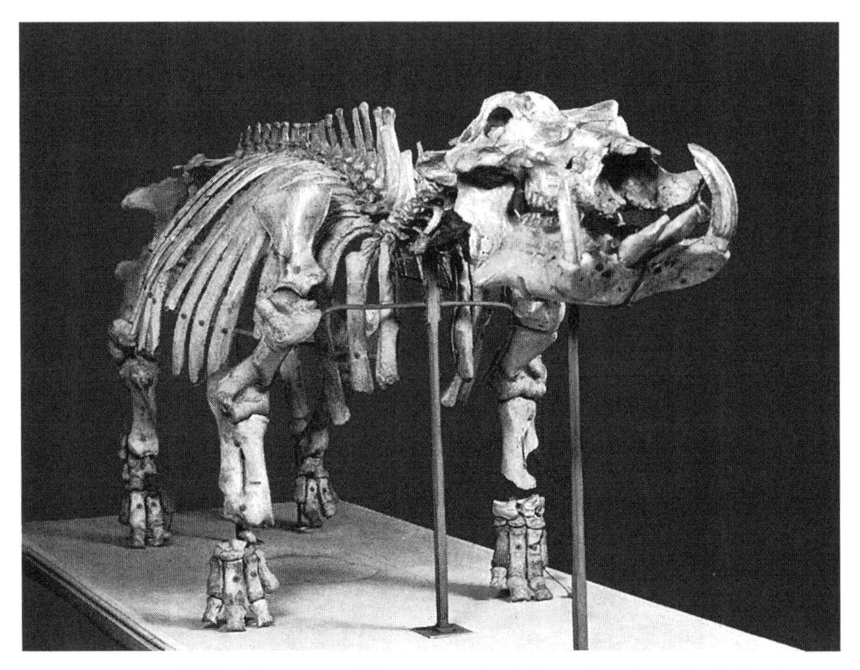

FIGURE 5.14. *Hippopotamus amphibius* composite skeleton, Last Interglacial, Barrington, Cambridge, UK. (Photo by the author, courtesy of Sedgwick Museum, UK.)

5.11 FOREST ELEPHANT, NARROW-NOSED RHINO, AND MERCK'S RHINO

Extinct species widespread in Europe during the Last Interglacial, in association with regional temperate and Mediterranean forests, included the huge straight-tusked or forest elephant (*Palaeoloxodon antiquus*)—up to 3.8 meters high and with a maximum estimated weight of 11.3 tonnes.[80] *P. antiquus* was accompanied by the narrow-nosed rhino (*Stephanorhinus hemitoechus*) and the related Merck's rhino (*Stephanorhinus kirchbergensis*), with estimated weights of 1,121 to 2,384 kilograms and 1,381 to 2,358 kilograms, respectively.[81] All three species seem to have retreated south of the Pyrenees and Alps in the early part of the Last Glacial, in response to cooler temperatures and spread of open vegetation, and were probably extinct before 50 kya.[82] Because this first wave of extinctions probably occurred mostly or entirely beyond the range of the radiocarbon method, the dating is very uncertain and unsatisfactory. Further research on this important issue is badly needed. However, two radiocarbon dates in the range of ca. 40 to 37 kya were obtained on *P. antiquus* molars from the Netherlands and North Sea,[83] suggesting much later survival of this species in northwest Europe and that it had adapted to the much colder climates and open vegetation of the Last Glacial. Ideally, further work would include independent dating of the same specimens by another laboratory. The possibility that Neanderthals (*H. neanderthalensis*) were involved in the disappearance of *P. antiquus*, *S. kirchbergensis*, and *S. hemitoechus* before ca. 50 kya shouldn't be ruled out, although it seems unlikely that they could have habitually hunted such huge formidable beasts.

5.12 NAUMANN'S ELEPHANT AND JAPANESE GIANT DEER

The youngest of forty-one accepted direct radiocarbon dates on molars and tusks of Naumann's elephant (*Palaeoloxodon naumanni*) from Japan is ca. 28.34 kya,[84] so it may have disappeared at around the beginning of the Last Glacial Maximum like cave bears and spotted hyena. Of the few available dates on the giant deer (*Sinomegaceros yabei*) from China and Japan, the youngest is about 44.6 kya (Japan).

5.13 STEPPE BISON: *BISON PRISCUS*

The extinct steppe bison (*Bison priscus*), another component of the mammoth steppe fauna, was widespread in northern Eurasia and North America in the Last Glacial. Its weight has been estimated at 363 to 1,930 kilograms.[85] Although similar in size to modern North American bison, *B. priscus* had larger horns and was distinct from both the living American and European species. There are rather few available dates, but it evidently survived into the early Holocene, ca. 9.8 kya in Taimyr (north-central Siberia) and ca. 8.9 kya in Chukotka (northeastern Siberia) (table 5.1). Research by Beth Shapiro and colleagues on the ancient DNA of steppe bison in Beringia (that is, northeastern Siberia plus Alaska/Yukon) revealed a dramatic decrease in genetic diversity that began around 37 kya.[86] There are many depictions of bison in Paleolithic art—for example, in Lascaux and Chauvet, and the outstandingly beautiful paintings from Altamira, Spain. A recent study based on aDNA and interpretation of cave paintings concluded that the extant European bison (*Bison bonasus*), which has only a meager fossil record, originated sometime before 120 kya as a hybrid between steppe bison and aurochs (*Bos primigenius*).[87] However, this interpretation has been contested by Grange and colleagues.[88]

5.14 SCIMITAR CAT: *HOMOTHERIUM LATIDENS*

The only putative Last Glacial record of scimitar cat (*Homotherium latidens*) from northern Eurasia is based on a single mandible trawled from the bed of the North Sea.[89] Six calibrated radiocarbon dates (Utrecht AMS Laboratory) gave a range from ca. 35.23 to 30.89 kya. This record is potentially very important as otherwise the youngest known record from the whole of Europe and Asia, about 300,000 years old, is from Schöningen, Germany (*Homotherium* survived much later in North America; see chapter 6). However, the lack of other Last Glacial records of this animal from Europe and Asia is difficult to explain if the North Sea find is correctly dated. We need more evidence in order to corroborate its presence in Europe during the Last Glacial, either by independent dating of the same specimen by another laboratory or by the discovery and dating of further material.

5.15 MUSK OX: *OVIBOS MOSCHATUS*

The last group of species to be considered disappeared either globally or regionally within the last few thousand or few hundred years.

During the Last Glacial, musk ox (*Ovibos moschatus*) was widely distributed across northern Eurasia (although its remains are never common), ranging from Iberia, Ireland, and Britain in the west, eastward across northern Asia into Alaska/Yukon, and as far south as the Great Lakes region.[90] Sometime after 2.8 kya (according to the youngest available radiocarbon dates), it was extirpated, perhaps by humans, from its last Eurasian foothold—the tundra of the Taimyr Peninsula, northern Siberia. Musk ox is now restricted to the tundra of Arctic North America and Greenland (ignoring recent human reintroductions to Siberia and Norway).

5.16 "EUROPEAN ASS": *EQUUS HYDRUNTINUS*

On the basis of archaeological associations, not direct radiocarbon dates, Jennifer Crees and Sam Turvey demonstrated that the Last Glacial range of *Equus hydruntinus* extended from southern Europe to southwest Asia.[91] In the Holocene, its area of distribution progressively shrank, with the latest known records approximately 3.5 to 3 kya from the Caucasus and Iran. They suggest that its extinction was due primarily to climate-driven reduction of open habitats and vulnerability of fragmented populations to human exploitation.

5.17 AUROCHS: *BOS PRIMIGENIUS*

During the Last Glacial, aurochs, or urus (*Bos primigenius*)—the wild ancestor of domestic cattle—ranged across much of Eurasia (except the north) and North Africa. It was a large animal, with bulls probably weighing as much as 2 tonnes (a range of 389 to 2,010 kilograms) and reaching 1.8 meters at the shoulder.[92] The large horns, twisted so that the tips point forward, could each reach 0.8 meters in length. Again, there are many depictions in the painted caves of Lascaux, Chauvet, and elsewhere. Aurochs were probably first domesticated in Turkey, around 10.5 kya. A study by Jennifer Crees and colleagues, based on numbers of archaeological sites with remains of aurochs, shows a progressive shrinkage of its European range from the Mesolithic (ca. 11.7 kya) to the late Medieval period.[93] Deforestation combined with hunting is thought to have progressively reduced their numbers. The last recorded individual died in 1627 in the Jaktorów Forest, Poland.[94]

5.18 OTHER LOSSES

Available information is inadequate to estimate the date of extinction of a large camel (*Camelus knoblochi*) and the poorly known spiral-horned antelope (*Spiroceros kiakhtensis*).[95]

5.19 PATTERNS OF EXTINCTION

As stated at the beginning of this chapter, since about 120,000 years ago some 41% of northern Eurasian megafauna went extinct. Significantly, these losses did not happen within a short time period, but on the contrary were conspicuously staggered over tens of thousands of years (fig 5.15). Moreover, for each species there were considerable differences in the timing of its disappearance from different areas of its geographical range so that each exhibited a unique and complex pattern of distributional shifts, culminating in extinction of some species and survival of others. As described above, there is ample evidence that climatic and vegetational changes were a major driver of range and population shifts of megafauna in northern Eurasia and are therefore strongly implicated in extinctions. Nevertheless, the possible role of humans has still to be fully explored. The extinction of one species of cave bear (*Ursus spelaeus*) at ca. 28.5 kya and another (*Ursus ingressus*) at ca. 25.3 kya, the demise of the extraordinary giant one-horned rhino (*Elasmotherium sibiricum*) ca. 38.48 to 35.06 kya, and

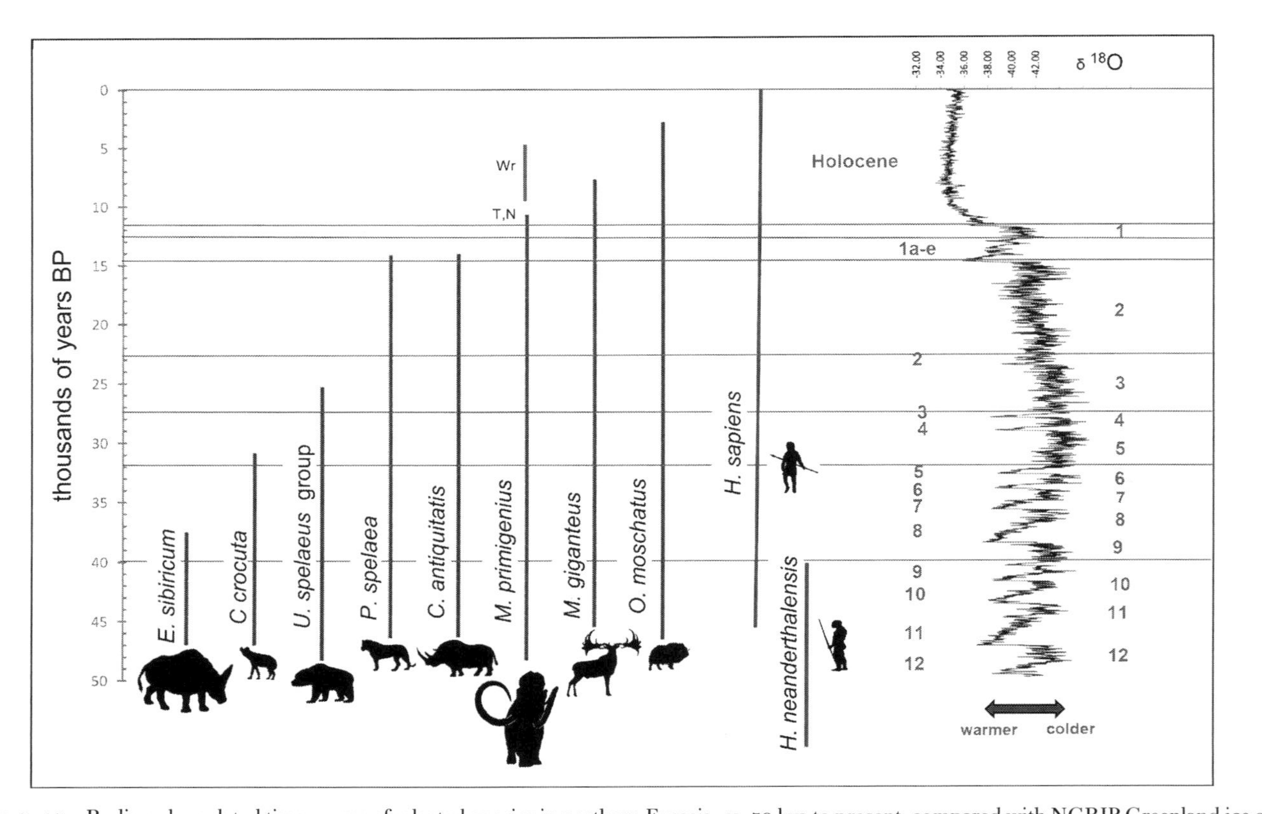

FIGURE 5.15. Radiocarbon-dated time ranges of selected species in northern Eurasia, ca. 50 kya to present, compared with NGRIP Greenland ice curve. Interstadials are numbered to the left, stadials to the right. T: Taimyr Peninsula. N: New Siberian Islands (then connected to the mainland). Wr: Wrangel Island. Time ranges of Neanderthals and modern humans are also shown on the right.

the extirpation from northern Eurasia of spotted hyena (*Crocuta crocuta*) ca. 31 kya, all occurred prior to the onset of the Last Glacial Maximum (Greenland Stadial GS-2), ca. 23 kya.

In marked contrast, both the cave lion and woolly rhino survived thousands of years longer, disappearing in the warmer and wetter Late Glacial, ca. 14 kya. Woolly mammoth survived substantially longer, only going extinct around 10.7 kya in the New Siberian Islands (then part of the mainland), and very much later (ca. 4 kya) on Wrangel Island off northeast Siberia. Remarkably, the giant deer also survived into the Holocene, to at least 7.7 kya in western Siberia and European Russia.

There is no evidence to indicate that people hunted woolly rhino, *Elasmotherium*, cave lion, or spotted hyena, and not very much to suggest that they hunted giant deer or cave bears either. Remains of the above-mentioned animals are generally scarce or absent on archaeological sites, whereas horse, reindeer, red deer, steppe bison, aurochs, and ibex are abundant, implying that they were favored prey items. However, paradoxically all of these (except steppe bison and aurochs) survive to the present day. Direct evidence for hunting in the form of spear- or arrowheads embedded in bone is very rare both for extinct and for surviving species. Again, as described above, only three such examples of woolly mammoth are known, and one possible cave bear.

An outstanding example a surviving species that was unequivocally hunted was provided by the 1970 find of the skeleton of a male Eurasian elk or moose (*Alces alces*) from Lancashire, England, radiocarbon-dated to 13.53 kya (Allerød Interstadial).[96] Two bone or antler spear points with neatly carved barbs were also associated with the skeleton.[97] One, excavated *in situ*, had been embedded in the hind foot, where it had remained for some weeks before the death of the animal, having eroded a groove in the lower end of the metatarsal bone (fig. 5.16). The bones also bore numerous injuries attributable to attacks by stone-tipped weapons soon before death. However, in spite of suffering this trauma, the animal evidently escaped the hunters, as its skeleton was recovered intact from undisturbed lake sediments, suggesting that it probably drowned after breaking through thin ice.

A clear pattern seen in most of the studied species (woolly mammoth, woolly rhino, giant deer, cave lion, spotted hyena, and cave bear) is the shrinkage and fragmentation of range prior to extinction, largely driven by climatic and vegetational changes and following a marked reduction in genetic diversity. However, such a process has not always resulted in extinction (well, not yet), as featured below.

FIGURE 5.16. Distal hind leg bones of European elk (moose) *Alces alces in situ* in lake sediments, with barbed projectile point resting in groove worn into the distal left metatarsal, Poulton-le Fylde, Lancashire, UK, 1970. The point had been snapped in half by the pressure of the overlying sediments. (Photo by John Hallam.)

5.20 SURVIVING MEGAFAUNA

Many megafaunal species in this region have managed to survive to the present day, although the future prospects of several are precarious. The survivors include the polar bear (*Ursus maritimus*), brown bear (*Ursus arctos*), tiger (*Panthera tigris*), leopard (*Panthera pardus*), Przewalski's or wild horse (*Equus ferus*), onager or wild ass (*Equus hemionus*), wild boar (*Sus scrofa*), Eurasian elk or moose (*Alces alces*), red deer (*Cervus elaphus*—western Eurasia), wapiti (*Cervus canadensis*—northeastern Eurasia and North America), sika deer (*Cervus nippon*), fallow deer (*Dama dama*), reindeer or caribou (*Rangifer tarandus*), saiga antelope (*Saiga tatarica*), argali or mountain sheep (*Ovis ammon*), Spanish ibex (*Capra pyrenaica*), European bison (*Bison bonasus*), Bactrian camel (*Camelus ferus*), and yak (*Bos mutus*). The 2017 Red List of Threatened Species of the International Union for Conservation of Nature (IUCN) classifies tiger and Przewalski's horse as endangered, and the onager, Bactrian camel, and saiga as critically endangered.

In the Last Glacial, saiga antelope (*Saiga tatarica*) occurred across much of Europe, northern Asia, and northwest North America.[98] However, it

TABLE 5.1 Youngest radiocarbon dates for extinct northern Eurasian megafauna

Species	Lab. no.	Cal BP	Median	Site	Source
Panthera spelaea	OxA-17268	14,075–14,915	14,413	Zigeunerfels Cave, Germany	5
Panthera spelaea	AA-41882	13,873–14,831	14,124	Le Closeau, France	5
Crocuta crocuta	OxA-11691	39,400–41,374	40,550	Duruitoarea Veche, Moldova	4
Crocuta crocuta	OxA-10523	30,328–31,221	30,813	Grotta Paglicci, Italy	4
Ursus spelaeus	OxA-12013	29,753–30,823	30,413	Sirgenstein Cave, Germany	2
Ursus spelaeus	GrA-52632	28,336–29,222	28,700	Rochedane, France	9
Ursus spelaeus	Beta-156100	28,055–29,054	28,540	Vindija Cave, Croatia	2
Ursus ingressus	GdA-3894	26,001–26,227	26,101	Stajnia Cave, Poland	16
Ursus ingressus	Poz-61719	24,807–25,648	25,267	Stajnia Cave, Poland	16
Bison priscus	Beta-148623	9684–10,148	**9845**	Taimyr Peninsula, North Siberia	10
Bison priscus	SPb-743	8643–9091	**8886**	Bilibino Region, Chukotka	17
Megaloceros giganteus	KIA-5669/OxA-13015	7615–7728	**7674**	Kamyshlov Mire, West Siberia	1
Megaloceros giganteus	OxA-19488/OxA-19487	7611–7697	**7662**	Maloarchangelsk, West Russia	12
Ovibos moschatus	OxA-17062	3067–3381	**3226**	Taimyr Peninsula, North Siberia	15
Coelodonta antiquitatis	OxA-10200/10201	16,658–17,016	16,835	Gönnersdorf, Germany	3
Coelodonta antiquitatis	AAR-11042/OxA-18604	14,589–15,192	14,933	Kama River, Urals	3
Coelodonta antiquitatis	OxA-20097	13,947–14,839	14,164	Ust'-Omolon, Northeast Siberia	3
Coelodonta antiquitatis	AAR-11027/OxA-18602	13,851–14,156	13,999	Lena-Amga, Northeast Siberia	3
Elasmotherium sibiricum	OxA-X-2677-52	36,270–38,690	37,501	Sarepta, Russia	11
E. sibiricum (phase analysis)	n/a	38,480–34,950	n/a	n/a	11
Palaeoloxodon naumanni	IAAA-53431	27,963–28,681	28,337	Aomori, Honshu, Japan	8
Mammuthus primigenius	GIN-11874	10,581–11,065	**10,713**	New Siberian Islands, Northeast Siberia	7
Mammuthus primigenius	GIN-1828	10,784–11,217	**11,072**	Taimyr Peninsula, North Siberia	6
Mammuthus primigenius	OxA-11181	4444–4813	**4602**	Wrangel Island, Northeast Siberia	13
Mammuthus primigenius	Ua-13366	3854–4225	**4024**	Wrangel Island, Northeast Siberia	14

Cal BP: calibrated date before present. Boldface dates: Holocene.

Sources: 1. Stuart et al. 2004. 2. Pacher and Stuart 2009. 3. Stuart and Lister 2012. 4. Stuart and Lister 2014. 5. Stuart and Lister 2011. 6. Sulerzhitsky and Romanenko 1997. 7. Nikolskiy, Sulerzhitsky, and V. V. Pitulko. 8. Iwase et al. 2012. 9. Bocherens et al. 2013. 10. MacPhee, Tikhonov, et al. 2002. 11. Kosintsev et al. 2018. 12. Lister and Stuart 2019. 13. Stuart and Lister 2012. 14. Vartanyan, Arslanov, et al. 2008. 15. Campos, Kristensen, et al. 2010. 16. Baca et al. 2016. 17. Kirillova et al. 2015.

subsequently experienced a decline in genetic diversity and has been extirpated from most of its former range, probably in the first instance largely due to the spread of forests at the expense of open vegetation. It survives today in the steppes and semideserts of Kazakhstan, southwestern Russia, and western Mongolia. However, the animals are hunted illegally both for their horns (used in traditional Chinese medicine) and for their meat, and alarmingly during the twentieth century the range was reduced to four disconnected populations. Moreover, saiga also experiences mysterious periodic mass die-offs, the most recent in 2015.

CHAPTER 6

North America: Mastodon, Ground Sloths, and Sabertooth Cats

Among the many noteworthy accomplishments of Thomas Jefferson (1743–1826), author of the Declaration of Independence and third president of the United States (1801–09), was his pioneering work on the fossil mammals of North America—the bones of which once filled a room in the White House. In 1796 he was sent remains of an extraordinary unknown animal with exceptionally large claws that had been unearthed from a cave in what is now West Virginia.[1] At first, he thought that the bones were of some giant carnivore, much larger than a modern lion, so he called it Megalonyx (Greek: "big claw"). However, it turned out to be a very different beast: a ground sloth, whose closest relatives are the living tree sloths of Central and South America. Subsequently, the animal was formally named *Megalonyx jeffersoni* in his honor. As Jefferson did not believe that any creature created by God could have become extinct, when in 1804 he commissioned Lewis and Clark to undertake their epic expedition across the continent to the Pacific Coast, he confidently expected that they would find living *Megalonyx*, together with mastodons and mammoths, in the unexplored American West. Of course, in this he was disappointed; we now know that these creatures had vanished many thousands of years earlier. A partial skeleton of *Megalonyx jeffersoni* found near Millersburg, Ohio, in 1890 has been dated recently by radiocarbon to about 13.1 kya,[2] the youngest known so far.

The North American continent has very many sites with spectacular megafaunal fossils dating from the Last Glacial (fig. 6.1).

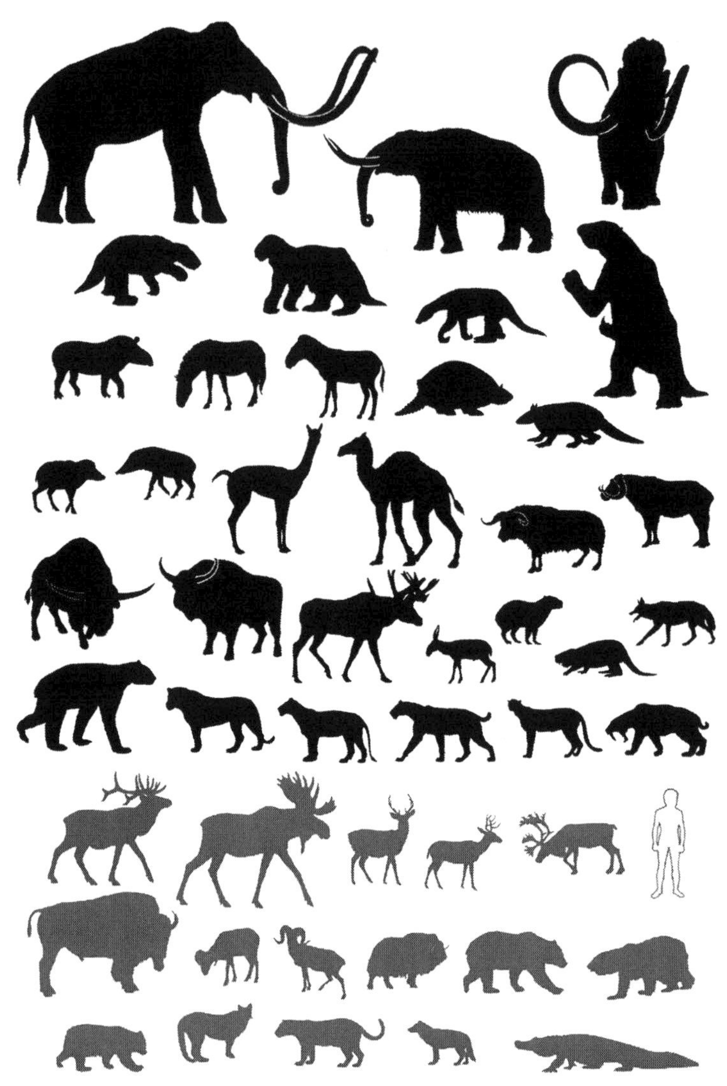

FIGURE 6.1. Extinct and extant megafauna from North America (Nearctic Ecoregion). From the top, row 1, left to right: *Mammuthus columbi*; *Mammut americanum*; *Mammuthus primigenius*. Row 2: *Glossotherium* (*Paramylodon*) *harlani*; *Megalonyx jeffersoni*; *Nothrotheriops shastensis*; *Eremotherium laurillardi* (*E. rusconii*). Row 3: *Tapirus veroensis*; *Equus occidentalis*/*Equus scotti*; *Haringtonhippus francisci* (stilt-legged horse); *Glyptotherium floridanum*; *Holmesina septentrionalis*. Row 4: *Platygonus compressus*; *Mylohyus nasutus*; *Hemiauchenia macrocephala*; *Camelops hesternus*; *Euceratherium collinum*; *Bootherium bombifrons*. Row 5: *Bison latifrons*; *Bison priscus*; *Cervalces scotti*; *Saiga tatarica*; *Neochoerus pinckneyi*; *Castoroides ohioensis*; *Canis dirus*. Row 6: *Arctodus simus*; *Panthera spelaea*; *Panthera atrox*; *Homotherium serum*; *Miracinonyx trumani*; *Smilodon fatalis*. Row 7: *Cervus canadensis*; *Alces alces*; *Odocoileus hemionus*; *Odocoileus virginianus*; *Rangifer tarandus*. Row 8: *Bison bison*; *Ovis canadensis*; *Ovis dalli*; *Ovibos moschatus*; *Ursus arctos*; *Ursus maritimus*. Row 9: *Ursus americanus*; *Puma concolor*; *Panthera onca*; *Canis lupus*; *Alligator mississippiensis*.
Black: selected extinct species. Gray: selected living species. The outlined *Homo sapiens* gives approximate scale.

6.1 RANCHO LA BREA

In 1792, when California was part of the Spanish Empire and the newly independent United States comprised just the thirteen colonies along the Atlantic seaboard, the naturalist and explorer José Longinos Martinez made some interesting observations on the area that was destined to become the city of Los Angeles:[3]

> Near the Pueblo de Los Angeles there are more than twenty springs of liquid petroleum, pitch, etc. Further to the west of the said town, in the middle of a great plain of more than fifteen leagues in circumference, there is a great lake of pitch, with many pools in which bubbles or blisters are continually forming and exploding. In hot weather animals have been seen to sink in it and when they tried to escape, they were unable to do so, because their feet were stuck, and the lake swallowed them. After many years their bones have come up through holes as if petrified.

It was many years before anyone realized that these "petrified" bones were actually fossils, not the remains of recently deceased animals. The first to do so was William Denton, an Englishman who taught at Wellesley College, Massachusetts. When he visited the La Brea area in 1875, Major Henry Hancock, who had purchased the land (now called Hancock Park), presented him with a large fossil tooth, which he correctly identified as the canine of a sabertooth cat. In his 1875 description of the site, Denton also reported other finds, including horse, deer, and "a large bovine animal" (presumably bison). However, at the time, these discoveries made little impression on the scientific community; major efforts to excavate the site did not take place until the early 1900s. Then, in 1908, Dr John C. Merriam, a vertebrate paleontologist at the University of California, published the first major account of the La Brea fossils, notably including the iconic sabertooth cat (*Smilodon fatalis*).

In 1913, in their laudable aim to protect the site for scientific study, the Hancock family awarded Los Angeles County Museum sole rights to excavate for two years and extensive excavations by the museum resulted in the recovery of many hundreds of thousands of fossils. The work was hazardous, as the pit sides were poorly shored up and there were frequent collapses, especially when it rained. Moreover, bones were cleaned on site by immersion and scrubbing in hot kerosene (heated by asphalt-fueled stoves), and inevitably the kerosene sometimes caught fire. The La Brea fossils were originally housed in the Los Angeles County Museum (now the Natural History Museum of Los

Angeles County) but were subsequently transferred to the purpose-built Page Museum, which opened in 1977.

Rancho La Brea, which translates as "tar (or pitch) ranch," is the type locality of the North American Rancholabrean Land Mammal Age, which began about 45 to 21 kya with the arrival of bison in North America.[4] Today it is the richest and best-known Quaternary fossil locality in North America, and indeed one of the richest and most famous in the world.[5] This fame owes much to the dedicated pioneer research and publications on the mammals by John Merriam (1869–1945) and Chester Stock (1892–1950), and on the birds by Loye Miller (1874–1970) and Hildegarde Howard (1901–1998).[6] The appeal of the La Brea Tar Pits and its spectacular extinct animals is such that, in addition to many scientific papers and museum booklets, it has featured in many popular publications. As a child I remember reading sensational accounts of doomed animals trapped in deep pools of tar, accompanied by appropriately lurid illustrations. These undoubtedly helped feed my early interest in paleontology.

In Hancock Park, crude oil seeps up along a geological fault from Miocene rocks below. On reaching the surface, the lighter petroleum fractions evaporate and various chemical changes take place resulting in semi-viscous asphalt, which eventually hardens due to oxidation. This process has continued over tens of thousands of years, with many separate seeps of extremely sticky asphalt trapping unwary animals—especially if the asphalt is effectively camouflaged by leaves, dust, or water. However, the traps did not operate continuously, as individual seeps that had been active for millennia ceased activity for long periods and subsequently resumed. It is probable that earthquakes have been responsible for "switching" the seeps on and off by opening or blocking the subterranean flow. As is well known, southern California is very much prone to earthquakes mainly due to periodic movement along the San Andreas Fault.

The wealth of fossil remains already excavated from La Brea is astonishing, and no doubt much more remains underground. Huge collections amounting to hundreds of thousands of fossils are stored in the Page Museum adjacent to the fossil localities in Hancock Park. In most "normal" Ice Age fossil assemblages, the remains of herbivores greatly outnumber carnivores, as would be expected from the ecological relationships of living animals. However, the La Brea assemblage is strikingly different. Chester Stock estimated that 90% of the mammal remains and 60% of the birds in the Hancock Collection (excavated 1913–1915) were of carnivores or scavengers. In 1960 a census by the paleontologist Leslie Marcus of the finds from fourteen productive pits showed that the proportions of species represented varied from one pit to another and that the preponderance of carnivores was less than noted by Stock, but still high.[7] For an explanation of this situation, we need to look to the highly unusual circumstances that prevailed at La Brea. It appears that several predators were attracted to feed on each dead or dying animal that had already become trapped in the asphalt and many of these became trapped in turn.[8]

The mammal fauna from the La Brea deposits includes such extinct megafaunal species as sabertooth cat (*Smilodon fatalis*—the California state fossil), scimitar cat (*Homotherium serum*—rare), American lion (*Panthera atrox*), dire wolf (*Canis dirus*), giant short-faced bear (*Arctodus simus*), Harlan's ground sloth (*Paramylodon harlani*), Jefferson's ground sloth (*Megalonyx jeffersoni*—rare), Shasta ground sloth (*Nothrotheriops shastensis*), Columbian mammoth (*Mammuthus columbi*), American mastodon (*Mammut americanum*), horses (*Equus occidentalis*, *Equus conversidens*), a tapir (*Tapirus californicus*—very rare), two species of bison (*Bison latifrons*, *Bison antiquus*), shrub ox (*Euceratherium* sp.), yesterday's camel (*Camelops hesternus*), large-headed llama (*Hemiauchenia macrocephala*), and flat-headed peccary (*Platygonus compressus*).[9] The larger extant mammals include jaguar, mountain lion, coyote, timber/gray wolf, black bear, pronghorn, and mule deer. In addition, numerous small mammals are recorded, together with many birds, lizards, snakes, turtles and amphibians. Dire wolves—more heavily built and with more robust teeth than the related living gray wolf—are the most abundant large mammal, followed by sabertooth cat. The Hancock Collection is estimated to contain remains of over four thousand dire wolves, about two thousand sabertooths, and about a thousand coyotes. Interestingly, no human remains or artifacts have been discovered in the Ice Age deposits at the site. The partial human skeleton known as "La Brea woman" has been radiocarbon dated to ca. 10 kya—much later than the other remains.

In the asphalt deposits, separate bones are jumbled together in a random manner; associated skeletons occur very rarely. The bones are usually packed tightly together in inverted cones, pipes, or fissures, the origins of which are not fully understood. Bones become progressively more tightly packed with depth, and clearly there has been extensive movement within the bone-rich asphalt, as demonstrated where bones of a single individual occur at different depths. Curious deeply incised grooves on some bones ("pit wear"), unique to La Brea, evidently result from the rubbing together of densely packed bones, possibly caused by trampling, subsidence, or earthquakes.

The La Brea Tar Pits continue to claim victims today, so the old excavations are fenced off to protect both people and animals. On a short visit to one of the seeps, I saw various small casualties, including a sparrow and a dragonfly trapped in asphalt that bubbled slowly with escaping methane gas. The asphalt becomes extraordinarily sticky in the warmth of the summer months, and even large animals can become trapped and immobilized in asphalt as little as 4 to 5 centimeters deep, demonstrated by unfortunate cattle that from time to time have perished in some of the hundreds of other active Californian asphalt seeps. However, since the asphalt becomes solid when cooled to below 18°C, animals would have fallen victim only during the warmer phases of the Last Glacial period. Moreover, the traps generally would have been less active at night and in winter, enabling predators and scavengers to reach mired animals with less risk of falling victim themselves.

Due to the fragility of their bones, birds are generally scarce in Ice Age fossil assemblages. However, La Brea is an outstanding exception to the rule, having yielded some 137 species, of which about 17 are extinct. The bird fauna comprises a wide range of species, including ducks and geese, storks, eagles, vultures, and the extraordinary extinct teratorns. With a wingspan of over 3.2 meters (similar to the living Andean condor) and estimated weight of 14 kilograms, the extinct teratorn (*Teratornis merriami*) is the largest bird known from La Brea, and was one of the largest birds that have ever flown.[10] It has no close living relatives. Previously believed to have scavenged carcasses like a condor, the structure of its skull and beak instead suggest that it actively stalked small mammals, birds, frogs and lizards, and swallowed them whole.

Although the total numbers of fossils recovered are impressive, there is no need to invoke mass deaths to explain them. Since each seep operated intermittently over a period of tens of millennia, it needed only one herbivore and several accompanying carnivores to fall victim every few years to account for the vast numbers of fossils in the deposit.

The climate was generally cooler and more humid than today—as shown, for example, by the occurrence at the site of land snails now found only above 1,500 meters in the mountains of California and Arizona, as well as by the plant fossils. Plant remains recovered from the asphalt represent the major habitats of the Los Angeles Basin during the Last Glacial: sagebrush scrub dotted with groves of oak and juniper; riparian woodland along the major stream courses; and chaparral vegetation on the surrounding hills. The predominantly grazing horses, mammoth, Harlan's ground sloth, and camel probably frequented the open sagebrush scrub habitat, whereas the predominantly browsing mastodon, tapir, Jefferson's ground sloth, and Shasta ground sloth are likely to have been more at home in the riparian woodland. Plant remains extracted from the teeth of *Equus occidentalis* show that, although largely grazers, these horses also ate leaves and shrubs.

The abundance of remains at La Brea opens up some interesting lines of research. For example, a study of the lower jaws of juvenile extinct bison (*Bison antiquus*) reveals discrete age classes—one year apart, showing that the migrating bison herds were present in the area only in late spring. On the other hand, horses were present all year. A recent analysis of stable isotopes reveals important information on the diet and local environment of several La Brea mammals.[11] For example, the ratio of nitrogen isotopes ^{15}N to ^{14}N changes for each species when moving higher in the food chain. From the nitrogen-isotope ratios in the La Brea bones, they found that the coyotes were omnivorous; that dire wolves and American lions consumed horses, ground sloths, bison, camels, and deer; and that sabertooth cats preferred bison and camels.

About 1% of the La Brea bones in the Page Museum collections show evidence of disease or injury, and these are kept in a separate paleopathology section.[12] Especially interesting is the abundant evidence of traumatic injuries to the bones of sabertooth cats and dire wolves, often with extensive secondary bone overgrowths (osteomyelitis). The injuries especially occur in the shoulder region and along the spinal column, probably resulting from bite wounds that became infected; there are also many examples of limb bones with healed fractures. Remarkable too are the sabertooth cats that survived for months or years with broken canine teeth (sometimes on both sides), including one instance where the broken tooth was driven into the nasal cavity. These observations provide some valuable insights into the social behavior of both species. Presumably, individual sabertooth cats and dire wolves whose ability to hunt was severely impaired were allowed to feed on kills made by other members of the group.

Pit 91

An asphaltic bone mass in Pit 91, discovered in 1915 and excavated only to a depth of 3 meters was left with the intention of displaying the exposed fossils as a permanent exhibition for public viewing (fig. 6.2). However, the pit was never roofed and eventually the sides caved in. Happily, though, this left a substantial body of intact, highly fossiliferous deposit available for future work. In 1969 excavations were resumed, for the first time recording the position and orientation of all fossils larger than 1 centimeter. Moreover and of particular importance, the enclosing asphalt matrix was treated with solvent to recover the smaller fossil elements, including rodents and other small mammals, reptiles and amphibians, insects, mollusks, and plant remains—most of which were also documented for the first time. The smaller fossils provide the best evidence on the habitats in the immediate vicinity of a fossil site. On the other hand, large mammals such as mammoths, bison, and horses, which may have traveled hundreds of kilometers during their lives, tell us about the wider regional environments. These efforts resulted in the recovery of more than twice the number of species recorded from the earlier excavations, enabling a much better reconstruction of the whole biota than was previously possible. Radiocarbon dating of mammal remains records at least two distinct entrapment episodes: ca. 45 to 35 kya, ca. 26.5 to 23 kya, and possibly also ca. 14 kya. Long-term flow and mixing within the asphalt has resulted in some bones recovered from deeper in the deposit dating younger than those above.

FIGURE 6.2. La Brea Pit 91 in 1915, showing a jumbled mass of large bones in asphalt. (Public domain.)

It is commonly stated that in the large-mammal finds from the 1913–15 excavations, carnivores outnumber herbivores by approximately 9:1. However, in the Pit 91 samples analyzed so far, it is clear that counting total bones greatly underestimates the number of herbivores because each individual carnivore is represented by more bones than is each individual herbivore. In other words, many of the herbivore bones, especially limb bones, are missing. The result is that although the ratio of carnivore to herbivore bones is 9:1, the ratio of carnivore to herbivore individuals is only 3:1. So why should this be the case? The likely answer provides a further interesting insight into scavenger activity at La Brea. Today, carnivores at a kill site often remove a limb from a carcass and carry it to a safer place to feed, thus minimizing unwelcome interference from other animals. At La Brea, scavengers could have successfully removed the uppermost limbs (of course, only when the trap was inactive), whereas those body parts mired in asphalt obviously would have been left behind. It is also apparent that not all species were equally attractive to scavengers: more than half of the bison limb bones had been removed, whereas only a quarter of horse limb bones were missing.

"Project 23"

In 2006 the Los Angeles County Museum of Art, also located in Hancock Park, was constructing a new underground parking lot. Workers came across the skull of what turned out to be the partly articulated, largely complete skeleton of an adult Columbian mammoth, nicknamed Zed, which unusually was preserved in a mixture of stream and asphalt deposits.[13] The challenge for the Page Museum staff was how best to handle this potentially hugely important fossil windfall. Time constraints meant that *in situ* excavation of the site was out of the question, so the problem was addressed by building a large wooden box around each fossil concentration (twenty-three in total—hence "Project 23"). These were lifted out by crane and removed by truck to another location in the park so that they could be carefully excavated—a task that has already taken several years. However, the potential scientific benefit is enormous. Here is a unique opportunity to dissect blocks of fossiliferous La Brea deposit under exceptionally well-controlled conditions and meticulously record and study everything found, from large mammals to rodent teeth, snake vertebrae, invertebrates, and plants, all contributing to a detailed reconstruction of the environment over a period of some thousands of years. Notable finds so far include entire millipedes and layers of oak leaf litter.

As with Pit 91, the aim of Project 23 is to recover all elements of the biota down to the smallest, not just the megafauna. This approach is yielding some very interesting results. In the 1913–15 excavations, most of the small vertebrates were washed away in the kerosene baths, whereas in the current project the recovered bones of hares, rabbits, and rodents greatly outnumber those of large mammals. For some years, bones were cleaned using trichloroethane, but its use was discontinued when it became known that this solvent contributes to atmospheric ozone depletion. It was replaced by more environmentally friendly solvents: n-propyl bromide in the late 1980s and biodiesel in 2012. At La Brea, even partial skeletons of any species are extremely rare, but a remarkable find from Project 23 is the partial associated skeleton (about 40% so far) of an American lion (*Panthera atrox*), nicknamed Fluffy. In addition, the remains of nine sabertooth cats (*Smilodon fatalis*), seven of which are juveniles, have been recovered from just 1 cubic meter of deposit. However, the limb bones and skull parts, packed tightly with thousands of other bones, have not yet been matched.

As epitomized by Rancho La Brea, North America during the Last Glacial featured a rich and diverse range of large mammals, most of which are now extinct. In part, this faunal wealth was a legacy of the Great American Biotic Interchange—a protracted and initially intermittent process that took place over several million years. As early as 9 mya (Miocene), tectonic uplift of the Isthmus of Panama provided limited opportunities for dispersal between South and North America, while closure of the Central American Seaway at about 2.8 mya was followed by the main interchange pulse that began ca. 2.6 mya (Pliocene).[14] This event allowed endemic South American animals such as ground sloths, glyptodonts, and capybaras to colonize North America, and many North American animals, such as proboscideans, camelids, horses, and sabertooth cats, to migrate in the opposite direction. Another factor was North America's lower rate of extinction throughout the earlier Quaternary period compared with northern Eurasia, so that by the Last Glacial it had many more megafaunal species and the impact of megafaunal extinctions was correspondingly greater. Out of about fifty-four species with a body weight greater than 45 kilograms, approximately thirty-seven went extinct (around 69%).

6.2 CLIMATE AND VEGETATIONAL HISTORY

During the Last Glacial, North America experienced rather similar climatic and vegetational changes to those of northern Eurasia, with the important difference that during the Last Glacial Maximum, the North American ice sheet was much more extensive. It covered most of the northern half of the continent (to roughly the Canada/US border) but left most of Alaska and the Yukon ice free.[15] For many millennia this ice sheet prevented free interchange of plants and animals with populations further south. However, with the Late Glacial warming, diminishing ice cover resulted in the opening of an ice-free corridor between the western Cordilleran and eastern Laurentide ice sheets, eventually allowing migrations in both directions.

Analyses of seven hundred pollen records from northern and eastern North America reveal a complex history of changing vegetational cover, although there are many difficulties and inherent inaccuracies in reconstructing such a history. Vegetational conditions were relatively stable during the Last Glacial Maximum ca. 21 to 17 kya, changed rapidly during the Late Glacial ca. 16 to 11.5 kya and early Holocene 11.5 to 8 kya, and thereafter showed little change until 500 years ago. As the ice sheets retreated after ca. 12 kya, boreal conifer forest (taiga) spread over vast areas in the north.[16]

The Last Glacial fauna of North America is extraordinarily rich and varied. In the following section I discuss most of the numerous extinct forms. The youngest available radiocarbon dates are listed in table 6.1.[17]

6.3 SABERTOOTH CAT: *SMILODON FATALIS*

With its enormous dagger-like canine teeth—the stuff of nightmares—the sabertooth cat (*Smilodon fatalis*) is, along with the woolly mammoth, probably the most easily recognized Ice Age animal, and one that has achieved worldwide fame (fig. 6.3).[18] It is recorded from many sites in the southern third of North America, most famously from Rancho La Brea. None would doubt that this was an active and ferocious carnivore, which potentially could have preyed on a variety of contemporary herbivores, ranging from deer and pronghorns to bison and horses, as well as the young of mammoths and mastodons. Isotope evidence indicates that the Rancho La Brea sabertooths targeted camels and horses.[19] It is the upper canines, up to 18 centimeters long, that are so impressive; the lowers are very much smaller. The upper canines, flattened side to side, had front and rear edges serrated like steak knives for slicing through flesh, as seen in several other mammalian and reptilian carnivores, living and extinct. With an estimated weight of 159 to 272 kilograms (males were larger than females)—twice that of an African lion—*Smilodon fatalis* was powerfully built, especially in the front limbs. The short, powerful limbs indicate that *Smilodon* was very likely an ambush predator that lay in wait for its victims before

FIGURE 6.3. Sabertooth cat *Smilodon fatalis* (cast) skull and mandible, La Brea, California. (Courtesy of Smithsonian Institution–National Museum of Natural History).

making a quick dash; it was not built for sustained running. The method that it used to kill prey has been much discussed. Modern big cats such as lions and tigers, having first brought down their victims, employ their large, conical canine teeth to cause asphyxiation either by means of a throat bite or by enclosing the nostrils and mouth. This process can take several minutes. Clearly a very different technique must have been used by *Smilodon*—one that minimized the risk of damage to its long, thin, and rather fragile canines, which would undoubtedly have broken if they hit bone (indeed, several skulls from Rancho La Brea do exhibit broken canines). So, one theory is that, having wrestled its victim to the ground using its powerful forelimbs, *Smilodon* then used its sabers to disembowel it. However, this action would not have killed sufficiently quickly, thereby risking injury to the attacker. The alternative and more likely technique is that, having immobilized the prey, *Smilodon* would have employed its sabers to inflict an accurate throat bite (avoiding the neck vertebrae), which would sever major blood vessels and the windpipe, resulting in rapid death by cutting off oxygen to the brain.[20]

The fact that some sabertooth cats at Rancho La Brea continued to live for long periods in spite of broken canine teeth or crippling injuries (which had time to heal) strongly implies that they lived in prides like modern lions. Only a social structure, in which disabled animals that could no longer hunt were permitted to share in kills, could have allowed them to survive for so long.

The youngest of forty-three radiocarbon dates on *Smilodon* (from La Brea) is ca. 13.18 kya.

6.4 SCIMITAR CAT: *HOMOTHERIUM SERUM*

The scimitar cat (*Homotherium serum*) was a second, rather less well-known, North American species of sabertooth cat.[21] Although recorded from far fewer sites than *Smilodon fatalis*, the distribution of the scimitar cat was very wide, extending from Texas and Florida to Alaska and the Yukon. Weighing some 190 kilograms (similar to a modern lion), it was built more lightly than *S. fatalis* and also had longer legs, suggesting that it could have run down its prey in short bursts, like a modern lion. However, its upper canines were even more flattened side to side, and were shorter and with coarser serrations—front and back—than in *S. fatalis*; and unlike *S. fatalis*, in *H. serum* all the teeth were serrated. Friesenhahn Cave, Texas, provides some very interesting evidence about the diet and mode of life of the scimitar cat. The remains of more than thirty-three individuals recovered from the site range from cubs to old individuals. The cave was evidently used as a den by scimitar cats that brought in remains of young Columbian mammoths, leaving their tooth marks on many of the bones. Like *Smilodon*, *Homotherium* probably specialized in hunting larger animals. In all cases, the likely method of killing prey, by throat bite, would have been as described above for *Smilodon*. The youngest available date (total eleven) for the scimitar cat, from Alaska/Yukon, is ca. 40.78 kya, while the only date so far from south of the ice sheets is ca. 26.48 kya from Tyson Spring Cave, Minnesota.[22]

6.5 OTHER EXTINCT LARGE CATS

The American cheetah (*Miracinonyx trumani*), at perhaps 90 kilograms, resembled a heftier version of a modern African or Asian cheetah, although the two species were not closely related.[23] As many have remarked, the phenomenal speed of the living pronghorn (*Antilocapra americana*)—about 56 to 88 kilometers per hour, depending on the distance—far exceeds that needed to outrun any of its potential present-day predators, such as wolves. To account for this extraordinary situation, it has been suggested (quite reasonably) that this grossly overengineered fleetness evolved in response to a now-vanished, very fast predator: the American cheetah. Its similarity to living cheetahs results from natural selection honing the pursuit of prey at high speeds in two distinct evolutionary lineages (an excellent example of convergent evolution). Presumably it would have also hunted a range of prey including deer. The dozen or so known localities for this animal are all from the western half of the contiguous United States.

The American lion (*Panthera atrox*), the largest of the lions from the Last Glacial with an estimated (male) weight of about 245 kilograms, is known from the southern half of North America and Mexico; Rancho La Brea has produced the remains of more than eighty individuals. The relationship of *P. atrox* to other lions has been discussed, with some authorities suggesting that it was not a lion at all, but an offshoot of an ancestor of the jaguar. However, ancient DNA studies suggest that it originated from a population of cave lion (*Panthera spelaea*) that dispersed into North America and subsequently became isolated.[24] Latest available dates include ca. 13.41 kya from California (La Brea) and ca. 13.2 kya from Alberta. Likely prey items included horses, deer, and bison.

The cave lion (*Panthera spelaea*) is described more fully in the chapter on northern Eurasia (chapter 5). Its vast range extended from Britain and Spain in the west for many thousands of kilometers across Siberia to the Bering region and into Alaska and the Yukon. Dramatic evidence that cave lions were capable of tackling large and formidable prey was provided by Dale Guthrie's meticulous forensic investigation of the naturally freeze-dried mummy of an extinct steppe bison (*Bison priscus*) discovered by Alaskan gold miners in the frozen "muck" deposits that overlie the gold-bearing placer deposits.[25] It was named Blue Babe, after the legendary Paul Bunyan's giant ox and in reference to the covering of crystals of the blue mineral vivianite. Puncture marks in the hide and a fragment of an embedded tooth matched in size and form with the cave lion. The bison carcass had been only partially eaten, suggesting that just a couple of lions were involved, as a large pride would have been able to consume the entire carcass before it froze. The mummy is now displayed in the Museum of the North in Fairbanks, Alaska. The youngest radiocarbon date for North American cave lion is ca. 13.29 kya from near Fairbanks, Alaska—nearly a thousand years younger than the youngest date for northern Eurasia.[26]

6.6 GIANT SHORT-FACED BEAR: *ARCTODUS SIMUS*

The giant "short-faced" bear (*Arctodus simus*) was the largest mammalian carnivore in Last Glacial North America, exceeding even the grizzly and polar bears

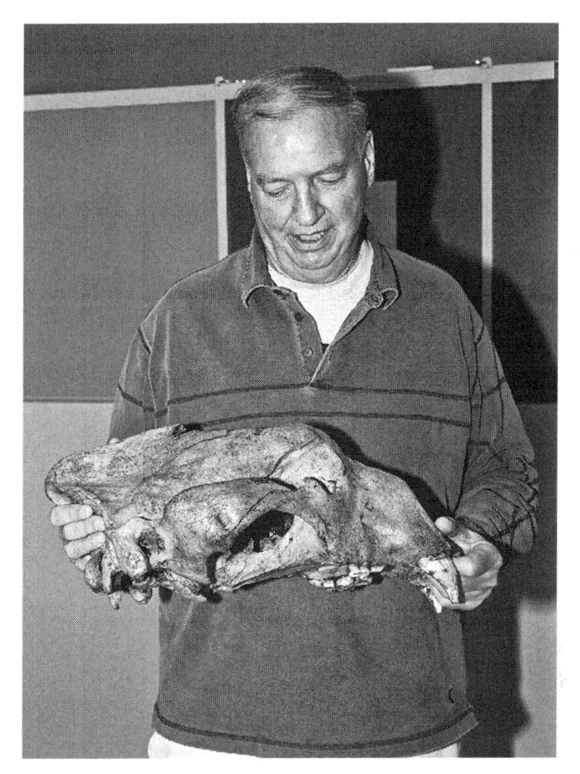

FIGURE 6.4. Dick Harington with giant bear *Arctodus simus* skull, Gold Run Creek, Yukon Territory. The specimen is in the Canadian Museum of Nature, Ottawa, cat. no. 7438. (Photo by Robert Young.)

in size (fig. 6.4). A detailed study concluded that its face was not unusually short, although the name has become well established.[27]

A. simus was widespread in Last Glacial North America and is recorded at over a hundred localities from Mexico to Alaska. As in living bears, it was sexually dimorphic in size, with males much larger than females; adult males weighed around 700 to 800 kilograms, with some individuals reaching 1 tonne—comparable to the distantly related Eurasian cave bears. The giant bear is generally believed to have been an active predator, and its long legs suggest that it could run fast in pursuit of prey.[28] It was probably omnivorous, consuming a range of animal and plant food as available, like the living brown bear (grizzly). However, isotope analyses of specimens from Alaska and the Yukon suggest that these populations, at least, were primarily carnivorous, contrasting with Eurasian cave bears, which are thought to have been primarily herbivorous (chapter 5).[29] Likely prey included most contemporary large herbivores such as bison, horse, and deer, and no doubt carrion also would have been eaten. Fox-Dobbs and colleagues concluded that giant bears living in Alaska and the Yukon specialized in caribou.[30] No doubt this formidable predator would have been able to see off competitors such as grizzly bears, wolves, sabertooth cats, American lions, and cave lions, and probably also steal their kills when the opportunity arose. Interestingly, the giant short-faced bear failed to colonize northern Eurasia, as for some reason (in common with several other species) it was unable to pass the "filter" of the Bering Land Bridge.[31]

The four available radiocarbon dates on *Arctodus simus* from Alaska/Yukon range from ca. 43.5 kya to ca. 24.73 kya, so it was probably extirpated from the region due to the increasing cold of the approaching Last Glacial Maximum and/or reduction in available prey. However, it survived very much longer in the southern half of the continent. A series of eight new dates made directly on *Arctodus simus* material, of which the youngest is ca. 12.78 kya (Kansas), indicates that the giant bear coexisted with humans of the Clovis culture and that it was one of the last megafaunal species to go extinct in North America.[32]

6.7 DIRE WOLF: *CANIS DIRUS*

The extinct dire wolf (*Canis dirus*) was about the same size as the living gray wolf (*Canis lupus*) but more heavily built, with an average weight of about 60 kilograms and height of about 0.8 meters at the shoulder. Its teeth were more robust and had larger attachments for the temporalis muscle on the skull, indicating that it could have inflicted a more powerful bite than the gray wolf. The abundance of remains from La Brea is consistent with living in packs, and the isotope evidence suggests that here it preyed on horses, ground sloths, bison, camels, and deer. The ultrafiltered radiocarbon dates from Rancho La Brea include a date of ca. 13.7 kya on dire wolf.[33]

6.8 SHASTA GROUND SLOTH: *NOTHROTHERIOPS SHASTENSIS*

No fewer than four species of ground sloth were present in North America during the Last Glacial, all of which ultimately originated from South America—a notable legacy of the Great American Biotic Interchange. The Shasta ground sloth (*Nothrotheriops shastensis*), the smallest of the North American ground sloths, has a key place in the history of Ice Age research in the American Southwest. Radiocarbon dating of dung deposits in several arid caves made a major contribution to understanding megafaunal extinctions in this region and

FIGURE 6.5. Paul Martin contemplating a dung bolus of the Shasta ground sloth *Nothrotheriops shastensis*, from Rampart Cave, Grand Canyon, Arizona. (Photo by Sandra L. Swift.)

FIGURE 6.6. *Nothrotheriops shastensis* dung bolus, Rampart Cave. Note the abundant undigested plant fragments. (Photo by Sandra L. Swift.)

in North America as a whole. Substantial well-preserved dung deposits, accompanied by some bones, are known from Rampart and Muav Caves in the Grand Canyon, Arizona (figs. 6.5, 6.6), Gypsum Cave, Nevada, and Williams Cave in the Guadalupe Mountains of Texas.[34] Moreover, abundant skeletal remains have been recovered from Rancho La Brea. The range of the Shasta sloth also extended as far south as central Mexico.

Rampart Cave is situated at the top of a long steep slope, high up in the south wall of the Grand Canyon, about 200 meters above the Colorado River and 535 meters above sea level. The cave preserves a remarkable record of use by Shasta ground sloths in the form of dung deposits several meters thick, together with sloth bones and hair. Remains of other animals, including the extinct Harrington's mountain goat (*Oreamnos harringtoni*), also occur. The lack of moisture in the caves inhibited the decay of organic remains by bacteria and fungi. The Muav Caves, about 1.7 kilometers upstream, also contain Shasta sloth dung. Radiocarbon dates of ca. 37 and 35 kya are available on dung from the lower sequence in Rampart Cave, while fifteen dates for the upper section range from 15.8 to 11.5 kya. There is a pronounced gap in the sequence, estimated to cover 23 to 16 kya, more or less corresponding to the Last Glacial Maximum, when only pack rat (*Neotoma*) middens are represented. At this time the climate was cooler and drier than now. Only by good fortune did the unique Rampart Cave fossil deposits escape total destruction in the disastrous fire of 1976 (which, sad to say, was started deliberately). More than thirty years earlier, a deep trench had been cut through the dung deposit from front to back in order to examine the stratigraphic sequence. This trench acted as a firebreak so that although the eastern part of the deposit and contained fossils were lost (about 70% of the whole), the western section survived.

By far the best-preserved skeleton of a Shasta ground sloth was discovered in 1928 in a fumarole (a vent from which volcanic gases escape to the surface) of the extinct Aden Crater volcano, New Mexico. Unlike other Shasta sloth specimens, which have been assembled from the bones of several animals, this one is from a single still-articulated individual and is virtually complete.[35] The skeleton (fig. 6.7) still has attached claw sheaths (composed of keratin), ligaments, muscle fibers, and adhering remnants of skin and fur; the rest of the carcass appears to have been eaten by rodents. The superb preservation results from both the peculiar mode of its death and the peculiar conditions in which it was entombed. The unfortunate animal had fallen into an open shaft 30 meters deep, from which there was no possible escape. Somehow it survived the fall, only to make its way to a blind-ending downward sloping gallery, which was destined to be its final resting place. The open shaft remains very dangerous

FIGURE 6.7. Skeleton of Shasta ground sloth *Nothrotheriops shastensis*, with adhering mummified tissues, Aden Crater, New Mexico. Yale Peabody Museum of Natural History, New Haven, Connecticut. (Photo by Brian Switek.)

today, with the only route in or out by means of a rope. The sloth's mummified corpse was preserved for millennia, because of the dry atmosphere and the accumulating blanket of dry droppings deposited by subsequent generations of unfeeling bats that roosted in the shaft. The skeleton was so tightly bound together by its ligaments that it presented considerable problems when arranging it into the desired pose for exhibition in the Yale Peabody Museum, where it may be seen today. An accompanying dung bolus contained twigs and roots of sagebrush (*Artemisia*).

Thanks largely to the Aden Crater find, more is known about the appearance of *Nothrotheriops shastensis* than of any other ground sloth. It was about 1 meter high at the shoulder, with an estimated weight of 180 kilograms and covered in coarse hair. In common with other ground sloths, it very likely used its long front claws to pull vegetation to within reach of its mouth and had a capacious gut that could digest large quantities of plant food over a long period of time. Its elongated face probably housed a long tongue that could curl around stems and strip off the leaves. The diet of the Shasta sloth is apparent from the wide range of plant remains preserved in dung from Rampart Cave, including: agave, yucca, Nevada jointfir, saltbush, catclaw acacia, common reed, desert globemallow, Freemont cottonwood, prickly pear cactus, Utah juniper, and various grasses, all of which still grow in the area today. Its cheek teeth were simple and peg-like, and from the conspicuous occurrence of large pieces of plant in the dung, R. M. Hansen concluded that it did not grind its food but instead "crunched and munched it," swallowing large pieces that were digested inefficiently (which is also true of living tree sloths).[36] It seems very likely that all ground sloths had slow metabolic rates similar to their living relatives. The dung boluses from Rampart Cave also preserve remains of internal parasites, including nematodes (roundworms). The presence of parasites is to be expected, but the fact of their preservation is remarkable.

Unfortunately, the only available radiocarbon date (ca. 13 kya) on the Aden Crater sloth was done as long ago as 1960 and may not be reliable. Of fifteen radiocarbon dates on Shasta sloth dung from Rampart Cave,[37] the youngest are 12.25 and 12.01 kya, whereas the youngest of four dates on dung from Gypsum Cave, Nevada, are 12.94 kya and 12.88 kya.[38] So, the available evidence indicates that the Shasta ground sloth survived into the early Younger Dryas phase of the Last Glacial (ca. 12.9 to 11.7 kya).

6.9 JEFFERSON'S GROUND SLOTH: *MEGALONYX JEFFERSONI*

The medium-size Jefferson's ground sloth (*Megalonyx jeffersoni*)—featured in the introduction to this chapter—weighed about 540 kilograms. It had the widest distribution of all the ground sloths, ranging from Central America to Alaska and the Yukon, where it is recorded only from the Last Interglacial, approximately 120 kya.[39] Radiocarbon dates of 13.34 kya (Illinois) and 13.1 kya (Ohio) show that south of the ice sheets it survived until near the end of the Last Glacial.

6.10 HARLAN'S GROUND SLOTH: *PARAMYLODON HARLANI*

Harlan's ground sloth (*Paramylodon harlani*, aka *Glossotherium harlani*) was about 1.2 meters at the shoulder (when walking on four feet) and weighed around 900 kilograms.[40] It is recorded from Mexico, Guatemala, California, Arizona, Texas, Florida, and South Carolina, and as far north as Oregon. It appears to have been a grazer like its close South American relative *Mylodon darwini* and had high-crowned cheek teeth. A strong correlation between occurrences of Harlan's ground sloth and the Columbian mammoth (*Mammuthus columbi*) implies that they shared similar habitat and diet.[41]

In 1882 scientists and public alike were much excited by a *New York Times* article with the dramatic headline "Footprints of Monster Men" describing giant human footprints (accompanied by prints of various animals) that had been uncovered by convicts quarrying sandstone at the Nevada State Prison, Carson City. The twists and turns of the subsequent history are entertainingly described by Don Grayson in his excellent book on the Great Basin.[42] The huge size of the footprints (46 to 51 centimeters long and 20 centimeters wide) was seen as confirmation of the much-quoted Biblical text "There were giants on the earth in those days" (Genesis 6:4). The curious rounded shape of the prints was no obstacle to this interpretation; evidently this giant was wearing wooden sandals! The truth is rather more prosaic. The geologist Joseph Le Conte deserves credit as the first person to suggest (in 1882) that the track maker was actually a large ground sloth, and Chester Stock demonstrated that the footprints exactly matched the assembled foot bones of *Paramylodon*.[43] The youngest available date on *P. harlani* is ca. 14.41 kya from Oregon.

6.11 LAURILLARD'S GROUND SLOTH: *EREMOTHERIUM LAURILLARDI*

With an estimated weight of more than 3 tonnes, Laurillard's ground sloth (*Eremotherium laurillardi*, aka *E. rusconii*) was the largest of the North American sloths. Predominantly South American (see chapter 7), it ranged northward to Central America and the southeastern United States, including Texas, Florida, and South Carolina. No dates are available.

6.12 SIMPSON'S GLYPTODONT: *GLYPTOTHERIUM FLORIDANUM*

Simpson's glyptodont (*Glyptotherium floridanum*) is recorded from Mexico, Texas, and Florida, with just a single known locality in South Carolina. This extraordinary animal was about 1.2 meters high and weighed about 1 tonne (fig. 6.8).[44] Other species of *Glyptotherium* are known from northern South America, Central America, and the southern United States.

Because of the strikingly shortened skull, Gillette and Ray suggested that *G. floridanum* and other glyptodonts featured a short trunk.[45] In common with its South American relatives, its entire body other than the belly was encased in

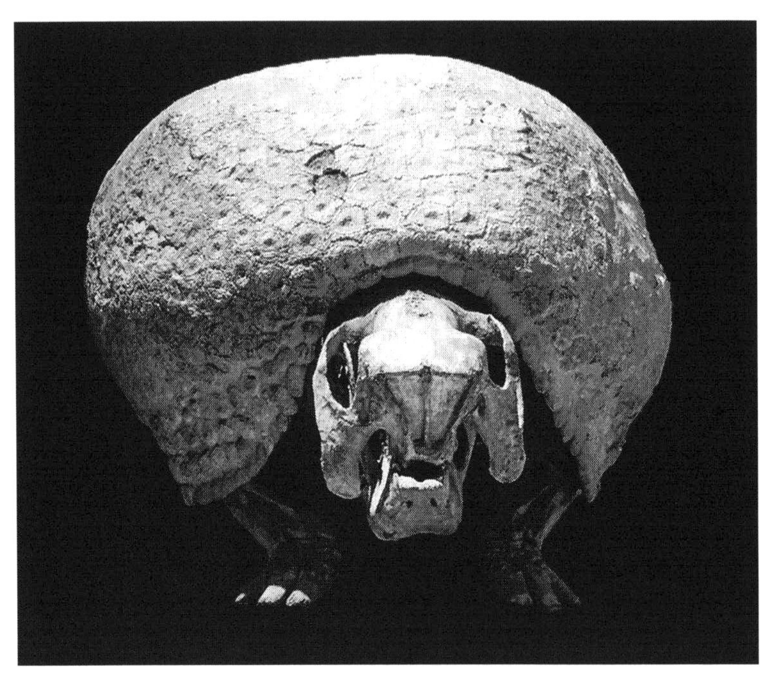

FIGURE 6.8. *Glyptotherium arizonae* skeleton. (Courtesy of Smithsonian Institution–National Museum of Natural History.)

a domed rigid protective shell made up of numerous bony interlocking plates, with additional plates forming a cap for the head. As in the South American *Glyptodon*, the flexible tail armor comprised a series of articulated, bony rings. However, this impressive armor might not always have been foolproof. A juvenile skull of *G. texanum* (an earlier species, ca. 2.5 million years old) from Dry Mountain, Arizona, features a pair of elliptical holes that correspond in size and spacing with the canine teeth of some kind of large cat, suggesting that this glyptodont was killed by a massive bite to the head. The specimen is on display at the American Museum of Natural History, New York. Since many *Glyptotherium floridanum* finds are from near lakes, marshes, and streams and are often associated with remains of an extinct capybara (a related species still lives in South American wetlands), it may have been semiaquatic, although some think it lived in grasslands. Gillette and Ray suggested that they "likely grazed on grasses and soft plants, perhaps even along shorelines or in water."[46] There are currently no available dates for this species.

6.13 NORTH AMERICAN MASTODON: *MAMMUT AMERICANUM*

The North American mastodon (*Mammut americanum*) has a special place in the history of North American paleontology. In 1705, a molar tooth said to be the size of a man's fist was discovered in the village of Claverack, New York.[47] The story goes that it was acquired (in exchange for a glass of rum) by Lord Cornbury, the eccentric governor of New York and New Jersey (his enemies said that he liked to cross-dress as his cousin Queen Anne). He sent the tooth to London labeled "tooth of a giant," and this unknown creature soon became celebrated as the "incognitum"—the first all-American prehistoric monster.[48]

Over the next decades, many finds of large bones, especially from the Midwest of the United States—were attributed to the "incognitum," and African American slaves are said to have noted the resemblance to the bones of elephants. In these early days, there was some confusion until it was realized that there were two distinct kinds of fossil elephant-like creatures: mastodon and mammoth, which are readily distinguished by their teeth. Mammoth molars are high-crowned with a series of transverse enamel ridges on a nearly flat grinding surface, whereas those of the mastodon are low-crowned with two parallel rows of rounded cusps. It was the renowned French anatomist Georges Cuvier who coined the very apt name *mastodon* ("breast-tooth") in 1806. However, previously in 1799 the German naturalist and polymath Johann Friedrich Blumenbach had called it *Mammut*, and this name has priority, while *mastodon* (or *mastodont*) continues to be universally used as a handy informal term.

For millennia, generations of large mammals appear to have been irresistibly drawn to the salt licks at the Big Bone Lick salt and mineral springs, in northern Kentucky.[49] Abundant bones of extinct and extant animals that littered the area were buried by accumulating sediments. Native Americans and European settlers alike marveled at the "big bones" that lay scattered around the lick. In 1765, the explorer Colonel George Croghan made a substantial collection of bones from Big Bone Lick. In his journal he described arriving at the site: "Early in the morning we went to the great lick, where those bones are only found, about four miles from the river."[50] Having followed a bison road to the Lick, Croghan and his men found several fossils, including "a tusk more than six feet long."[51] Although this collection of bones was lost as the result of an Indian attack (in which several of his companions were killed and he was almost brained by a tomahawk), an undaunted Croghan returned the next year and obtained many more fossils.

He sent some of these to Benjamin Franklin, who was then in London. Franklin's response includes some perceptive observations: "I return you many thanks for the box of elephants' tusks and grinders. They are extremely curious on many accounts; no living elephants having been seen in any part of America by any of the Europeans settled there or remembered in any traditions of the Indians. The tusks agree with those of the African and Asiatic elephant in being nearly of the same form and texture. But the grinders differ, being full of knobs like the grinders of a carnivorous animal: when those of the elephant, who eats only vegetables, are almost smooth."[52] Undoubtedly, he was describing the molars of a mastodon. However, Franklin was a lot wider of the mark with another observation: because he thought elephants could live only in the tropics, he was led to suppose that the "earth had anciently been in another position, and the climates differently placed than what they are at present."[53]

The American mastodon *Mammut americanum* is recorded from hundreds of localities in North America extending from Alaska to Mexico and Honduras, with finds especially concentrated in the formerly glaciated parts of Illinois, Indiana, Michigan, and Ohio, south of the Great Lakes. It was confined entirely to North America, neither making it through the Isthmus of Panama to reach South America, nor managing to cross the Bering Land Bridge into Eurasia.

The American mastodon was a "long, low and stocky" animal weighing around 3.5 to 5.4 tonnes, with a shoulder height of 2.5 to 3 meters—considerably smaller than both woolly and Columbian mammoths.[54] The limbs were robust and the skull large compared with mammoths and elephants, broad and flat-topped. The skull carried large tusks up to 2.5 meters long that projected forward and curved gently inward and upward toward the tip, and, unlike both mammoths and elephants, the lower jaw in young animals often also had small vestigial tusks. The cheek teeth with their blunt conical cusps were well adapted to browsing woody vegetation, and there are many records from the Midwest dating from after the Last Glacial Maximum in association with plant fossils indicating open spruce and pine parkland. Plant remains found in association with mastodon skeletons and presumed to be gut contents were reported to include twigs of spruce and pine. In addition to this evidence, microwear scratches on teeth can tell us a great deal about the mastodon diet. The presence of coarse and very coarse scratches, large pits, and puncture pits is characteristic of browsing on leaves and bark. This contrasts strongly with mammoths whose diet comprised grass and browse. The feet of mastodons were wide and squat compared with mammoths and elephants—useful for walking on soft boggy ground.

The "Warren mastodon" was recovered in 1845 from a bog in Newburgh, New York, in the course of digging for peat fuel.[55] In this case, putative gut contents consisted of crushed leaves and twigs in a convoluted column about 10 centimeters in diameter that went through the pelvic opening. The near-complete skeleton is now on display at the American Museum of Natural History, New York.

In 1989 an exceptionally well-preserved skeleton was unearthed from a peat bog during the construction of a water hazard on Burning Tree Golf Course in Newark, Ohio.[56] The animal is estimated to have been 2.7 meters at the shoulder and weighed as much as 4 tonnes. A discrete, pungent cylindrical mass, interpreted as intestinal contents, comprised swamp grass, leaves, moss, sedges, water lilies, and other wetland vegetation, but no conifer twigs. From the growth rings in its tusks, Dan Fisher estimated that the mastodon was about 30 years old at time of death.[57] Both the tusk growth rings and the gut contents suggest that it died in late autumn. The Burning Tree Mastodon is notable for the claim that live gut bacteria *Enterobacter cloacae* had remained dormant for more than 12,000 years in the ground, preserved by the cold and lack of oxygen, allowing a sample to be successfully cultured in the laboratory. However, others have suggested that the bacteria were more likely recent contaminants. A radiocarbon date of ca. 12.75 kya (Pitt-0830) was obtained on collagen from a mastodon bone, whereas twigs from its gut contents gave a date of ca. 13.49 kya (Beta-38421/ETH-6758).

The records of the Smithsonian Museum for 1915 gives an account of the finding of the "Indiana mastodon": "Each year the Museum receives reports of many finds of mastodon and mammoth remains, especially from different localities in those states bordering the Great Lakes. These finds, which come for the most part from swamp deposits of the Pleistocene, usually consist of a few isolated bones or teeth, but they give evidence of the great abundance of these larger creatures which roamed over the continent during the geological age just preceding the present."[58] It goes on to say that complete skeletons are rarely recovered because generally the finders lack the necessary expertise. Fortunately, however, in the spring of 1914, when bones were unearthed by workmen digging a drainage canal on the Pattison Farm near Winamac in northwestern Indiana, the landowners notified the Smithsonian Museum. In 1915 James W. Gidley of the Smithsonian made two trips to excavate or otherwise retrieve the remaining bones, many of which were found on the spoil heap where they had been dumped by the steam shovel. Interestingly, some of the bones were found by probing the bog with a long iron rod—the very same method that was often used to find giant deer remains in Ireland. The happy outcome was

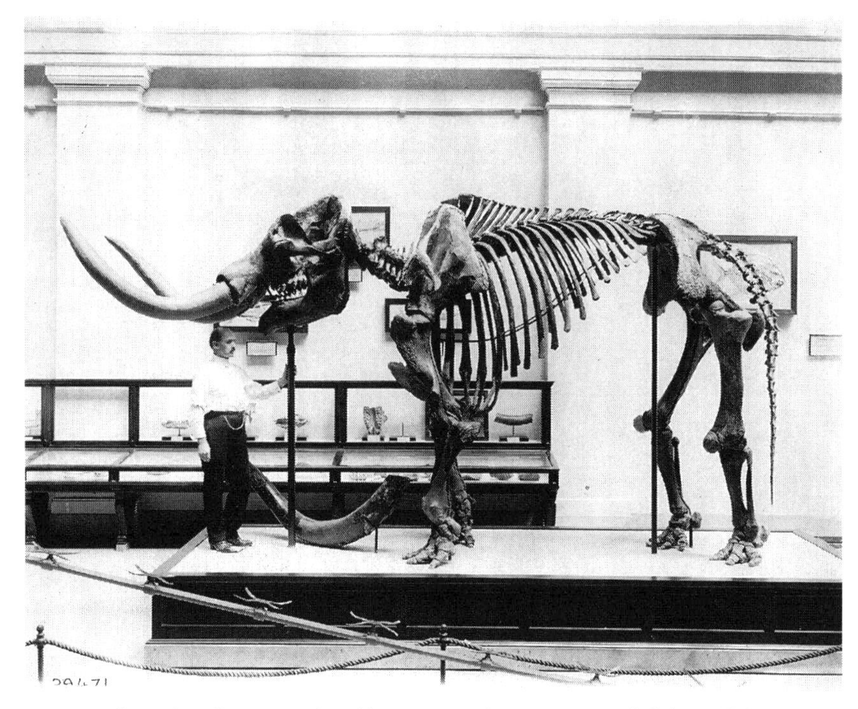

FIGURE 6.9. American mastodon, *Mammut americanum*, mounted skeleton, Winamac, Indiana. (Courtesy of Smithsonian Institution–National Museum of Natural History.)

that a near-complete, excellently preserved adult male skeleton was recovered and donated by the landowners to the Smithsonian. In 1916 the reconstructed skeleton went on display (fig. 6.9). It has now been remounted and redisplayed in the new Deep Time exhibition, which opened in 2019.

The remains of the "Overmyer mastodon" were discovered in 1976 when an agricultural drainage ditch was dug through a former peat bog near Rochester, northern Indiana. Two years later, the recovered skeleton, estimated at 41 to 48% complete, was excavated from a layer of organic marl 1.8 meters thick, containing abundant mollusk shells and overlying glacial outwash.[59] The bone-bearing deposit was in turn overlain by about 2 meters of peaty marl and fibrous peat. This sequence is similar to other bog sites that have yielded mastodon and other megafaunal remains in the region. Moreover, there are also interesting parallels with the circumstances of giant deer (*Megaloceros*) finds from Ireland (chapter 5). In both cases, near to the end of the Last Glacial the retreating ice sheets left lime-rich freshwater pools, in which megafaunal remains were beautifully preserved, to be succeeded by fen, peat bog, and forest with the transition to the Holocene. The skull and dentition indicate a large mature female estimated at about 32 to 36 years old (in equivalent African elephant years). The lack of weathering, carnivore chewing, and rodent gnawing indicates that the carcass sank quickly into the pond—again, similar to several other mastodon sites in Indiana and Ohio.

The Kimmswick Site (now part of Mastodon State Historic Site, Missouri) has a long and colorful history. In 1839 the showman Albert Koch was drawn there by reports of large bones eroding from the banks of Rock Creek, one of two creeks that flow from Kimmswick into the Mississippi. Among the many bones that he excavated was a large skull, which, along with various other fossil bones, he cobbled together to construct a composite skeleton of much exaggerated size: the fantastic "Missouri Leviathan," which he exhibited in his St. Louis museum. In 1843, after a successful tour of Europe, Koch sold the

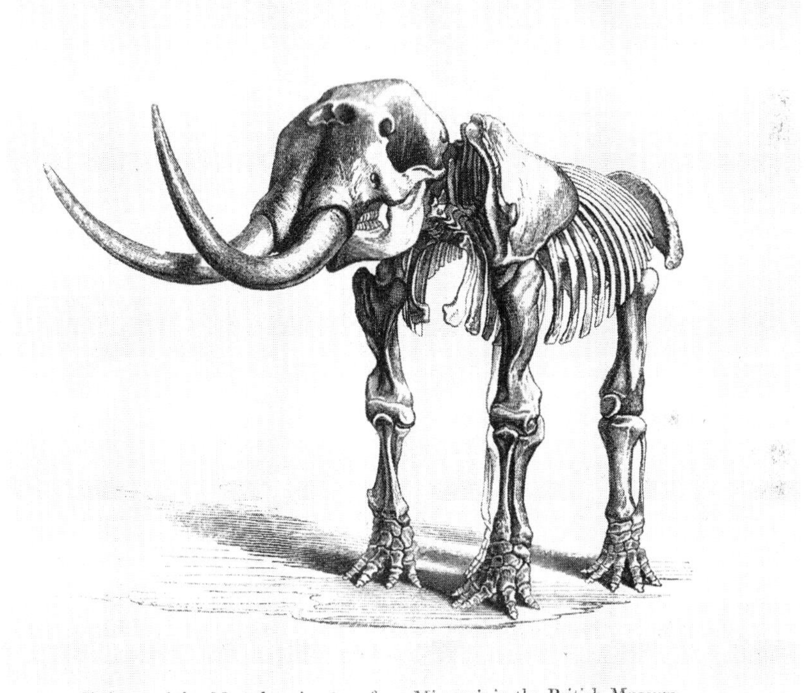

FIGURE 6.10. The supposed "Missouri Leviathan," in actuality American mastodon *Mammut americanum*, mounted skeleton, Kimmswick, Missouri. This engraving is from the nineteenth century; the specimen is exhibited at the Natural History Museum, London.

"Leviathan" to what is now the Natural History Museum in London, where, putting aside the bones that didn't belong, the celebrated but controversial comparative anatomist and museum's founder Sir Richard Owen (1804–1892), reassembled the skeleton correctly as an American mastodon (fig. 6.10).

Much more recently, Kimmswick has provided clear evidence that an American mastodon was killed by Clovis hunters.[60] Excavations in river terrace deposits yielded two characteristic Clovis projectile points and other stone tools in close direct association with mastodon and other faunal remains. The two projectile points were found 1.25 meters apart horizontally and were vertically separated by less than 1.5 centimeters. The larger steel-gray point was found among the disarticulated foot bones of an adult mastodon. It shows minor damage to the tip—presumably sustained on impact when it was used in killing the animal. The second projectile, made of olive-green chert and extensively reworked, was discovered directly beneath and in contact with a large mastodon bone fragment. It too shows minor damage to the tip and may have been embedded in the animal's flesh. Other artifacts were probably used in butchering the mastodon. Associated small-mammal fossils—tree squirrel, marmot, ground squirrel, pocket gopher, meadow vole, and bog lemming—indicate the presence of deciduous woodland with open grassy areas.

Six dates on mastodon range from 13.6 kya (Heisler, Michigan) to 12.7 kya (Deerfield, Wisconsin) and 12.6 kya (Hiscock, New York).[61] Radiocarbon dates published in 2009 on the "Overmyer mastodon" (Indiana), including ca. 11.58 kya on filtered collagen (probably the most reliable), fall just within the earliest Holocene or alternatively at the very end of the Last Glacial (Younger Dryas cold phase).[62] Notably, in view of the proposal in chapter 4 that Clovis hunters were responsible for the demise of the North American megafauna, the Overmyer dates are considerably younger than the range given for the Clovis culture (see below).

6.14 CUVIER'S GOMPHOTHERE: *CUVIERONIUS HYODON*

As described in chapter 7 on South America, the gomphothere (*Cuvieronius hyodon*) stood about 2.5 meters at the shoulder and probably weighed about 3.5 tonnes. Each of its low-crowned molars had three transverse rows of rounded cusps—consistent with a browsing diet—and unique spirally twisted tusks. Predominantly South American, it only marginally ranged into the United States, with just one record from South Carolina and another from Texas. Of particular interest, however, are finds of two gomphothere mastodonts that were described in 2014 from El Fin del Mundo in northwestern Sonora, Mexico.[63] The bones, accompanied by stone flakes and four Clovis points, were indirectly dated (associated dates on charcoal) to about 13.4 kya, which is at the older end of Clovis dates from elsewhere. Unfortunately, since the bones are badly weathered and lack collagen, it was not possible to obtain a direct radiocarbon date.

6.15 COLUMBIAN MAMMOTH: *MAMMUTHUS COLUMBI*

Two species of mammoth were present in continental North America during the Last Glacial: woolly mammoth (*Mammuthus primigenius*), extending from northern Eurasia into the northern half of the continent, and the much larger Columbian mammoth (*Mammuthus columbi*) in the south. Their ranges overlapped to a limited extent at around the Canada/U.S. border. A putative third species (*Mammuthus jeffersoni*) might have been a hybrid resulting from interbreeding between these two.[64]

The Columbian mammoth (*Mammuthus columbi*) was huge, reaching up to 4 meters at the shoulder and weighing around 10 tonnes (fig. 6.11).[65] It had much more widely spaced enamel plates than the molars of the woolly mammoth and was very similar, possibly identical, to *Mammuthus trogontherii*, from the Middle Pleistocene of northern Eurasia, which migrated into North America via the Bering region.[66] *M. columbi* ranged from the northern United States south to Mexico. The Mammoth Site of Hot Springs, South Dakota, features an extraordinary and unique concentration of mammoth skeletons: currently fifty-eight Columbian and three woolly mammoths, many of which are exhibited still in the sediment as they were found (fig. 6.12).[67]

The collapse of underlying limestone caused a sinkhole about 18 meters deep to open in the Spearfish Shale (of Permian–Triassic age), which was filled by a warm spring. Attracted to the resulting pond, over a period of at least 50,000 years from time to time mammoths were trapped and perished, unable to climb up the slippery, steep-sided walls. Other animals, including the giant bear, met a similar fate; all were eventually entombed in the silts that gradually filled the pond. Remarkably, the sediment "plug" infilling the sinkhole proved more resistant to erosion than the surrounding shale bedrock, so that the infill now stands proud of the bedrock. Unfortunately, there are no radiocarbon dates on the mammoths or other animals, as collagen is not preserved at this site, presumably due to degradation in warm water. However, very recently, dates of ca. 190 kya were obtained via optically simulated luminescence from

FIGURE 6.11. Columbian mammoth *Mammuthus columbi* skull *in situ*, Hot Springs Mammoth Site, South Dakota. (Courtesy of the Mammoth Site.)

FIGURE 6.12. Columbian mammoth *Mammuthus columbi* skeletal remains *in situ*, Hot Springs Mammoth Site, South Dakota. (Courtesy of the Mammoth Site.)

the base of the exposed sediments, and ca. 140 kya at the top.[68] In addition, some 15 meters of sediment have still to be excavated and dated.

About a dozen sites, including Naco, Lehner Ranch, Murray Springs (all in Arizona), Blackwater Draw (New Mexico), Domebo (Oklahoma), Lubbock Lake (Texas), Colby (Wyoming), and Dent (Colorado), have produced Clovis spear points in association with Columbian mammoth remains.[69] These provide good evidence that Columbian mammoths were hunted by Clovis people. In 1951 excavations at Naco, in the San Pedro Valley, southern Arizona, revealed no fewer than eight Clovis spear points in intimate association with the bones of a Columbian mammoth; one at the base of the skull, one close to a scapula (shoulder blade), and five among ribs and vertebrae.[70] Presumably the points originally had been embedded in the flesh of the mammoth but shifted down to the positions in which they were found as the carcass rotted and collapsed. Some of the bones and *in situ* Clovis points are displayed, as excavated, in the Arizona State Museum, Tucson.

Bechan Cave, Utah, a dry cave in the arid Southwest of the United States, boasts a blanket of dung 40 centimeters thick that uniquely is attributed to Columbian mammoth.[71] The name *Bechan* in the indigenous Navajo language translates (politely) as "large feces." Measuring about 31 meters wide, 9 meters high, and 52 meters from front to back, the cave is easily large enough to allow mammoths to come and go and is well lit in the daytime. Mammoths, shrub ox, and other animals probably made use of the cave to shelter in bad weather. Perfectly preserved by the dry atmosphere, dung boluses measuring approximately 230 by 170 by 85 millimeters are nearly identical in size and form to those of extant elephants. The largest constituents of the dung are stalks measuring 60 by 4.5 millimeters. Grasses and sedges dominated the diet, although woody species were also commonly eaten. Nineteen radiocarbon dates on plant remains from dung boluses range from ca. 16.22 kya to 13.59 kya.[72] The youngest dates on bone collagen from Columbian mammoth are ca. 12.86 kya from Hacienda de Hornos, Coahuila, Mexico, and ca. 12.48 kya from Dent, Colorado.[73] The latter is the most recently obtained, and probably the most reliable of several dates from this site.

6.16 WOOLLY MAMMOTH: *MAMMUTHUS PRIMIGENIUS*

The woolly mammoth *M. primigenius* is described in chapter 5 on northern Eurasia. Many well-preserved specimens have been recovered during placer mining in Alaska and the Yukon (fig. 6.13). The youngest available date for mainland North America is ca. 13.34 kya from Alaska.[74]

FIGURE 6.13. Woolly mammoth *Mammuthus primigenius*, skull and tusks (incorrectly labeled "mastodon") found 1904 in the course of placer mining, Quartz Creek, Yukon Territory. (Dawson City Museum 1990.54.35, courtesy of Grant Zazula.)

Saint Paul Island, the largest of the Pribilof group of islands, in the Bering Sea off Alaska, is notable for its late-surviving population of woolly mammoth—an interesting parallel with the late-surviving mammoths of Wrangel Island off northeast Siberia (chapter 5). Saint Paul is more than 450 kilometers from the Alaskan mainland, and has low relief with a few freshwater lakes but no springs or streams. It is free of permafrost; the vegetation is moderately productive moss-herb tundra. Saint Paul became isolated by rising sea levels between ca. 14.7 and 13.5 kya, was rapidly reduced in area until 9 kya, and then slowly shrank until 6 kya. A series of radiocarbon dates made directly on mammoth remains, ancient DNA, and marked decline in the spore counts of three species of dung fungus in the sediments of a small lake, all concur that woolly mammoths on Saint Paul went extinct within about a century of 5.6 kya,[75] making this one of the best studied and dated prehistoric extinctions. The vegetation composition remained stable over the period of extinction. Additionally, overkill can be ruled out as a cause, since there is no record of the arrival of people (in the form of Russian whalers) until 1787 CE. The critical factor driving local mammoth extinction here appears to have been a shortage of freshwater as the climate became drier and the island shrank in area. Intensified mammoth activity around the lake is thought to have exacerbated the decline in water quality and grazing. Clearly, this study has implications for the future of terrestrial vertebrate populations on islands in the face of increasing temperatures, sea level rise, and decreasing freshwater availability

6.17 YESTERDAY'S CAMEL: *CAMELOPS HESTERNUS*

Yesterday's camel (*Camelops hesternus*, aka western camel) probably would have looked rather similar to a modern Arabian camel, although with 20% longer legs. Whether or not it also had a hump is not known. It reached around 2.3 meters at the shoulder and probably weighed about 1 tonne.[76] A recent aDNA study demonstrated that *Camelops* is much more closely related to extant Arabian and central Asian camels (dromedary and Bactrian camel, respectively) than to New World camelids, such as guanacos.[77] Evidence from tooth-wear studies and plant remains extracted from teeth indicate that it was a mixed feeder, probably consuming grass and browse according to what was available.[78] There are many records from the middle and western part of the United States, extending south into Mexico and north into Canada. Several finds from Alaska and the Yukon date from the Last Interglacial. A toe bone from Last Interglacial sediments of the White River, Yukon, was associated with plant fossils indicating birch shrub-tundra vegetation and a cool climate dating to between approximately 115 and 87 kya.[79] The latest available date on *Camelops* from south of the ice sheets (Casper, Wyoming) is ca. 13.06 kya. A butchered *Camelops* carcass from Wally's Beach, Alberta, with median calibrated dates of 13.27 and 13.31 kya, provides unique plausible evidence that this species was hunted.[80]

6.18 GIANT BEAVER: *CASTOROIDES OHIOENSIS*

Weighing about 150 to 200 kilograms, the giant beaver *Castoroides ohioensis* was about the same size as a black bear and was the largest rodent to have inhabited Last Glacial North America. Most finds are from the eastern United States, but there are also a few records from the Yukon and Alaska.[81] From the eight (southern) sites that they considered, McDonald and Bryson concluded that *Castoroides* could tolerate colder temperatures than exist at these sites today.[82] It was much larger than the extant beavers *Castor canadensis* or *Castor fiber* (at only 25 kilograms) and evidently had a very different mode of life; more like the very much smaller South American coypu or North American muskrat. There is a close resemblance between *Castoroides* and the smaller *Trogontherium cuvieri* from the Middle Pleistocene of Europe, and they had a common ancestor in North America in the Miocene.[83] The overall shape of the skull (flattened in comparison with *Castor*) is similar; both also had large incisor teeth, rounded in cross section and multigrooved, contrasting with those of beaver, which are flattened and smooth. Neither *Castoroides* nor *Trogontherium* would have been able to fell trees with these teeth, and the tail was rather narrower than in the beaver. The youngest available dates are ca. 12.74 kya from Ohio and ca. 11.82 kya from New York State.

6.19 EXTINCT MUSK OXEN: *EUCERATHERIUM* AND *BOOTHERIUM*

The shrub ox (*Euceratherium collinum*) is an imperfectly known extinct relative of the living musk ox, intermediate in size between a musk ox and a bison, that is recorded from several sites in the American Southwest and Mexico.[84] Its distinctive dung pellets have been likened in shape to Hershey's Kisses (a kind of chocolate shaped like a flat-bottomed teardrop), and both analyses of the contents and studies of the skull and teeth show that it was a browser. The youngest available date (on *E. collinum* dung from Bechan Cave) is 13.46 kya.

Another extinct musk ox relative, the helmeted or Harlan's musk ox (*Bootherium bombifrons*), had a much wider distribution taking in much of the contiguous United States, southern Canada, and Alaska/Yukon. It is better known than *E. collinum* because there are many more finds of skeletal material, as well as a near-complete frozen carcass discovered in Alaska. *Bootherium* was taller and more slenderly built than the extant musk ox and had larger horns.[85] There are significant differences in form between the skulls of males and females, to the extent that in former times male skulls were assigned to a different genus (*Symbos*). As pointed out by Dale Guthrie, these sexual differences likely reflect the predominant use of the horns for combat between males, whereas female horns had a purely defensive function.[86] The latest available date is ca. 12.86 kya from Alberta.

6.20 HARRINGTON'S MOUNTAIN GOAT: *OREAMNOS HARRINGTONI*

The living mountain goat (*Oreamnos americanus*) is native to southern Alaska, western Canada, and the adjacent United States, whereas the extinct

Harrington's mountain goat (*Oreamnos harringtoni*) was confined to the American Southwest, the Great Basin, and Mexico. *O. harringtoni* has been especially well studied because of the occurrence of well-preserved skeletal remains and dung pellets from dry caves in the Grand Canyon and elsewhere on the Colorado Plateau.[87] *O. harringtoni* was about 30% smaller than its extant relative, with a slenderer and more elongated skull together with cannon bones that were shorter and wider. Ancient DNA studies confirm that, while closely related, *O. harringtoni* and *O. americanus* are indeed separate species. Preserved hair shows that their coats were as white as the living animal; and no doubt they would have been equally skilled at negotiating steep rock faces, making them just as effective in avoiding predators. Major studies by Jim Mead and colleagues included extensive dating of dung pellets, the youngest of which are ca. 12.79 and 12.75 kya.[88] However, it should be noted that the dates, which were done more than 30 years ago, have large errors.

6.21 NORTH AMERICAN HORSES: *EQUUS* AND *HARINGTONHIPPUS*

The number of valid horse species in Last Glacial North America remains uncertain at present, but two groups can clearly be distinguished: caballine horses (*Equus* species) and slender-limbed, stilt-legged, "hemione-like" asses. Using aDNA, Weinstock and colleagues went some way to clarifying the confusion; suggesting that only two species may be represented.[89] Moreover, the New World stilt-legged horses (NWSL), previously regarded as related to Old World "hemionid" asses that originated in Eurasia, are in fact endemic to North America. A recent study based on mitochondrial and partial nuclear genomes from across their geographical range, concluded that the NWSL equid is distinct from *Equus* and that it merits its own genus.[90] They proposed a new genus, *Haringtonhippus* (species: *H. francisci*), named in honor of the eminent Canadian paleontologist Dick Harington, who first described the fossils.

Remarkably, all species of horse in both North and South America died out at the end of the Last Glacial, whereas they survived into the Holocene in northern Eurasia and of course were domesticated there. Even more remarkably, horses—descended from domesticated animals introduced by the Spanish in the sixteenth century—subsequently thrived and still thrive as feral populations (mustangs) in large areas of North America. The youngest date for the "stilt-legged" horse from Alaska is ca. 35.7 kya, suggesting that it disappeared from there before the Last Glacial Maximum.[91] Records of the same species from Wyoming and Nevada have yet to be dated. The latest dates on caballine horses are ca. 13.04 kya from Alberta, ca. 12.64 kya from California (La Brea), and ca. 14.50 kya (on the basis of many dates) from Alaska. Evidently caballine horses were extirpated significantly earlier in Alaska, coinciding with the time that the mammoth steppe was giving way to more mesic shrub vegetation.[92]

6.22 EXTINCT BISON: *BISON PRISCUS*, *B. LATIFRONS*, AND *B. ANTIQUUS*

Three extinct species of bison are known from North America. The steppe bison (*Bison priscus*), which was also widespread in northern Eurasia (chapter 5), was restricted to the northwest, from Alaska to British Columbia. Weighing about a tonne, like the modern American bison (*Bison bison*), it had larger horns that spanned about a meter. Mummified remains of this species are known from Alaska, notably including Blue Babe, discussed earlier in the context of predation by cave lions.[93]

The giant longhorn bison (*Bison latifrons*) carried much larger outspread horns, of which the bone horn cores alone (in males) spanned more than 2 meters; adding on the keratin horn sheaths, they would have been considerably larger.[94] Males are estimated to have weighed about 2 tonnes, although females were probably only half as heavy. The giant bison is recorded from much of the contiguous United States, with sparse records from southern Canada and from Mexico. In contrast to the grazing steppe bison, it is thought to have been a mixed feeder.

The third species (*Bison antiquus*), which probably weighed up to 1.5 tonnes, has many records from the southern half of the continent, notably at La Brea. It appears to have given rise to *Bison bison*, which was the largest North American species to survive extinction in the Last Glacial. A major study of the genetics of fossil *Bison* includes 201 radiocarbon dates on North American material but mostly does not distinguish species.[95] However, their aDNA analyses show that a large diverse population of bison existed in Beringia until ca. 37 kya, when its genetic diversity began to decline sharply.

6.23 EXTINCT PRONGHORNS: *STOCKOCEROS ONUSROSAGRIS* AND *S. CONCKLINGI*

There were probably four extinct species of pronghorn in North America during the Last Glacial, all of which sported four horns, contrasting with the sole surviving species (*Antilocapra americana*), which has just two.[96] The extinct *Stockoceros onusrosagris* and *S. conckling*i had bone horn cores that divided into two branches of approximately equal length (the outer horn sheaths, common to all pronghorns, are not preserved). *S. onusrosagris* was of similar size to the extant species, whereas *S. concklingi* was somewhat smaller. Both were rather stockier than *A. americana*, so presumably were not as fast. They are known from only a few localities—mainly from Mexico and the extreme south of the United States. However, San Josecito Cave, Nuevo León, and Papago Springs Cave, Arizona, have each produced remains of more than fifty individuals, providing compelling testimony that they were herd animals. Shuler's pronghorn (*Tetrameryx shuleri*) is known only from a handful of localities in Texas, Utah, and Mexico. Of its four horns, the two rear ones were much longer than those at the front. Lastly, there was the diminutive pronghorn (*Capromeryx furcifer*), which has the largest number of localities, mainly in Mexico, California, Texas, and New Mexico. Weighing only around 10 kilograms, it was decidedly not megafauna, but nevertheless went extinct.

6.24 OTHER EXTINCT SPECIES

Pampatheres—larger extinct relatives of the armadillo—are known from South and Central America and the southern United States. Both the southern pampathere *Pampatherium* and northern pampathere *Holmesina septentrionalis* were encased in flexible armor plates and probably each weighed about 250 kilograms. *Holmesina* is recorded from a number of sites, mainly in Texas and Florida, while *Pampatherium* is rarer. Like armadillos, pampatheres probably fed on insects and other invertebrates and were similarly restricted to the south of the continent, where such food was more plentiful.[97]

Today, peccaries are typical of South and Central America, although one species, the collared peccary (*Pecari tajacu*), is also common in southern Arizona and southern Texas. Two extinct peccaries are known from the Last Glacial in North America. The long-nosed peccary (*Mylohyus nasutus*) had long, slender legs, implying that it was a better runner than most extant species.[98] The flat-headed peccary (*Platygonus compressus*) also had long legs, and its skull structure suggests that it had a highly developed sense of smell. Like extant peccaries both would have been omnivorous, taking a wide range of food such as roots, fruits, cactus, insects, and small vertebrates. *Mylohyus* is recorded from the eastern half of the United States westward to Oklahoma and Texas, while *Platygonus* was much more widespread, ranging across most of the contiguous United States, with an outlying record (of Last Interglacial age) from the Yukon. Sheriden Cave, Ohio, has produced the latest dates for both species: ca. 13.67 kya (*Mylohyus*) and ca. 12.92 kya (*Platygonus*).[99]

At least two species of tapir have been recognized from the Last Glacial of North America: *Tapirus californicus* in the southwest and *Tapirus veroensis* in the east, extending into Mexico. Both were rather larger than extant South American tapirs, and like them probably browsed in wooded areas with ready access to water.

The stag moose or elk moose (*Cervalces scotti*) was a large deer, in which the males bore antlers larger and more elaborate than those of the extant moose, to which it was closely related.[100] It was mainly distributed to the south of the Great Lakes, but there are also several records from Alaska and the Yukon. A similar species, *Cervalces latifrons*, is known from the Middle Pleistocene of northern Eurasia. The youngest date on *C. scotti* is 13.24 kya from Illinois.

The mountain deer (*Navahoceros fricki* aka *Odocoileus lucasi*) is known from only a few localities, all in the central region of the United States and Mexico. Intermediate in size between white-tailed deer and elk, it bore simple three-tined antlers, while its robust limb bones and very short broad cannon bones suggest that it was well adapted to rock climbing, like mountain goats and the European chamois.[101] The youngest available date is ca. 42.08 kya from Nevada.

Llamas and guanacos—relatives of the Old World camels—are seen as typical South American animals. However, two extinct species of llama were present in Last Glacial North America as well.[102] The stout-legged llama (*Palaeolama mirifica*), weighing about 200 kilograms, had relatively low-crowned cheek teeth and was probably a browser or mixed feeder. Known from much of South and Central America, it is also recorded from Texas, Florida, and South Carolina. The longer and more slender limbs of the large-headed llama (*Hemiauchenia macrocephala*) indicate that it was a faster runner, which, with its high-crowned cheek teeth embedded in abundant cementum (helping to resist pressure when chewing), suggests that it was better adapted to grazing and open environments than was *Palaeolama*.[103] At about 300 kilograms, it was also larger. *H. macrocephala* was widespread in South and Central America and also across much of the United States, especially in the South and West. The youngest date for *Hemiauchenia* is ca. 40.97 kya from Nevada.

The capybara (*Hydrochoerus hydrochaeris*), which averages about 50 kilograms, is by far the largest rodent alive today. It is semiaquatic, inhabiting large areas of South America, including dense forest and savannas. At more than 100 kilograms, the extinct *Neochoerus pinckneyi* was considerably larger but was probably otherwise similar in appearance and ecology, feeding largely on grasses and aquatic plants.[104]

The saiga antelope (*Saiga tatarica*) is now restricted to the steppes and semideserts of Kazakhstan, southwestern Russia, and western Mongolia (chapter 5), but during the Last Glacial its range extended via northeastern Siberia into Alaska, the Yukon, and just into the Northwest Territories.[105] It was probably extirpated from North America soon after its last recorded date of ca. 14.61 kya, in the face of increased snowfall and the replacement of dry, open mammoth steppe with shrubby tundra vegetation

With a very different ecology, the extinct giant tortoise *Geochelone crassiscutata* is recorded as far north as Illinois during the Last Interglacial (Sangamonian), indicating a climate significantly warmer than today.[106] It presumably disappeared in the face of decreasing temperatures at the onset of the Last Glacial, as did *Hippopotamus* in Europe.

6.25 THE PEOPLING OF NORTH AMERICA

In contrast to Africa and Eurasia, only modern humans (*Homo sapiens*) are known to have colonized North and South America and did not do so until late in the Last Glacial period, almost certainly entering via Beringia from northeastern Siberia.[107] Earlier studies concluded that people had probably reached Alaska by ca. 15 ka.[108] However, recent work on the Bluefish Caves, Yukon Territory, indicates human arrival by ca. 24 ka (see chapter 4).[109] During the Last Glacial Maximum, most of the northern half of North America was covered by a huge ice sheet that cut off Alaska and the Yukon, which were largely ice free, from the rest of the continent. For several decades, the "Clovis first" theory held sway, in which ancestors of the Clovis people were believed to have been the first to penetrate south of the ice sheets to what is now the contiguous United States. Previously it was believed that they were blocked from moving southward until ca. 13 to 12 kya, when deglaciation opened a 1,500 kilometer-long ice-free corridor between the western (Cordilleran) and eastern (Laurentide) ice sheets.[110] The highly distinctive Clovis stone spear points (named after Clovis, New Mexico, from where they were first described) are known only from south of the ice sheets. The beautifully flaked points were fluted at the base, which enabled them to be readily attached to a shaft. The Clovis culture was estimated to range from ca. 13.2 kya to 12.8 kya.[111] However, more recently, stone flakes, Clovis points and gomphothere remains from El Fin del Mundo in northwestern Sonora, Mexico, were dated (associated dates on charcoal) to about 13.4 kya (see above).[112] Although there is substantial evidence for earlier sites, Clovis is the most widespread, abundant, and readily recognizable, and there is compelling evidence that Clovis people hunted Columbian mammoths and mastodons with the aid of their signature spear points.

The "Clovis first" theory is closely entwined with Paul Martin's "Blitzkrieg" hypothesis envisaging "big-game" hunters emerging from the southern exit of the ice-free corridor into virgin territory and rapidly exterminating most of the megafauna, which until then had never encountered humans ("naive prey"). However, there are problems with this scenario, in the shape of older archaeological sites. In particular, Monte Verde in southern Chile, dated to ca. 14.8 kya, strongly argues for earlier human colonization of North America, as it is significantly older than the opening of ice-free corridor.[113] It seems highly unlikely that people could have colonized South America other than via North America, and the far-fetched theory that Clovis ancestors had crossed the Atlantic from Europe was effectively laid to rest by aDNA analysis of a Clovis burial demonstrating close similarities both with other indigenous people of the Americas and with people from northeastern Siberia.[114] Other researchers have promoted the hypothesis that people utilized an earlier route along the islands of the Pacific coasts of North, Central, and South America, which were deglaciated from ca. 15 kya. Attracted by rich marine resources and almost certainly using boats or rafts, it is plausible that in a few generations at most, people could have made their way from island to island all the way from Alaska to Chile, a distance of some 14,000 kilometers. But finding archaeological evidence for such a coastal migration route is challenging since most of the likely sites would have been drowned by the Holocene global sea level rise resulting from melting of the ice sheets. However, recently the coastal route hypothesis has received powerful indirect support from multidisciplinary studies of lake sediments from within the ice-free corridor region, which indicate that this would not have provided a viable route until after 12.6 kya—several centuries later than the earliest Clovis.[115]

Moreover, there are now several sites with good evidence for the presence of people south of the ice sheets substantially earlier than the Clovis culture. Especially notable are the Paisley Caves, Oregon, one of which has produced the earliest directly dated evidence of people in North America in the form of human coprolites (identified by aDNA) preserved by the arid environment.[116] Due to the likely significance of the finds for the timing of human arrival in North America, the oldest coprolite was independently radiocarbon dated by the Oxford and Beta labs, which were in good agreement with a calibrated date of 14.27 to 14 kya. The earlier levels of pre-Clovis age yielded only small undiagnostic stone artifacts. However, the younger Clovis-age levels contained Western stemmed projectile points—a technology distinct from Clovis that existed in parallel in the western half of North America.

On the basis of mitochondrial aDNA analyses of pre-Columbian human remains from South America, Llamas and colleagues inferred a population burst at ca. 16 kya in North America south of the ice sheets.[117] As this considerably predates the opening of the inland ice-free corridor from ca. 11.5 to 11.0 kya, it provides further support to the coastal route hypothesis.

A highly controversial claim for a very much older presence of hominins (perhaps Neanderthals or Denisovans), in North America has been made on the basis of the smashed bones and teeth (presumably for marrow extraction) of an individual mastodon, *Mammut americanum*, with associated large cobbles from the Cerutti Mastodon Site, California. Thorium-uranium (^{230}Th/U) analyses of multiple bone specimens from the site indicate a burial date of 130.7 ± 9.4 kya, which is very broadly of Last Interglacial age.[118] The deposits yielded "spiral-fractured bone and molar fragments, indicating that breakage occurred while fresh," and "the occurrence and distribution of bone, molar and stone refits suggest that breakage occurred at the site of burial. Five large cobbles (hammerstones and anvils) in the CM bone bed display use-wear and impact marks and are hydraulically anomalous relative to the low-energy context of the enclosing sandy silt stratum."[119] However, corroborative evidence, including distinctive artifacts, from other sites would be required before most authorities would accept that hominins had arrived in North America some 100,000 years earlier than previously accepted.

6.26 EXTINCTION PATTERNS

The richness of the North American fossil record offers huge potential toward answering the key question: what caused megafaunal extinctions? Detailed studies on northern Eurasia have shown marked differences in the timing of megafaunal extinctions from one region to another, as described in chapter 5; so we might reasonably expect to find similar patterns for North America (for those species that have wide geographical ranges). However, at present we are far short of being able to do this, as we don't even have good radiocarbon coverage for most of the extinct species (including reasonable estimates of last appearance dates, or LADs), let alone sufficient data to explore geographical variations adequately. We need very many more radiocarbon dates made directly on megafaunal remains (securely identified to species) from a range of geographical regions across North America. Of course, this would be a huge task, but one that is vital to understanding the process of megafaunal extinctions on that continent. Realistically, in the first instance it would probably be best to focus on dating several of the more abundant species from across their geographical ranges, and to substantially improve the date coverage for all the species.

However, there are many available radiocarbon dates from Alaska/Yukon (eastern Beringia) based on large quantities of well-preserved remains recovered from frozen ground in the course of placer gold mining, providing good radiocarbon-date coverage for several species. Some geographical variation in LADs is apparent when compared with the rest of North America. The latest of eight dates for giant short-faced bear (*Arctodus simus*) from Alaska is ca. 24.7 kya, compared with ca. 12.8 kya (Kansas) and ca. 12.7 kya (Texas), suggesting that it survived very much longer south of the ice sheets.[120] The youngest of eleven dates for scimitar cat (*Homotherium serum*) from Alaska/Yukon is ca. 40.78 kya, while the only date so far from south of the ice sheets, in Minnesota, is ca. 26.48 kya.[121] The latest Alaska/Yukon date for stilt-legged horse (*Haringtonhippus francisci*) is ca. 35.7 kya. It seems likely that it lasted much later in the south, but currently we have no dates to confirm this. The available evidence suggests that all three species were extirpated from Alaska/Yukon before the Last Glacial Maximum, probably in response to decreasing temperatures. As far as they go, these results are consistent with a staggered pattern of losses (much as seen in northern Eurasia) and imply that significant megafaunal disappearances occurred here well before the arrival of humans (ca. 24 kya).

The latest known date of ca. 14.67 kya on a caballine horse (*Equus* species) from Alaska/Yukon compares with ca. 12.74 kya for Alberta, strongly suggesting significantly later survival south of the ice sheets. In mainland Alaska/Yukon, woolly mammoth (*Mammuthus primigenius*) survived to at least ca. 13.34 kya, cave lion (*Panthera spelaea*) to ca. 13.75 kya, and saiga antelope (*Saiga tatarica*) to ca. 14.61 kya. Today saiga is entirely absent from North America but still survives in a limited area of northern Eurasia (mainly Kazakhstan; chapter 5). There is no evidence from Alaska/Yukon for an extinction event at ca. 12.9 kya, as predicted by the impact hypothesis.

The significantly warmer climate of the Last Interglacial (broadly 130 to 117 kya) evidently allowed several species to extend their ranges as far north as Alaska/Yukon: yesterday's camel (*Camelops hesternus*), Jefferson's ground sloth (*Megalonyx jeffersonii*), American mastodon (*Mammut americanum*; fig. 6.14), flat-headed peccary (*Platygonus compressus*), and giant beaver (*Castoroides ohioensis*).[122] Presumably they were subsequently extirpated from this region because of deteriorating temperatures at around the beginning of the Last Glacial.

All these species survived much later in the contiguous United States, with the following LADs: *Camelops*, ca. 13.06 kya; *Megalonyx*, ca. 13.1 kya; *Mammut*, ca. 11.58 kya; *Platygonus*, ca. 12.92 kya; and *Castoroides*, ca. 11.82 kya (table 6.1). The available radiocarbon dates from the contiguous United States and the extreme south of Canada (which were south of the ice sheets) record

FIGURE 6.14. Partial skeleton of a Last Interglacial American mastodon, *Mammut americanum*, Yukon Territory. (Courtesy of Grant Zazula.)

the loss of about thirty megafaunal species between ca. 43 kya and ca. 11.6 kya. Of these, eighteen have LADs between 13.7 and 11.6 kya. Several other species of likely Last Glacial age currently have no dates at all. Tyler Faith and Todd Surovell proposed that megafaunal extinction in North America was a sudden event, consistent with catastrophic hunting by humans, as advocated so passionately by the late Paul Martin.[123]

They recognized the LADs of sixteen out of thirty-five genera of mammals as securely falling between ca. 13.8 kya and ca. 11.4 kya. Although on the face of it the latest dates on the other nineteen genera might lend to the argument that these animals disappeared earlier, the authors believe that this is an artifact due to small sample size ("Signor-Lipps effect"[124]) obscuring the "true" picture of sudden extinction of all the genera. They claim that the evidence is consistent with "an extinction mechanism that is capable of wiping out up to 35 genera across a continent in a geologic instant," and "North American late Pleistocene extinctions are best characterized as a synchronous event"—a "sudden surge in extinction rates consistent with overkill or extra-terrestrial impact." (This issue is discussed in chapter 13.)

Many researchers (including the author) would argue that we require vastly more evidence, in the form of many more radiocarbon dates, to determine the chronological pattern of extinctions much more accurately before we can

TABLE 6.1 Youngest direct radiocarbon dates for extinct North American megafauna

Species	Lab. no.	Cal BP	Median	Site	Source
Megalonyx jeffersoni	UCIAMS-116401	13,034–13,180	13,099	Millersburg (OH)	1
Nothrotheriops shastensis **[D]**	Ua-12506	12,270–13,061	12,883	Gypsum Cave (NV)	3
Paramylodon harlani	UCIAMS78125, AA87426 US	14,603–14,213	14,408	Willamette Valley (OR)	13
Arctodus simus	AA-17511	24,237–25,246	24,727	Fairbanks area (AK)	4
Arctodus simus	NZA-28889	12,700–12,917	12,776	Bonner Springs (KS)	5
Canis dirus	LACMHC H1872	13,560–13,741	13,652	La Brea (CA)	6
Smilodon fatalis	LACMHC 35915	13,093–13,272	13,179	La Brea (CA)	6
Homotherium serum	CAMS-131351	38,390–43,225	40,783	Fairbanks area (AK)	4
Homotherium serum	NZA-30409	26,121–26,956	26,481	Tyson Spring Cave (MN)	15
Panthera atrox	LACMHC X7113	13,302–13,487	13,413	La Brea (CA)	7
Panthera atrox	OxA-12900	13,087–13,300	13,198	Edmonton (AB)	7
Panthera spelaea	OxA-10080	13,094–13,644	13,290	Fairbanks Creek (AK)	7
Castoroides ohioensis	CAMS-26783	12,673–12,838	12,738	Sheriden Cave (OH)	18
Castoroides ohioensis	NOSAMS OS-73632	11,501–12,050	11,819	Clyde (NY)	17
Equus sp. (caballine)	CAMS-82411	12,693–12,813	12,743	Pashley Gravel Pit (AB)	8
Equus sp. (caballine)	AA-26830	14,238–15,075	14,666	Upper Cleary (AK)	9
Harringtonhippus francisci	AA-26780	33,480–38,883	35,696	Ester Creek (AK)	9
Mylohyus nasutus	Beta-139687	13,571–13,762	13,667	Sheriden Cave (OH)	18
Platygonus compressus	CAMS-10349	12,775–13,070	12,923	Sheriden Cave (OH)	18
Camelops hesternus	SR-5893	13,043–13,235	13,119	Pine Springs (WY)	8
Camelops hesternus	CAMS-61899	12,933–13,159	13,063	Casper (WY)	20
Hemiauchenia macrocephala	Beta-134248	40,264–41,606	40,967	Mineral Hill Cave (NV)	2
Navahoceros fricki	Beta-134241	41,465–42,700	42,083	Mineral Hill Cave (NV)	16
Cervalces scotti	Beta-127907	14,103–15,286	14,720	Sheriden Cave (OH)	18

Cervalces scotti	CAMS-82932	13,123–13,364	13,241	Lang Farm (IL)	19
Saiga tatarica	AA-3075	14,242–14,995	14,612*	Lost Chicken Creek (AK)	9
Euceratherium collinum **[D]**	Beta-18269	13,165–13,760	13,464	Bechan Cave (UT)	8
Oreamnos harringtoni **[D]**	R-1134	10,428–13,049	11,739	Rampart Cave (AZ)	8
Bootherium bombifrons	TO-7691	12,716–13,030	12,856	Wally's Beach (AB)	8
Mammut americanum	NZA- 29627	11,345–11,795	**11,576**	Overmyer (IN)	10
Mammuthus columbi	AA-2941	12,124–12,705	12,479	Dent (CO)	2
Mammuthus columbi **[D]**	A-3212	12,825–14,350	13,548	Bechan Cave (UT)	11
Mammuthus exilis	CAMS-24429	13,954–16,524	15,260	Santa Rosa Island (CA)	12
Mammuthus exilis	CAMS-71697	12,750–13,035	12,895	Santa Rosa Island (CA)	12
Mammuthus primigenius	AA-17601	13,096–13,706	13,374	Delta (AK)	9
Mammuthus primigenius	AA-22573	13,064–13,709	13,341	Galena (AK)	9
Mammuthus primigenius	UCIAMS-149817	5333–5585	**5530**	Saint Paul Island (AK)	14

D: date on dung. Cal BP: calibrated date before present. Boldface dates: Holocene.

AB: Alberta. AK Alaska. AZ: Arizona. CA: California. CO: Colorado. IL: Illinois. IN: Indiana. KS: Kansas. MN: Minnesota. NV: Nevada. NY: New York. OH: Ohio. OR: Oregon. UT: Utah. WY: Wyoming.

Sources: 1. McDonald, Stafford, and Gnidovec 2015. 2. Waters and Stafford 2007. 3. Hofreiter et al. 2000. 4. Fox-Dobbs, Leonard, and Koch 2008. 5. Schubert 2010. 6. Fuller et al. 2014. 7. Barnett, Yamaguchi, et al. 2006; Stuart and Lister 2011. 8. Faith and Surovell 2009. 9. Guthrie 2006. The "stilt-legged horse" was named *Harringtonhippus francisci* by Heintzman, Zazula, and R. D. E. MacPhee 2017. 10. Woodman and Athfield 2009. 11. Mead and Agenbroad 1992. 12. Agenbroad, Johnson, et al. 2005. 13. Gilmour et al. 2015. 14. Graham, Belmecheria, et al. 2016. 15. Widga et al. 2012. 16. Hockett and Dillingham 2004. 17. Feranec and Kozlowski 2010. 18. Tankersley 2011. 19. Schubert et al. 2004. 20. Frison 2000.

adequately test possible causes. The dates available at present from North America, with the exception of Alaska/Yukon, are inadequate for constructing a reliable chronology of megafaunal extinctions. There is an overwhelming need for an extensive program of direct radiocarbon dating of a wide range of North American megafaunal species. Only then will we have a sound basis for solving the problem of cause or causes of megafaunal extinctions in this region. On similar lines, Don Grayson argued that "the continuing debate over the causes of North American losses is not likely to be resolved unless the history of each species is analyzed individually. . . . : Recent advances in understanding Eurasian extinctions provide a research guide for extracting ourselves from the explanatory morass that now characterizes the North American situation."[125]

6.27 TESTIMONY OF THE DUNG FUNGUS

Further evidence bearing on megafaunal extinctions is provided by studies of the changing abundance of the dung fungus *Sporormiella* in lake sediments.[126] Marked declines in the abundance of *Sporormiella* at Appleman Lake, Indiana, and several New York sites have been interpreted as reflecting a megafaunal decline ca. 14.8 to 13.7 kya, during the Bølling–Allerød warm period, before both the Younger Dryas cooling and the conjectured extraterrestrial impact event at ca. 12.9 kya. This inferred decline in megafaunal populations occurred at least several centuries before their final extinction.

6.28 SURVIVING MEGAFAUNA

The larger surviving mammals of the region are illustrated in figure 6.1: polar bear (*Ursus maritimus*), grizzly or brown bear (*Ursus arctos*), black bear (*Ursus americanus*), mountain lion (*Puma concolor*), jaguar (*Panthera onca*), pronghorn (*Antilocapra americana*), wapiti or North American elk (*Cervus canadensis*), caribou (*Rangifer tarandus*), mule deer (*Odocoileus hemionus*), white-tailed deer (*Odocoileus virginianus*), moose (*Alces alces*), musk ox (*Ovibos moschatus*), bighorn sheep (*Ovis canadensis*), Dall sheep (*Ovis dalli*), mountain goat (*Oreamnos americanus*), and American bison (*Bison bison*). Happily, none of these is listed as endangered in the IUCN Red List. The American bison was brought close to extinction during the nineteenth century as a result of indiscriminate and commercial hunting, but following active conservation efforts in the twentieth century, their numbers have increased considerably, although far short of their original numbers.

CHAPTER 7

South America: Ground Sloths and Glyptodonts

Finds of huge bones in South America have long been the object of wonder and imaginative speculation. The eighteenth-century English Jesuit Thomas Falkner wrote:[1]

> On the banks of the River Carcarania, or Tercero (*Patagonia*), about three or four leagues before it enters into the Parana, are found great numbers of bones, of an extraordinary bigness, which seem human. There are some greater and some less, as if they were of persons of different ages. I have seen thigh-bones, ribs, breast-bones, and pieces of skulls. I have also seen teeth, and particularly some grinders which were three inches in diameter at the base. These bones (as I have been informed) are likewise found on the banks of the Rivers Parana and Paraguay, as likewise in Peru. The Indian Historian, Garcilasso de la Vega Inga, makes mention of these bones in Peru, and tells us that the Indians have a tradition, that giants formerly inhabited those countries, and were destroyed by God for the crime of sodomy.

Later investigations revealed that such finds were the remains not of giant human sinners, but of giant extinct animals. In 1787, on the banks of the Luján River in Argentina, Father Manuel Torres discovered the first known skeleton of the iconic giant ground sloth *Megatherium* (see below). Subsequently in the course of the *Beagle* voyage, Charles Darwin discovered and collected remains of *Toxodon*, *Macrauchenia*, glyptodonts, ground sloths, and other extraordinary animals from Uruguay and Argentina. The South American Lujanian Land Mammal Age takes its name from the Luján River. From the association

FIGURE 7.1. Extinct and extant megafauna from South America (Neotropic Ecoregion). From the top, row 1, left to right: *Notiomastodon platensis*; *Cuvieronius hyodon*; *Toxodon platensis*. Row 2: *Lestodon armatus*; *Glossotherium robustum*; *Megatherium americanum*; *Eremotherium laurillardi*. Row 3: *Catonyx cuvieri*; *Mylodon darwini*; *Scelidotherium leptocephalum*; *Holmesina septentrionalis*. Row 4: *Glyptodon clavipes*; *Doedicurus clavicaudatus*; *Panochthus* sp.; *Neosclerocalyptus* sp. Row 5: *Hippidion saldiasi*; *Equus* sp.; *Hemiauchenia* sp.; *Macrauchenia patachonica*. Row 6: *Neochoerus* sp.; *Arctotherium tarijense*; *Smilodon populator*; *Canis dirus*. Row 7: *Tapirus bairdii*; *Tapirus terrestris*; *Blastocerus dichotomus*; *Lama guanicoe*; *Myrmecophaga tridactyla*. Row 8: *Hydrochoerus hydrochaeris*; *Puma concolor*; *Panthera onca*; *Tremarctos ornatus*; *Crocodylus acutus*; *Eunectes murinus*.
Black: selected extinct species. Gray: selected living species. The outlined *Homo sapiens* gives approximate scale.

of their remains with mollusk shells of modern species, he correctly inferred that these giants had still existed in geologically recent times.[2] On the basis of Darwin's material, between 1837 and 1845 Richard Owen described eleven species or genera that were then new to science.

During the Quaternary, South America exhibited a hugely rich and varied fauna (fig. 7.1), resulting from the admixture of extraordinary endemic animals that had evolved during millions of years when it was an isolated island continent, and the immigration of North American forms as part of the Great American Biotic Interchange (chapter 6).

Closure of the Central American Seaway at about 2.8 mya was followed by the main interchange pulse that began ca. 2.6 mya (Pliocene).[3] In the late Quaternary, most South American megafauna was wiped out, amounting to a loss of at least 70% of species. The severity of these extinctions is strikingly illustrated by the fact that today, by far the largest native mammals are Baird's tapir and the lowland tapir, which average only about 350 kilograms and 250 kilograms respectively. Interestingly, tapirs are predominantly forest animals, while most losses were of open grassland (pampas) species. As in North America, all species of horse disappeared.

7.1 GIANT GROUND SLOTH: *MEGATHERIUM AMERICANUM*

Ground sloths are among the most extraordinary and bizarre mammals to have walked the earth. Their nearest living relatives are the much smaller, famously slow-moving tree sloths of South and Central America. At least eleven species of ground sloth were present in Last Glacial South America, and several sloths made it into North America as part of the Great American Biotic Interchange. Ground sloths came in a range of sizes from the giants *Megatherium* and *Eremotherium*, at around 3 to 4 tonnes—as big as some elephants—down to the smallest *Nothrotherium* at about 150 kilograms. Large quantities of vegetation would have been dealt with by high-crowned, peg-like teeth, used for cutting plant material rather than chewing, and a large body cavity that evidently housed a capacious gut for digestion of cellulose by bacterially assisted fermentation. Ground sloths are generally thought to have had an unhurried life style comparable to their modern relatives' that would have enabled them to process large amounts of plant food at leisure. Nevertheless, the abundance of recognizable plant remains in the dung of several species indicates that they were not efficient digesters.

From detailed reconstructions of muzzle shape in five species of ground sloth, Bargo and colleagues were able to reconstruct their likely diets and methods of feeding.[4] The wide-muzzled *Glossotherium robustum* and *Lestodon armatus*, which had square, non-prehensile upper lips were interpreted as mostly bulk feeders, with lips and tongue used to pull up grass and herbaceous plants. On the other hand, the narrow-muzzled *Megatherium americanum*, *Mylodon darwini*, and *Scelidotherium leptocephalum* had conical prehensile lips, indicating mixed or selective feeders. The prehensile lip would have been used to select particular plants or plant parts.

The giant ground sloth is arguably the best-known extinct animal of Ice Age South America. The first *Megatherium* remains to come to scientific attention were found in 1787 on the banks of the Luján River about 65 kilometers west of Buenos Aires, in the Provincias Unidas del Río de la Plata (Argentina).[5] In 1789, the bones—of what proved to be a near-complete skeleton, although lacking the tail—were sent to Madrid by the viceroy of Río de la Plata, the Marques de Loreto. The skeleton, some 4 meters long, was assembled and mounted in more or less correct anatomical order by Juan Bautista Bru, although (understandably, given the novelty of the find) in an awkward unnatural pose, and installed "on a grandiose pedestal in the hall of petrified fossils in the Royal Cabinet."[6] Bru also proceeded to make a detailed study including drawings of the skeleton and individual bones. On the basis of Bru's drawings and descriptions, which he reproduced (fig. 7.2), the celebrated French comparative anatomist Baron Georges Cuvier named the creature *Megatherium americanum* ("giant beast of America").[7] Crucially, he recognized that the animal was extinct, although with similarities to the much smaller tree sloths that still live in Central and South America. Standing on its hind legs, *M. americanum* reached about 4 meters high and probably weighed about 4 tonnes (fig. 7.3).[8] A trackway attributed to *Megatherium* at Pehuén-Có, Argentina, clearly demonstrates that the animal walked bipedally with its body weight mainly supported by its massive hind legs, touching the ground more lightly with its smaller front feet.[9] The average speed would have been about 1.21 meters per second and a top speed of 1.68 meters per second (compared with a normal human walking speed of 1.4 meters per second). Several smaller species of *Megatherium* have been proposed, mainly from western South America, and some of these may turn out to be valid.[10]

The teeth of ground sloths, tree sloths, armadillos, and their other relatives (Xenarthrans) are rootless and continuously growing. Lacking enamel, they comprise dentine with or without an outer layer of cementum (due to their specialized method of feeding, anteaters lack teeth entirely). Much interest

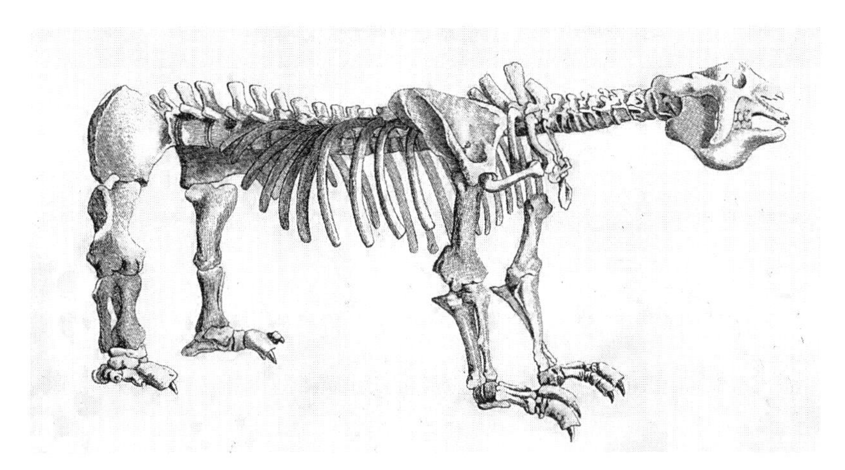

FIGURE 7.2. Earliest (eighteenth-century) reconstruction of *Megatherium americanum* skeleton by Bautista Bru, as reproduced by Georges Cuvier.

has been shown in the likely diet of *Megatherium americanum*, including the suggestion that it may have been a scavenger, or sensationally even an active predator. Fariña and Blanco proposed that this huge ground sloth had morphological features consistent with some carnivorous habits.[11] In particular, they drew attention to the forearms, which they thought were better suited to the aggressive use of its large claws than for pulling down branches to feed. They further suggested that "the high mechanical advantage of the biceps might have made it possible for the animal to have lifted and carried heavy weights. This, in turn, suggests the possibility that the animal could have manipulated large prey (for instance, turning dorsally armored prey or carcasses upside down to expose softer parts) and cached large food pieces in a safer place. By this view, *Megatherium americanum* would be the largest land mammal hunter to have existed."[12] This scenario conjures up a rather comical mental picture of *M. americanum* feeding on a glyptodont carcass, with the latter's inverted carapace providing a ready-made bowl. However, more prosaically, microwear patterns on its teeth indicate that the giant ground sloth's diet comprised plants of "low to moderate intrinsic toughness."[13] The bilophodont teeth, intermeshing between the upper and lower jaws, are notable for their strongly pronounced transverse sharp cutting edges maintained by a self-sharpening process.[14] Its narrow muzzle and inferred prehensile lip[15] indicate selective browsing rather than grazing; moreover, a sample of

FIGURE 7.3. Giant ground sloth *Megatherium americanum* skeleton (cast). This engraving is from the nineteenth century; the cast is on display at the Natural History Museum, London.

coprolite material from Argentina, probably from *Megatherium*, contained mainly woody plant remains.[16]

At first sight, ground sloths—slow-moving and lacking obvious armor, tusks, or horns—would appear to have been almost defenseless against large predators such as sabertooth cats, jaguars, and giant short-faced bears; in fact, they possessed a range of formidable defensive features, which no doubt is a major reason why they were so successful for so long. Ground sloths exhibited a wide range of form, but all had long claws that could be used for digging and would also have been very effective for combating predators. A broad pelvis and massive tail, for example, in *Megatherium* and *Eremotherium* not only allowed these animals to stand upright and feed on leaves and branches of trees, which were beyond the reach of other herbivores, but also would have enabled them to employ their claws to ward off would-be attackers. Like their living relative the giant anteater (*Myrmecophaga tridactyla*), they walked on the sides of their feet, thus protecting their claws. This gait, however, would have precluded rapid locomotion. In mylodont ground sloths (*Mylodon* and *Glossotherium*), the deeper layer of skin (dermis) was studded with bony nodules (osteoderms), which would have offered substantial protection against the teeth of carnivores such as *Smilodon* and *Arctotherium*, although perhaps would have been rather less effective against human hunters using long-range projectile weapons: spear throwers (atlatls) or bows and arrows. The youngest dates for *M. americanum* are ca. 9 kya and 8.38 kya (both early Holocene) from Campo Laborde, Argentina.

7.2 LAURILLARD'S GROUND SLOTH, *EREMOTHERIUM LAURILLARDI*

Weighing approximately 3.5 to 4 tonnes, *E. laurillardi* (aka *E. rusconii*) was another elephant-size ground sloth, equal in size to or perhaps even larger than *Megatherium americanum*.[17] Although anatomically broadly similar, there were several differences between the two: the skull was less robust and the limb bones longer in *E. laurillardi* than in *M. americanum*, and there were only three digits in the hand of the former, compared with four in the latter (fig. 7.4). The upper and lower jaw (maxilla and dentary) are deeper in *M. americanum* than in *E. laurillardi*, accommodating higher-crowned cheek teeth. *M. americanum* was "apparently better suited for consuming a variety of turgid or moderate to soft tough food items."[18] Additionally, "*E. laurillardi* inhabited more tropical to subtropical, closed or forested environments, and is considerably less hypsodont than *M. americanum*, which inhabited a more temperate, arid to semiarid environment."[19] *E. laurillardi* has been called the "Panamerican ground sloth" because of its exceptionally wide distribution, extending from the extreme south of Brazil (Rio Grande do Sul) through the tropics of South and Central America, and as far north as the southern United States. On the other hand, *M. americanum* was restricted to the southern, more temperate regions of South America, including Argentina and Uruguay, but not reaching as far north as Brazil. Although so far no actual preserved skin

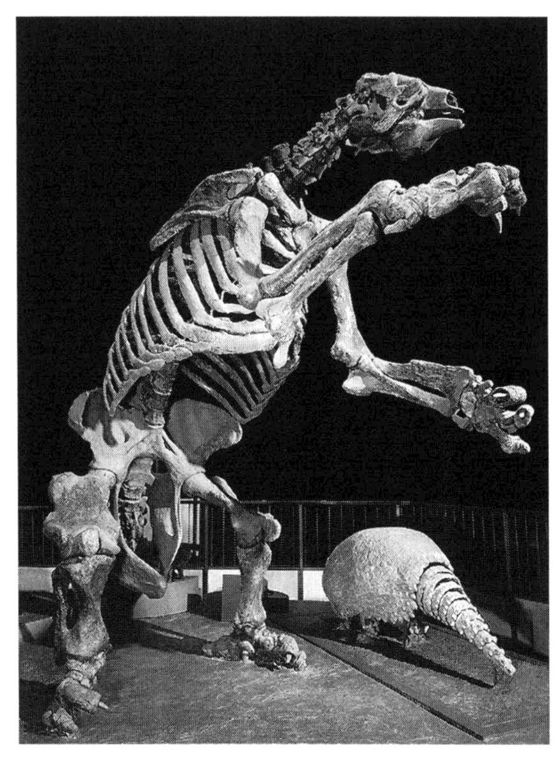

FIGURE 7.4. Mounted skeleton of a Laurillard's ground sloth *Eremotherium laurillardi* (aka *E. rusconii*), from Panama. Note the deeper mandible compared with *M. americanum*, and only three digits on the forelimbs. On the right is a skeleton of *Glyptotherium arizonae* from North America. (Courtesy of Smithsonian Institution–National Museum of Natural History.)

has been discovered, finds in association with its skeletal remains strongly suggest that, like *Mylodon*, *Eremotherium* had osteoderms embedded in its skin.[20]

In 1950 and 1951, expeditions from the Smithsonian Museum in Washington DC excavated 9 tons of fossil material, including many bones of Laurillard's ground sloth from El Hatillo, southern Panama. Dr. C. Lewis Gazin, then curator of the Division of Vertebrate Paleontology who led the expedition, wrote that the fossils were found "in a mud deposit in the vicinity of a large and swampy spring."[21]

The excavated bones were encased in jackets of burlap dipped in wet plaster that was allowed to harden for safe lifting—a time-honored method, almost universally still used today. The excavators had also applied tissue paper soaked in gum acacia, partly to harden the fossils and partly to lessen adhesion of plaster to the bone surface. (Nowadays we generally use aluminum foil as a separator and reinforce the jackets, for example, with metal mesh; if necessary, we employ a range of consolidants, such as Paraloid.) The continual seepage of water presented problems for the excavators, as it interfered with the setting of the plaster jackets. An additional complication was the necessity to avoid leaving a hole that would fill with stagnant water and create an ideal breeding ground for mosquitoes. However, despite the difficulties, the partial skeletons of between eight and twelve ground sloths were collected. With the aid of the US Air Force—who provided a truck and crane—two composite skeletons were shipped to the Smithsonian's National Museum of Natural History, where they were assembled. An entry in the museum's 1953 annual report, published shortly after the discovery, asserted that "the skeleton of this huge, uncouth creature promises to be a spectacular addition to the hall of vertebrate paleontology."[22] Until recently, the two giant ground sloth skeletons were displayed in the Ice Age Hall of the museum, with the larger specimen about 4 meters tall when mounted in a standing position (fig. 7.4). The latter also features in *David H. Koch Hall of Fossils—Deep Time*, which opened in June 2019.

The youngest available dates on *E. laurillardi* are ca. 14.79 kya from Abismo do Fossil and ca. 13.19 kya from Itaituba, both in Brazil.

7.3 DARWIN'S GROUND SLOTH: *MYLODON DARWINII*

The huge Mylodon Cave (Cueva del Milodón), approximately 250 meters long, 145 meters wide, and 30 meters high, is situated in Ultima Esperanza ("Last Hope") Province near the southern tip of South America in Chilean Patagonia. This remote area, formerly visited by very few, has now become a popular tourist attraction. The enduring interest and fascination with this site began in 1895, when the German explorer Hermann Eberhard, who had recently settled in the area, discovered a piece of hairy hide of an unknown creature in a cave on his land.[23] The find was all the more puzzling due to the numerous bony nodules embedded on the inside of the skin. Subsequent discoveries included several more pieces of skin, long claws (with keratin sheath), and well-preserved skeletal remains of a Darwin's ground sloth (*Mylodon darwinii*), to which the hide was eventually attributed. *M. darwini* was a large beast, up to about 1.2 meters at the shoulder when standing on all fours, probably weighing as much as 2.5 tonnes and with high-crowned but simple prismatic cheek teeth. In places the cave floor was covered with a thick blanket of dried sloth dung, indicating that sloths visited the site extensively. Analyses of plant remains in the dung showed that *M. darwinii* fed on grasses, sedges, and herbs in an open, moist, cool, boggy sedge-grassland. Moore concluded that it selectively grazed a diverse grassland-forest zone.[24] Other remains from the cave included an extinct horse (*Hippidion saldiasi*); the extraordinary extinct ungulate (*Macrauchenia patachonica*); and the extant guanaco (*Lama guanicoe*), puma (*Puma concolor*), and jaguar (*Panthera once*).[25] Although evidence of humans was very sparse (apparently confined to the younger layers), some suggested that the ground sloths had been hunted by humans and even (highly improbably) that they were semidomesticated and corralled within the cave—thus accounting for all that dung.

In *Mylodon darwinii* as in other related ground sloths, the dermis (immediately below the hairy epidermis) was closely studded with numerous rounded bony "pebbles," or osteoderms, up to approximately 0.8 centimeters in diameter. These probably would have functioned much like chain mail, affording significant protection against the claws and teeth of predators. Examples of skin are preserved in the collections of several museums around the world, including the Museo de La Plata in Argentina, the Natural History Museum, London, and the Museum für Naturkunde, Berlin. Because of the remarkably good preservation of the ground sloth material, in the few years following its discovery, it was widely supposed that these animals might still exist somewhere in Patagonia. In 1901 the English *Daily Express* newspaper went so far as to send a reporter to the region to determine whether any "antediluvian monsters" still existed there, but he found none. The apparent freshness of the remains turned out to be deceptive, instead resulting from the cold and exceptionally dry environment in the cave and enhanced by the presence of magnesium sulphate—a most effective preservative, as it readily absorbs the least trace of moisture. Many years later, when the question of age was addressed by radiocarbon dating, the remains proved to be very much older than had been supposed. The dates (on dung) place the finds toward the end of the Last Glacial.[26] In 1993, together with several other caves in the Ultima Esperanza area, the Mylodon Cave was designated a natural monument. A life-size *Mylodon* statue stands at the entrance.

Paul Martin proposed that *Mylodon darwinii* would have been easy prey for human hunters who therefore could have wiped them out very readily.[27] However, Luis Borrero and Fabiana Martin suggested that they "might have been easy to find but not easy to hunt" and that "they were probably dangerous animals to approach due to their strong claws" and even using stone-tipped projectile weapons propelled by atlatls (spear-throwers) they would have been very difficult to kill. "Ground sloth hides were not only thick, but also included thousands of osteoderms that acted as dermal armor. While the economic returns of sloths for hunter-gatherers were surely high, the costs were probably equally high."[28] Their study found no cut marks on sloth bones from the Mylodon Cave, and only two from one other cave (Cueva Fell) in southern Fuego–Patagonia, on the basis of which they stated that there was only very sparse evidence for intensive skinning of or butchering of sloths. They concluded that "there is little to no evidence for active human hunting of ground sloths and that scavenging is the only form of interaction that can be defended."[29]

The youngest available dates for Darwin's ground sloth are ca. 12.82 kya and 12.65 kya, on dung from Cueva del Milodón, Chile

7.4 *GLOSSOTHERIUM ROBUSTUM*

The ground sloth *Glossotherium* was similar to *Mylodon* but rather larger, with a shorter, wider rostrum. The first remains were found in 1833 by Charles Darwin at Arroyo Sarandi on the Rio Negro, Uruguay (the same locality from which he acquired the first *Toxodon*; see below). Back in England, working with just part of the rear of a skull, Richard Owen thought that the 2.5-centimeter-diameter, approximately circular depression on the underside close to the ear region was for the attachment of the hyoid bones, which he believed would have supported a very large tongue. Accordingly, he named the animal *Glossotherium* ("tongue beast").[30] Christiansen and Fariña estimated its weight at 500 to 1,000 kilograms.[31] The youngest known date for *G. robustum* is ca. 12.48 kya from Arroyo Seco, Argentina. The closely related *Paramylodon harlani* ranged widely across North America, reaching as far north as Alaska in the Last Interglacial (chapter 6).

7.5 *LESTODON ARMATUS*

With an estimated weight of over 3 tonnes, *Lestodon* was surpassed in size only by the other giant ground sloths *Megatherium* and *Eremotherium*.[32] Although lacking incisors and canines, the first remaining tooth in *Lestodon* (Greek: "robber tooth") resembles a canine and protrudes forward as a small tusk. The animal was described and named by the nineteenth-century French paleontologist Paul Gervais from its fancied resemblance to a carnivore. The fact that the remains of individuals of varying ages have been found together in the Arroyo del Vizcaíno suggests that they lived in herds; moreover, their massive ear ossicles suggest sensitivity to low frequencies for long-distance communication, comparable to elephant herds today.[33]

7.6 "NARROW-HEADED" GROUND SLOTH: *SCELIDOTHERIUM LEPTOCEPHALUM*

A partial skeleton of this medium-size ground sloth was collected by Charles Darwin from "hard consolidated gravel" in sea cliffs at Punta Alta, Patagonia, which, as with other discoveries from the *Beagle* voyage, was described and named by Richard Owen.[34] Subsequently, a near-complete skull was collected from near Buenos Aires by Auguste Bravard, a French paleontologist, living in South America. Owen was especially struck by its unusually long slender skull, hence the specific name *leptocephalum* (Greek: "narrow head"; fig. 7.5).

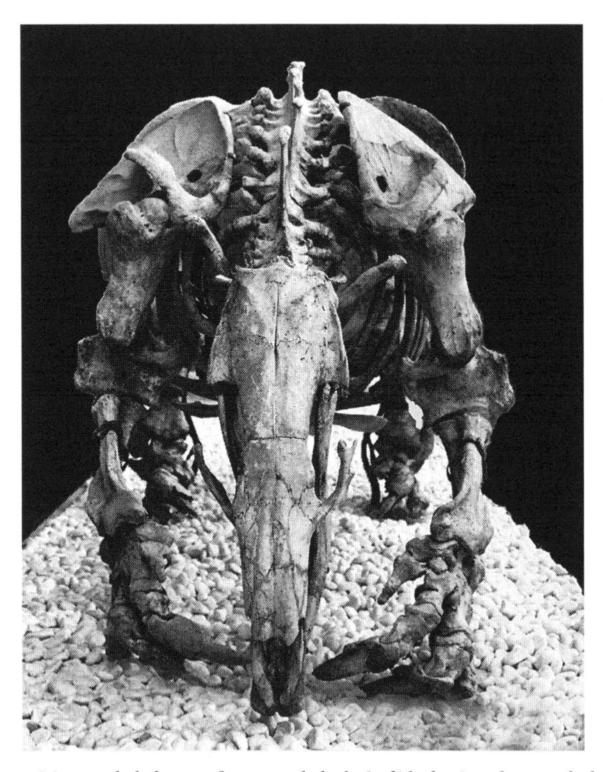

FIGURE 7.5. Mounted skeleton of a ground sloth *Scelidotherium leptocephalum*, shoulder height about 1 meter. Note the massive front limbs and claws, and the narrow, elongated skull. (Courtesy of the Photographic Archive of the Museum of Natural Sciences of Valencia.)

The skull and mandible together contained only eighteen simple prismatic unrooted (that is, permanently growing) cheek teeth. As described above, the long, narrow skull and conical prehensile lip indicate a selective browser or mixed feeder.[35] Another interesting feature noted by Owen was the relatively small size of the brain, suggesting rather low intelligence. Its weight has been estimated at about 600 kilograms to 1 tonne.[36]

Other Last Glacial relatives of scelidotheres include *Catonyx cuvieri*, *Valgipes*, and *Nothrotherium*.[37] All had narrow skulls and consequently also laterally compressed tooth rows. The youngest date for *S. leptocephalum* is ca. 8.44 kya (early Holocene) from Arroyo Tapalqué, Argentina; for *C. cuvieri*, ca. 9.99 kya (early Holocene) from Pampa de Fosiles, Peru; for *Valgipes* sp., ca., 12.88 kya from Cuvieri Cave, Brazil; and for *Nothrotherium* sp., ca. 14.09 kya from Gruta de Brejões, Brazil.

7.7 BURROWING SLOTHS

Improbable though it might seem, there is convincing evidence that some ground sloths were active burrowers—probably the largest animals known to have done so.[38] Their powerful forelimbs armed with impressive claws appear to have been superbly adapted for digging. Preserved burrows occur in the sea cliffs and river banks in Argentina and Brazil; some are infilled with sediment while others remain nearly empty, and in the case of the larger, burrows can be entered by people standing upright. Good examples are known from within the city of Mar del Plata near Buenos Aires in Argentina on the Atlantic coast. The largest burrows—more than 1.4 meters high, and up to 2.1 meters wide and 15 meters long, with large and deeply incised two-digit scratch marks—are thought to have been made by the claws of the mylodontid sloths *Glossotherium*, *Mylodon*, or *Scelidotherium*. No doubt they would also have been highly effective for defense against predators, especially when the animal was holed up in its burrow and could use its claws to best advantage. Burrowing obviously would also have offered effective protection from the weather, and this adaptation could explain how some species of ground sloths survived in cold climates; fossils of Jefferson's ground sloth (dating to the Last Interglacial) have been found as far north as Alaska (chapter 6), and, as described above, *Mylodon darwini* is known from caves in the extreme south of Chile.

Other burrows were not dug by ground sloths but by armadillos, rodents, or other animals. A burrow 1 meter wide and 0.75 meters high with distinctive three-digit scratch marks was attributed to *Pampatherium*, a large extinct close relative of the armadillo.

7.8 GLYPTODONTS

"I myself found the shell of an animal, composed of little hexagonal bones, each bone an inch in diameter at least; and the shell was near three yards over. It seemed in all respects, except its size, to be the upper part of the shell of the armadillo; which, in these times, is not above a span in breadth."[39] Of all the weird and wonderful animals that inhabited Last Glacial South America, the glyptodonts were perhaps the weirdest (fig. 7.6). During the late Quaternary, this continent supported an extraordinary diversity of glyptodonts—around twelve species—which a recent aDNA study has shown were closely related to living armadillos.[40] Glyptodonts took armored protection to extremes not seen in any other mammal living or extinct.[41] Unlike the armadillo body, which has flexible armor, the glyptodont body (other than the belly) was entirely encased in a rigid carapace or shell, comprising thousands of interlocking thick hexagonal bony plates. The head, which couldn't be withdrawn into the shell, was protected by a bony cap, and even the tail was heavily armored. These four-legged mammalian tanks invite comparisons with giant tortoises and particularly with the long-extinct ankylosaurs—armored dinosaurs that disappeared at the end of the Cretaceous period around 66 mya. Different species

FIGURE 7.6. Top: Mounted carapace and skeleton, glyptodont *Glyptodon clavipes*, about 1.45 meters high, external and cutaway views. Bottom: Mounted carapace and skeleton, club-tailed glyptodont *Doedicurus clavicaudatus*, about 1.5 meters high, external view. (Artwork by Rob Stuart, based on various sources.)

of glyptodont had evolved differently shaped carapaces and a variety of tail armor. Their body shape and dentition indicate an herbivorous diet, but their preferred habitats are uncertain.

According to Gillette and Ray, *Glyptotherium* in North America was associated with swampy areas.[42]

7.9 *GLYPTODON CLAVIPES*

With an estimated weight of 2 tonnes, *Glyptodon clavipes* (fig. 7.6, top), frequently compared in size and shape to a Volkswagen Beetle car, is the best-known species of glyptodont.[43] A smaller species, *G. reticulatus*, weighed nearly 1 tonne.[44] Its tail was well protected by a series of hefty articulated bony rings, which would have been able to move one upon the other, allowing it to flex. As in other glyptodonts, its high-crowned cheek teeth were simple, three-lobed prisms lacking enamel, on the basis of which Richard Owen in 1839 named the genus *Glyptodon* (Greek: "grooved or carved tooth"). The greatly foreshortened skull, which suggests that it (and other glyptodonts) may have possessed a short trunk like a tapir, had no canine or incisor teeth. The youngest available date for *Glyptodon* sp. is ca. 25.55 kya from Inciarte, Venezuela.

7.10 CLUB-TAILED GLYPTODONT: *DOEDICURUS CLAVICAUDATUS*

Perhaps the most extraordinary of all the glyptodonts was *Doedicurus clavicaudatus*, whose tail ended in a 1-meter-long rigid bony tube bearing a formidable club or mace (fig. 7.6, bottom).[45] On the basis of the massive posterior part of the rigid bony tube, which was all that he had available in 1846, Owen included it in the genus *Glyptodon* and coined the specific name *clavicaudatus* ("club-tailed"). In 1874, working with much more complete material, Hermann Burmeister reclassified it on the basis that the carapace was strongly domed or humped, and because of its very distinctive osteoderms, which are each pierced by two or more large holes. He named the genus *Doedicurus* ("pestle-tailed"). The proximal part of the tail comprised a series of articulated bony rings, allowing some flexibility, whereas the 1-meter-long distal section consisted of a rigid bony tube. The far end of this tube bore polygonal scars to which keratin pads or spines were very probably attached. In most artistic reconstructions, the tail is depicted armed with fearsome spikes resembling the mace of a medieval knight. Since the rear limb bones of large glyptodonts were remarkably strong, Fariña considered the possibility that they were capable of

FIGURE 7.7. Carapace, skull, and tail sheath of *Sclerocalyptus ornatus*, one of the smaller species of glyptodont. Shoulder height about 1 meter. Note the rigid bony tail sheath. (Courtesy of the Photographic Archive of the Museum of Natural Sciences of Valencia.)

walking on their hind legs.[46] The youngest available date for *D. clavicaudatus* is ca. 7.84 kya (early Holocene) from La Moderna, Argentina.

7.11 OTHER GLYPTODONTS

Panochthus was generally similar to *Doedicurus*, although its carapace was less strongly domed and the tail sheath bore elliptical scars along its entire length, to which keratin pads or spines were probably attached.[47] *Sclerocalyptus* (fig. 7.7) and *Lomaphorus* also featured a flexible proximal tail and distal rigid bony tube, probably used as a club for combat with others of the same species (see below) or to ward off predators. *Hoplophorus euphractus* was a medium-sized glyptodont, named by the Danish paleontologist Lund in 1839 on the basis of osteoderms and carapace fragments from caves in the Lagoa Santa Region of Brazil. The youngest known date for *Sclerocalyptus* is ca. 16.09 kya from Chascomús, Argentina.

7.12 FIGHTING GLYPTODONTS

In 1894 Richard Lydekker figured a *Doedicurus* carapace in the Museo de La Plata, which toward the top exhibited a circular dent about 35 centimeters across and surrounded by a bony growth—evidently a healed fracture.[48] Unfortunately, due to repeated repairs of the specimen over many years, this feature is no longer available for study. Other, more minor evidence is also seen on several other glyptodont specimens in museum collections—notably, a carapace of *Glyptodon reticulatus* in the Museo Argentino de Ciencias Naturales, in which the damaged area, about 15 centimeters in diameter, is not dented, but has fused osteoderms and is near the front on the lower right side. As discussed above, in many cases the muscular tails of glyptodonts terminated in a rigid bony caudal sheath; in both *Doedicurus* and *Panochthus*, this sheath formed a massive club (aptly termed a "biological hammer"). Although *Glyptodon* lacked a caudal sheath, its heavily armored tail would also have constituted a formidable weapon. Alexander and colleagues interpreted the observed fractures on some carapaces as resulting from intraspecific fights (most likely between males), and their estimates of the energy generated by the tail muscles are of the same order of magnitude as the energy required to fracture a rival's carapace.[49] They also speculated that the space between the thoracic and lumbar vertebrae and the carapace may have been occupied by a fatty pad which could have helped cushion the force of blows. Interestingly, the pronounced dorsal hump in the carapace of *Doedicurus* might have served to accommodate such a pad. *Panochthus* also had a hump, although rather smaller.

Blanco and colleagues compared the function of glyptodont tails to humans wielding tennis rackets and baseball bats, where in order to avoid injuries it is important that powerful blows are sustained close to the center of percussion.[50] Indeed, several species had tail clubs that can be modeled as rigid beams, like baseball bats. They found that, for example, *Doedicurus clavicaudatus* had the "sweet spot" very close to the inferred location of keratin spikes or pads, supporting the inference that they were adapted to inflicting powerful blows at this point and that the tail clubs were used combatively. In species such as *Glyptodon clavipes* where the tail was articulated and flexible along its entire length, the force exerted would have been moderate. On the other hand, the combination of a flexible proximal tail with a distal rigid tube of bone (more than 1 meter long in species such as *Doedicurus clavicaudatus* and *Panochthus tuberculatus*) would have delivered much heavier blows. The authors concluded that, for the larger glyptodonts, because of the need for precision in delivering the blow, such weapons would have been ineffective against fast-moving predators such as sabertooth cats. On the other hand, these tails seem better suited to intraspecific combat, which may have been ritualized, as seen in several living species—for example, bighorn sheep and wapiti.

A prime example of convergent evolution is seen in comparing glyptodonts with ankylosaurids, which were heavily armored (non-avian) dinosaurs from the late Cretaceous of western North America, and much larger than any glyptodont. Several ankylosaurids are known to have possessed a tail club. The functional resemblance to *Doedicurus* and *Panochthus* is remarkably close, with the ankylosaurid tail comprising a flexible proximal and a rigid distal section terminating in a bony club.[51] In this case, a "biological hammer" was formed by fusing the caudal vertebrae into a solid rod. Presumably, the ankylosaurid club was similarly used for defense against predators and/or combat with others of its own species.

7.13 *MACRAUCHENIA PATACHONICA*

Macrauchenia and *Toxodon*—bizarre ungulates (hoofed mammals) unique to South America—were both discovered by Charles Darwin on the *Beagle* voyage.[52] Attempts using aDNA to determine their phylogenetic relationships have proven unsuccessful. However, by means of a different technique—ancient protein sequences—Welker and colleagues were able to demonstrate that *Macrauchenia* and *Toxodon* were closely related, and that their sister group comprised horses, tapirs, and rhinos (perissodactyls).[53]

One of the most extraordinary of all the Last Glacial megafauna from South America (and a personal favorite of mine) was the litoptern *Macrauchenia patachonica*, the last of its evolutionary line (fig. 7.8). Litopterns are an extinct

FIGURE 7.8. Artist's impression of *Macrauchenia patachonica* in life. (Painting by Kate Scott.)

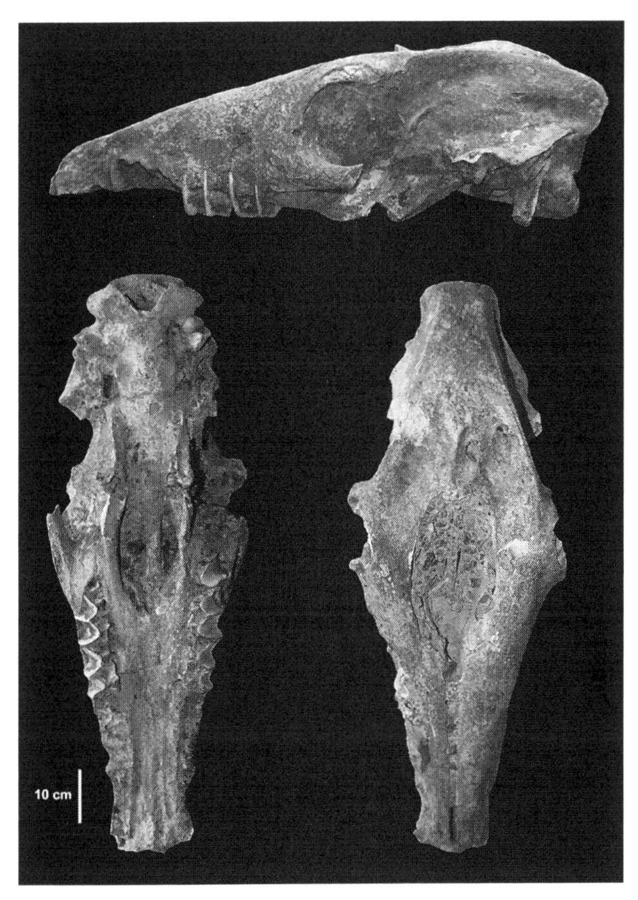

FIGURE 7.9. Skull of *Macrauchenia patachonica*, Kamak Mayu, Calama, northern Chile. Top: lateral view. Left: view of palate. Right: dorsal view. Note the large central nasal opening, suggesting the presence of a trunk. (Photos by Patricio López Mendoza and Ismael Martinez.)

group of ungulate mammals with origins dating back to the early Cenozoic. They were confined almost entirely to South America, although there is a single record (several teeth) from the Eocene of the Antarctic Peninsula.[54] In 1834, Charles Darwin collected some fragmentary remains of a most unusual beast—including vertebrae, a radius, ulna, forefoot, and femur—from a bed of sandy soil on the south side of Puerto San Julián, Patagonia.[55] In 1837, these were described by Richard Owen, who named it *Macrauchenia* (Greek: "big neck"), for its greatly elongated cervical vertebrae. He was later able to describe more complete material.[56] *Macrauchenia* was a large herbivore, with an estimated weight of 1 to 1.2 tonnes, with a long neck and long legs.[57] It has been aptly described as resembling a humpless camel with a short trunk, rather like a modern tapir; the existence of a trunk was suggested by the presence of elliptical nasal openings situated on the central top part of the skull (fig. 7.9, right). The skull was long and narrow, with forty-four moderately low-crowned teeth (the full complement for a placental mammal) that resembled those of a rhino. It was probably a mixed feeder, eating both leaves and grass. Its feet each had three hooves—again, like a rhino, and quite unlike the single hoof of a horse. However, like a horse, its long limbs were clearly designed for speed which would have helped it to escape predators such as sabertooth cats, giant short-faced bear, and jaguar. More important, the construction of the limb bones, which were stronger laterally than front to back, is thought to have made it adept at swerving to elude attackers.[58] It also seems likely that, as in horses, moose, llamas, and many other ungulates, the hooves were used as effective defensive weapons. The youngest available date for *M. patachonica* is ca. 11.28 kya (early Holocene) from Centinela del Mar, Argentina.

7.14 *TOXODON PLATENSIS*

In marked contrast to the agile *Macrauchenia*, *Toxodon platensis*—the last representative of the endemic South American group of mammals known as notoungulates—was a bulky, ponderous herbivore, weighing about 2 tonnes—the size of a white rhino.[59] It had a large head, capacious body, and short legs, somewhat similar to a hippo or rhino (fig. 7.10). The massive broad incisors—for cutting vegetation—and exceptionally high-crowned permanently growing (unrooted) cheek teeth indicate adaptation to eating silica-rich grasses and/or accidentally ingesting grit from low-growing plants. Stable isotope $\delta^{13}C$ values for *Toxodon platensis* from Brazil indicate that it was a mixed feeder, mostly eating C4 plants, while carbon isotope values on *Toxodon* from elsewhere in South America show increased consumption of C3 plants further south.[60]

As with other very large animals, the massive bulk of *Toxodon* would no doubt have rendered adults almost immune to predation, although the young are likely to have been more vulnerable. Judging by the abundance of its remains, it seems to have been a common animal in the Last Glacial. In his journal account of the *Beagle* voyage, Charles Darwin describes how he acquired a *Toxodon* skull from a local farmer in Uruguay: "On November 26th—I set out on my return in a direct line for Monte Video. Having heard of some giant's bones at a neighbouring farm-house on the Sarandis, a small stream entering the Rio Negro, I rode there accompanied by my host, and purchased for the value of eighteen pence the head of the Toxodon. When found it was quite

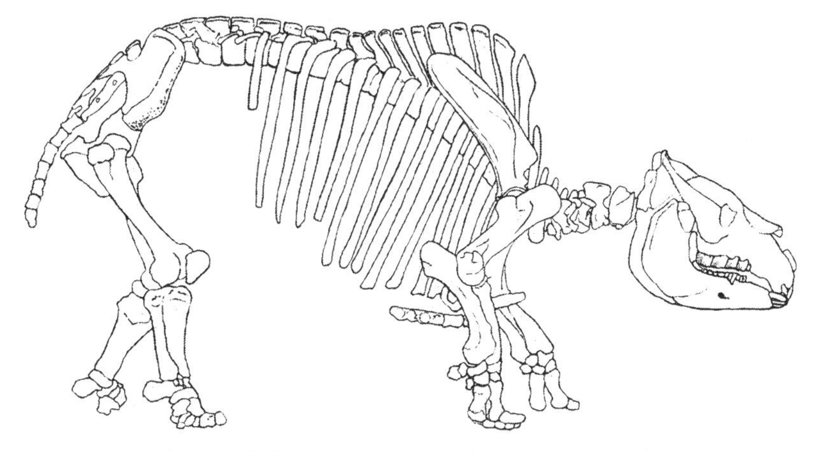

FIGURE 7.10. Skeleton of *Toxodon platensis*, shoulder height about 1.5 meters. (Drawn by the author from a photograph.)

perfect; but the boys knocked out some of the teeth with stones, and then set up the head as a mark to throw at."[61] This skull, the first to be scientifically studied, was described and illustrated by Owen in 1840, who named it *Toxodon* ("bow-tooth") for its prominent, strongly arched upper incisors.[62] In contrast, the lower incisors were horizontally arranged, giving a spade-like appearance to the front of the lower jaw. Suggestions that it had an amphibious mode of life like a hippo are not supported by the isotopic evidence.[63] The youngest available date for *Toxodon* sp. is ca. 11.9 kya (late Last Glacial) from Arroyo Tapalqué, Argentina.

7.15 SOUTH AMERICAN HORSES: *HIPPIDION PRINCIPALE*, *H. SALDIASI*, AND *EQUUS NEOGEUS*

the genetic (aDNA) relationships of New World horses were analyzed by Weinstock and colleagues, who proposed a simple classification with just two South American genera *Equus* (*Amerhippus*) and *Hippidion*.[64] *Hippidion* (Greek: "little horse") species were shorter than other South American horses, weighing about 500 kilograms, and generally more stoutly built and less clearly adapted to open environments.[65] A unique feature of *Hippidion* is that the nasal bones are separated from the rest of the skull for most of their length. As in *Equus*, the limb bones in *Hippidion* are stronger antero-posteriorly—an adaptation for escaping predators by fast running, not swerving as in *Macrauchenia*, whose limb bones are stronger transversely (see above). The cheek teeth of *Hippidion* were less high-crowned than in *Equus* and had relatively less enamel folding suggesting a less abrasive diet. The youngest known date for *Hippidion* sp. is: ca. 10.28 kya (early Holocene) from Barro Negro, Argentina, and for *Equus* (*Amerhippus*) *neogeus*, ca. 11.84 kya (late Last Glacial) from Río Quequén Salado, Argentina.

7.16 GOMPHOTHERES: *CUVIERONIUS* AND *NOTIOMASTODON*

Proboscideans (elephants, mammoths, mastodonts, and relatives) originated in Africa, whence they migrated into Asia and thence to North America via the Bering region. Mammoths never reached South America, but in the Last Glacial this continent supported two species of elephant-like gomphotheres, descendants of ancestral forms that had emigrated from North America in the Great Biotic Interchange. Unfortunately, the literature on South American gomphotheres is confusing, as different classifications have been made by different authors. For example, Prado and colleagues distinguished *Cuvieronius hyodon* (Cuvier's gomphothere), *Stegomastodon waringi* and *S. platensis*, whereas Ferretti recognized *C. hyodon*, *Haplomastodon chimborazi*, and '*Stegomastodon*' *platensis* (distinct from the true *Stegomastodon* of North America, Pliocene to Middle Pleistocene).[66] However, Mothé and colleagues recognized only two genera and species: *C. hyodon* and *Notiomastodon platensis* (aka "*Stegomastodon*" *platensis*).[67] All had low-crowned cheek teeth with rounded cusps, suggesting that browse was a major part of their diet. Cuvier's gomphothere had characteristic spirally twisted tusks, stood about 2.3 meters at the shoulder, and weighed some 3.5 tonnes, compared with *N. platensis* at about 2.5 meters and 4.4 tonnes.[68] According to Mothé and colleagues, *C. hyodon* was confined to the Andes region of Ecuador, Peru, and Bolivia, whereas Sánchez and colleagues show it extending as far as south-central Chile.[69] It also ranged northward to Mexico and marginally into the southern United States. On the other hand, *N. platensis* is recorded from every South America country except Guyana, Suriname, and French Guiana. Sánchez and colleagues analyzed carbon and oxygen isotopes from sixty-eight gomphothere bone and tooth samples from twenty-four South American localities. They found that the sampled *Cuvieronius* from Chile fed exclusively on C3 plants, whereas those from further north (Bolivia and Ecuador) had a mixed C3/C4 diet. There was a similar trend for *Stegomastodon*: analyzed samples from Quequén Salado, Argentina, indicated that it ate only C3 plants, contrasting with those from La Carolina Peninsula, Ecuador, which were exclusively C4 feeders. The authors suggested that the reason mammoth (*Mammuthus*) and mastodon (*Mammut*) were unable to reach South America is likely due to the lack of suitable habitat for these specialized feeders along the Central American land bridge.

The youngest available date for *Notiomastodon platensis* is ca. 12.67 kya from Arroyo Chasicó, Argentina, and for *Cuvieronius* sp., ca. 13.87 kya from Monte Verde, Chile.

7.17 CARNIVORES: *SMILODON POPULATOR*, *PANTHERA ATROX*, *CANIS DIRUS*, AND *ARCTOTHERIUM* SPECIES

large mammalian carnivores, all of which originated as emigrants from North America, were well represented in the South America megafauna.[70] The South American sabertooth *Smilodon populator*, with an estimated weight of up to 405 kilograms, was more heavily built and presumably even more powerful than its well-known North America cousin *S. fatalis*.[71] *S. fatalis* also reached South America but was confined to west of the Andes. Both species apparently were ambush predators that used their powerful front limbs to overpower their victims and then inflicted a fatal bite to the throat with their enlarged canines (chapter 6).

The huge American lion *Panthera atrox* ranged from the south of North America to the northern part of South America (chapter 6). The dire wolf *Canis dirus* is better known from North America, especially Rancho La Brea. Much the same size as the living gray wolf (*Canis lupus*) but more heavily built, it very likely operated in packs to hunt large prey (chapter 6). Five species of the South American giant short-faced bear *Arctotherium* have been recognized, of which *A. tarijense* (about 400 kilograms), *A. bonariense* (perhaps about 300 kilograms), and *A. wingei* (about 105 kilograms) are known from the Last Glacial.[72] None occurred west of the Andes. They were all probably omnivorous; in addition to consuming a range of plant foods, they likely acquired meat by active hunting, scavenging, and stealing prey from other carnivores.

Morphological studies suggest that *Arctotherium tarijense* was mainly an omnivore that may have also scavenged and occasionally hunted medium-large mammals like camelids and horses. On the other hand, analyses of stable isotopes ($d^{13}C$ and $d^{15}N$) indicates a highly carnivorous diet, probably resulting from scavenging.[73]

In addition, the jaguar *Panthera onca*, puma *Puma concolor*, and spectacled bear *Tremarctos ornatus* survive to the present day.

S. populator, dated to ca. 13.14 kya, is recorded from Ultra Esperanza Cave, Chile. The youngest known date on *Smilodon* sp. (presumably *S. populator*) is ca. 10.3 kya (early Holocene) from Lapa da Escrivânia 5, Brazil.

7.18 GIANT CAPYBARA: *NEOCHOERUS*

South America today is home to a diverse range of endemic rodents, the caviomorphs—for example, guinea pigs, agoutis, maras, New World porcupines, and the capybara (*Hydrochoerus hydrochaeris*), which is the largest rodent alive today. Known only from fragmentary remains, the extinct giant capybara was similar to its living relative, although at least twice as large, with an estimated mass of 100 to 150 kilograms, making it one of the largest rodents of all time.[74] It was very probably amphibious, feeding on aquatic and waterside vegetation, as does *H. hydrochaeris* today.

7.19 LLAMAS: *PALAEOLAMA* AND *HEMIAUCHENIA*

llamas and guanacos—camel relatives whose ancestors had emigrated from North America—are now restricted to the Andes and Patagonia, but were much more widespread in South America during the Last Glacial and occurred as far north as the southern United States (chapter 6). Fariña and colleagues give an estimated weight of 200 kilograms for *Palaeolama*. The youngest known date for South American *Palaeolama* sp. is ca. 11.21 kya from Sitio Arroyo Seco 2, Argentina.

7.20 HUMAN ARRIVAL

According to Prado and colleagues, in the Pampas region there is evidence for widespread human presence at ca. 13 kya, and sparse indications of people around a thousand years earlier.[75] Most early South American archaeological sites are dated to ca. 13 kya or younger. However, there is one major site—Monte Verde in southern Chile—dated to about 14.8 kya, with claims of possible occupation a few thousand years earlier.[76] The site is widely recognized as the oldest evidence of humans in South America, although some archaeologists remain unconvinced. But there is now additional evidence for human presence of broadly comparable age. Arroyo Seco 2, in the Argentinian Pampas, is a rich multicomponent archaeological site. Radiocarbon dates (calibrated means) on the oldest level, containing stone tools, humanly fractured animal bones, and remains of extinct megafauna, range from 14.064 kya to 13.068 kya.[77] According to these authors, this evidence for hunter-gatherers in the Southern Cone of South America at ca. 14 kya adds to "the growing list of American sites that indicate a human occupation earlier than the Clovis dispersal episode, but posterior to the onset of the deglaciation of the Last Glacial Maximum (LGM) in the North America."[78]

Excluding the highly unlikely idea that Paleolithic people somehow crossed the Atlantic from Africa or Europe or the Pacific from Asia, the only possible route for human entry to South America is from the north. The previous puzzling absence of archaeological sites of comparable age en route can be explained if people largely followed the western coastal migration routes (chapter 6).

7.21 ENVIRONMENTAL CHANGES

Unlike the northern continents, South America was not very extensively glaciated in the Quaternary, although mountain glaciers in the Andes expanded during the Last Glacial Maximum (LGM) and covered the western part of Patagonia.[79]

The cyclical advance and retreat of Andean glaciers was accompanied by alternating arid and humid conditions in the Pampas; a pattern that, according to Prado and Alberdi favored alternate floral and faunal assemblages and also higher extinction rates.[80] During the LGM, around 22 kya, arid, cold faunal elements predominated. The pollen record also suggests extremely arid to arid conditions associated with more continental environments, as well as lowered sea level exposing an extensive area of Atlantic continental shelf. A shift to warmer and more humid conditions occurred at the Pleistocene–Holocene transition (around 11.7 kya). Mayle and colleagues found that there were no large-scale differences in vegetational distributions in tropical South America between the LGM (21 kya) and the present; in the Andes, though, expanding montane grasslands replaced montane forest, and at the Amazon basin margins, savanna increased at the expense of rainforest and gallery forest.[81] However, the species composition and structure of these LGM forests was very different than those of today, with low atmospheric CO_2 thought to have resulted in forest communities with substantially reduced canopy density.

7.22 PATTERNS OF EXTINCTION

South America, with its wealth of wonderful vanished creatures that survived to comfortably within radiocarbon range, is an especially exciting region for researching megafaunal extinctions. The potential of obtaining large numbers of radiocarbon dates also offers a prime opportunity to track extinctions geographically as well as chronologically. Prado and Alberti estimated that fifty genera (about 83%) of South American megafauna disappeared during the late Quaternary. Three entire mammalian orders (Notoungulata, Proboscidea, Litopterna) were lost, together with all the large xenarthrans. As elsewhere it was the largest animals most at risk of extinction, so that all species of 1 tonne or more disappeared, as did most over 45 kilograms.[82] It is notable that most of the species that disappeared lived in more open habitats, whereas forest animals were much less affected.

Our knowledge of these extinctions is growing rapidly as more radiocarbon dates (made directly on megafaunal material) continue to be published, and there are already sufficient dates to establish an outline chronology. It is now clear that a substantial number of megafaunal species survived into the Last Glacial, and twenty recently published dates on extinct megafaunal species are younger than 14 kya (table 7.1). Especially interesting and significant is that the youngest available dates for nine species (*Megatherium americanum*, *Catonyx cuvieri*, *Scelidotherium leptocephalum*, *Doedicurus clavicaudatus*, *Eutatus seguini* [a large extinct armadillo], *Macrauchenia patachonica*, *Palaeolama* sp., *Hippidion* sp., and a *Smilodon* species) fall within the early Holocene—that is, later than ca. 11.7 kya. So far, the youngest record (*D. clavicaudatus*) is ca. 7.84 kya. This evidence for Holocene survival contrasts sharply with the continental North American record, in which all extinctions apparently occurred significantly earlier (chapter 6). Luis Borrero observed: "The age difference between Last Dated Appearances of Pleistocene megafauna and First Contact with humans is around 1,000 radiocarbon years at some regions and was perhaps longer at places where Holocene survival is suggested. Thus, the length of coexistence can be measured in thousands of years."[83] Since this was written, substantially more evidence of species surviving well into the Holocene has accumulated. Crucially, this would appear to rule out a South American "Blitzkrieg," as humans were evidently well established on this continent by ca. 14 kya, several thousand years before the extinction of most of the megafauna. Similarly, Prado commented that humans did not coexist with megafauna throughout South America and therefore could not have been the sole driver of megafaunal extinctions.[84]

A final intriguing question is why South America suffered major extinctions whereas sub-Saharan Africa had very few (chapter 12), although both have a broadly similar range of climate and vegetation at the present day, and apparently also did during the Last Glacial.

7.23 SURVIVORS

The surviving megafauna of the region include the jaguar (*Panthera onca*), mountain lion (*Puma concolor*), spectacled bear (*Tremarctos ornatus*), Baird's tapir (*Tapirus bairdii*), lowland or Brazilian tapir (*Tapirus terrestris*), mountain tapir (*Tapirus pinchaque*), giant anteater (*Myrmecophaga tridactyla*),

TABLE 7.1 Youngest direct radiocarbon dates for extinct South American megafauna

Species	Lab number	Cal BP	Median	Site	Source
Glossotherium robustum	OxA-4591	13,827–14,683	14,145	Arroyo Seco (Ar)	1
Glossotherium robustum	AA-9049	12,132–12,769	12,478	Arroyo Seco (Ar)	1
Megatherium americanum	GrA-49130	13,348–13,473	13,411	Río Salado (Ar)	6
Megatherium americanum	AA-55118	8522–9480	**8988**	Campo Laborde (Ar)	1
Megatherium americanum	AA-55117	7871–9010	**8381**	Campo Laborde (Ar)	1
Eremotherium sp.	Beta-237348	14,430–15,141	14,786	Abismo do Fossil (Br)	5
Eremotherium sp.	Beta	13,087–13,286	13,186	Itaituba, lower Rio Tapajos (Br)	7
Mylodon darwinii	GX-6243	11,998–13,406	12,820	Cueva del Milodón (Ch)	1
Mylodon darwinii	GX-6248	11,254–13,216	12,365	Cueva del Milodón (Ch)	1
Catonyx cuvieri	Beta-165398	11,252–11,606	**11,375**	Gruta Cuvieri, Lagoa Santa (Br)	1
Catonyx cuvieri	GIF-4116	9535–10,497	**9988**	Pampa de Fosiles (Pe)	1
Valgipes sp.	Beta-248057	12,749–13,010	12,880	Cuvieri Cave (Br)	4
Scelidotherium leptocephalum	GrA-48388	8353–8535	**8444**	Arroyo Tapalqué (Ar)	6
Nothrotherium sp.	NZA-6984	13,786–14,640	14,091	Gruta de Brejões (Br)	2
Glyptodon sp.	AA-33646B	27,272–28,757	27,997	Inciarte (Ve)	1
Glyptodon sp.	AA-33646A	24,378–26,805	25,546	Inciarte (Ve)	1
Doedicurus clavicaudatus	GrA-48961	15,220–15,435	15,328	Río Salado (Ar)	1
Doedicurus clavicaudatus	TO-1507-1	7663–8014	**7838**	La Moderna (Ar)	1
Neuryurus sp.	GrA-48668	12,249–12,547	12,398	Río Quequén Salado (Ar)	6
Sclerocalyptus ornatus	GrA-48955	15,994–16,192	16,093	Chascomús (Ar)	6
Holmesina sp.	AA-33647C	38,413–51,857	43,587	Inciarte (Ve)	1
Eutatus seguini	AA-90117	8010–8339	**8159**	Arroyo Seco 2 (Ar)	9
Smilodon populator	UCIAMS-142836	13,065–13,235	13,137	Ultima Esperanza (Ch)	11
Smilodon populator	GrA-49131	11,630–11,961	11,796	Arroyo Tapalqué (Ar)	6
Smilodon sp.	Beta-174722	9792–10,712	**10,320**	Lapa da Escrivânia 5 (Br)	1

TABLE 7.1 (*continued*)

Species	Lab number	Cal BP	Median	Site	Source
Toxodon platensis	Beta- 218193	12,916–13,093	13,004	Ponta de Flecha (Br)	6
Toxodon sp.	UCIAMS-143035	11,970–12,110	12,040	Arroyo Tapalqué (Ar)	1
Toxodon sp.	UCIAMS-143034	11,865–11,935	11,900	Arroyo Tapalqué (Ar)	1
Macrauchenia patachonica	OxA-25840	12,130–12,295	12,185	Luján (Ar)	6
Macrauchenia patachonica	GrA-49321	11,231–11,331	**11,281**	Centinela del Mar (Ar)	6
Hippidion principale	GrA-48962	16,082–16,281	16,182	Arroyo La Carolina (Ar)	6
Hippidion principale	GrA-49323	17,166–17,396	17,281	Río Salado (Ar)	6
Hippidion saldiasi	OxA-9247	12,533–12,734	12,643	Tres Arroyos (Ar)	8
Hippidion saldiasi	OxA-9504	11,405–12,587	12,098	Cueva Lago Sofía 1 (Ch)	8
Hippidion saldiasi	LP-1528	9665–10,195	**9939**	Cerro Bombero, Santa Cruz (Ar)	10
Hippidion sp.	AC-0969(1)	9908–10,655	**10,282**	Barro Negro (Ar)	3
Equus (Amerhippus) neogeus	GrA-47178	16,255–16,483	16,369	Arroyo Tapalqué (Ar)	6
Equus (Amerhippus) neogeus	GrA-49125	11,709–11,967	11,838	Río Quequén Salado (Ar)	6
Equus sp.	OxA-4590	12,829–13,110	12,950	Arroyo Seco 2 (Ar)	1
Equus sp.	LP-1235	11,411–12,656	12,088	Río Quequén Grande (Ar)	1
Palaeolama sp.	GrA-47340	11,187–11,233	**11,210**	Sitio Arroyo Seco 2 (Ar)	6
Cuvieronius hyodon	OxA-105	13,330–14,765	13,914	Monte Verde (Ch)	1
Cuvieronius hyodon	TX-3760	13,389–14,574	13,874	Monte Verde (Ch)	1
Notiomastodon platensis	GrA-47267	12,636–12,709	12,673	Arroyo Chasicó (Ar)	6

Cal BP: calibrated date before present. Boldface dates: Holocene.
Ar: Argentina. Br: Brazil. Ch: Chile. Pe: Peru. Ve: Venezuela.
Sources: 1. Barnosky and Lindsey 2010; Politis and Messineo 2007. 2. Czaplewski and Cartelle 1998. 3. Fernández et al. 1991. 4. Hubbe, Hubbe, and Neves 2009. 5. Hubbe, Hubbe, and Neves 2013. 6. Prado, Martinez-Maza, and Alberdi 2015. 7. Rossetti et al. 2004. 8. Borrero 2008. 9. Politis, Gutiérrez, et al. 2016. 10. Paunero et al. 2008. 11. Paijmans et al. 2017.

capybara (*Hydrochoerus hydrochaeris*), vicuña (*Vicugna vicugna*), guanaco (*Lama guanicoe*), marsh deer (*Blastocerus dichotomus*), taruca or North Andean deer (*Hippocamelus antisensis*), huemul or South Andean deer (*Hippocamelus bisulcus*), American crocodile (*Crocodylus acutus*), black caiman (*Melanosuchus niger*), Orinoco crocodile (*Crocodylus intermedius*), and green anaconda (*Eunectes murinus*). The greater rhea *Rhea americana*—a ratite like the emu and ostrich—reaches only about 40 kilograms, so it is rather too small to qualify as megafauna. The 2017 IUCN Red List classifies mountain and Baird's tapirs as endangered and the Orinoco crocodile as critically endangered. In view of so many losses, actual and potential, it is good to note the remarkable discovery of a "new" tapir species, *Tapirus kabomani*, the smallest of the genus (at 110 kilograms), announced in 2013.[85] However, it is also listed as endangered.

CHAPTER 8

Sahul: Giant Marsupials, a Thunderbird, and a Huge Lizard

During the cold phases of the Quaternary, at times of lowered sea level when global water was tied up in expanded ice sheets, Australia, Tasmania, New Guinea, and some smaller offshore islands were united into a single large continental land mass, known as Sahul. In July 1993, I spent several memorable days in the main chamber in Victoria Fossil Cave, near Naracoorte in South Australia, helping to excavate the skeletal remains of extinct and living marsupials (fig. 8.1).

I especially recall unearthing an exquisitely preserved skull of one of the smaller extinct short-faced kangaroos (*Sthenurus*). Judging by the huge quantities already recovered, this vast deposit of red cave earth (estimated at 5,000 tonnes) must contain hundreds of thousands (quite possibly millions) of vertebrate remains, making it one of the richest Quaternary fossil vertebrate sites in the world.[1] Along with twenty-five other nearby bone caves, all formed when water gradually dissolved the Oligocene and Miocene limestone bedrock, the status of the Naracoorte Caves as a World Heritage Site is amply justified. The excellent preservation of skulls and bones and the high proportion of articulated remains indicate that postmortem disturbance was moderate in spite of episodic flowing water and the futile struggles of trapped animals.

Although providing such a wonderful opportunity for paleontologists many thousands of years later, Victoria Fossil Cave was undoubtedly very bad news for the animals that once lived in the surrounding area. For hundreds of millennia the cave acted as a huge pitfall trap. From time to time, mammals, reptiles, birds and frogs fell to their deaths down an open shaft in the cave roof, accompanied by soil and stones dislodged or washed in from the surface, gradually forming a huge debris cone of sediment and bones. The Victoria Fossil Cave chamber

FIGURE 8.1. *Sthenurus* bones *in situ* on cave floor, Naracoorte, South Australia. (Photo by Steve Bourne.)

deposits span several hundred-thousand years of the Middle Pleistocene, with ESR (electron spin resonance) dates (made directly on tooth enamel) ranging from ca. 500 kya to ca. 280 kya.[2] Interestingly, ESR dates of about 280 to 170 kya for Cathedral Cave and ca. 125 kya (Last Interglacial) for Grant Hall in Victoria Fossil Cave, record little or no change in the megafaunal assemblage from the early Middle to the early Late Pleistocene.[3] However, there appears to have been a dramatic change "after the last interglacial, when a large proportion of the megafauna suddenly disappeared."[4] As discussed later in this chapter, the timing and speed of this extinction event is much debated.

I also found it most evocative and rather poignant on emerging from Victoria Fossil Chamber to see such characteristic Australian animals as kangaroos, wallabies, and brushtail possums that are present as fossils inside the cave, alive and running around outside.

The wonderful and unique mammal fauna of Australia and New Guinea today—the result of tens of millions of years of evolution in virtual isolation from the rest of the world—is dominated by marsupials such as kangaroos, wombats, possums, and koalas, together with the extraordinary egg-laying echidnas and platypus (monotremes). The Ice Age down under also boasted very many spectacular species of megafauna that are now totally extinct (fig. 8.2).[5]

FIGURE 8.2. Extinct and extant megafauna from Sahul (Australasia Ecoregion). From the top, row 1, left to right: *Diprotodon optatum*; *Zygomaturus trilobus*; *Palorchestes azael*. Row 2: *Procoptodon goliah*; *Sthenurus stirlingi*; *Procoptodon raphe*; *Sthenurus tindalei*. Row 3: *Sthenurus atlas*; *Sthenurus andersoni*; *Procoptodon gilli*; *Macropus ferragus*. Row 4: *Simosthenurus occidentalis*; *Protemnodon anak*; *Protemnodon brehus*; *Propleopus oscillans*. Row 5: *Thylacoleo carnifex*; *Ramsayia magna*; *Phascolonus gigas*; *Thylacinus cynocephalus*.[E] Row 6: *Varanus priscus* (aka *Megalania prisca*); *Genyornis newtoni*; *Pallimnarchus pollens*; *Wonambi naracoortensis*. Row 7: *Zaglossus hacketti*; *Macropus rufus*; *Macropus fuliginosus*. Row 8: *Casuarius casuarius*; *Dromaius novaehollandiae*; *Crocodylus porosus*; *Crocodylus johnsoni*.
Black: selected extinct species. Gray: selected living species. [E]Extinct following European settlement—not megafauna. The outlined *Homo sapiens* gives approximate scale.

As pointed out by several authorities, a curious and unique characteristic of Australian Ice Age mammal faunas, in comparison with those from other continents, was that as a whole the size spectrum was shifted toward the smaller end. This phenomenon probably relates to the generally sparse available nutrients together with the restricted area habitable by large mammals because of the extensive arid continental interior. The largest-known Australian mammal of all time was *Diprotodon optatum*, which was roughly comparable in size to a white rhinoceros; but there was no marsupial giant to compare with an elephant or mastodon.

Although the timing is highly controversial, probably all would concur that most megafaunal extinctions in Sahul occurred much earlier than on other continents. A widely cited paper by Roberts and colleagues concluded that extinction of the megafauna happened over a short period at ca. 46 kya.[6] However, both the date and suddenness of extinctions have been hotly disputed by other researchers and are closely tied up with the highly contentious issue of what caused them. In Australia, as in North America, there is much lively and even fierce debate between the supporters of the overkill hypothesis and the advocates of climate change. "Overkill" enthusiasts perceive a close correspondence in time between the arrival of humans and the extinction of the megafauna, believed to result from hunting and the burning of vegetation. However, others argue that extinctions resulted from increasing aridity and were staggered over a much longer period.[7] Such radically contrasting opinions are a symptom of the difficulty of establishing a reliable extinction chronology. Moreover, the dating problem is compounded by a patchy fossil record across the continent, so that we don't know the stratigraphic range of many species nor when they disappeared.

It is clear, however, that in the Late Pleistocene Australia suffered spectacular extinctions of its megafauna, although at the present state of knowledge this figure cannot be estimated accurately. However, the extent of this loss is starkly illustrated by the fact that there are only three living mammal species that qualify as megafauna as customarily defined (that is, weighing 45 kilograms or more; see below).

8.1 LATE QUATERNARY CLIMATIC AND VEGETATIONAL CHANGES

Unlike the northern continents, Sahul was glaciated only to a very limited extent during the Quaternary, although the climate fluctuated between more arid and wetter periods. Determining the occurrence and timing of arid phases has an important bearing on whether or not climate change caused or contributed to extinctions. As is apparent from the examples described below, there is ample evidence for such episodes in the Australian record. A recent study of the now-arid interior of Australia demonstrates the crucial role of climate change in the history of this continent. Optical luminescence dating of a sequence of ancient shorelines and river deposits of the vast Lake Eyre Basin through the Last Glacial cycle has revealed major hydrological changes due to a major increase in aridity. Mega-lakes that had been previously overflowing were drastically reduced in extent, culminating in "a final and catastrophic drying phase at 48±2 ka."[8]

Pollen and charcoal analysis on an offshore marine sediment core (Fr10/95, GC-17) provides a record of vegetation, fire, and climate change covering most of the last 100,000 years for the Cape Range Peninsula, Western Australia.[9] Drier conditions after 46 kya, compared with 100 to 64 kya, are indicated. In the period 46 to 40 kya, vegetation changed from open *Eucalyptus* woodlands rich in grasses to open *Eucalyptus* and *Gyrostemon* shrublands.

The long terrestrial pollen record from the volcanic crater lake of Lynch's Crater in northeastern Queensland is crucial for understanding Ice Age climatic and vegetational change in tropical Australia.[10] Tropical rain forest plants that predominated from about 130 to 78 kya (broadly Last Interglacial), were replaced by drier (sclerophyll) vegetation with eucalypts and acacia from about 46 to 11 kya in the later part of the Last Glacial, followed by the reestablishment of rain forest plants in the Holocene. The dung fungus *Sporormiella* shows a sharp decline at ca. 40 kya, which was interpreted as due to megafaunal loss (although no megafauna has been recorded from the region), whereas three later peaks are attributed to surviving kangaroos and/or ratites.

8.2 HUMAN ARRIVAL

The first humans to reach Sahul (*Homo sapiens*—ancestors of the modern Aboriginal people) came via southeast Asia. Although sea levels were substantially lower at this time—so that many islands became connected—nevertheless, boats or rafts would have been needed to cross several intervening water gaps. On the basis of radiocarbon dates from Nawarla Gabarnmang Rock Shelter (Northern Territory), O'Connell and Allen estimated that Sahul was first colonized by humans at ca. 47 kya, and cast doubt on claims for earlier arrival.[11] Several other sites in New Guinea and Australia have dates of ca. 45 kya. However, shortly afterward, Clarkson and colleagues published OSL (optically stimulated luminescence) dates on sediment enclosing abundant stone artifacts at the Madjedbebe Rock Shelter (also Northern Territory) suggesting that colonization occurred as early as ca. 60 to 50 kya; and more recently, new excavations at the site have pushed this date back considerably further.[12] A series of OSL dates were obtained through a sequence of three separate archaeological layers. The oldest level, at a depth of 2.6 to 2.15 meters, was dated by OSL to ca. 65 kya—the earliest putative evidence of human presence in Sahul. In addition to an *in situ* hearth, this level yielded a distinctive stone tool assemblage fashioned from quartzite, silcrete, mudstone, dolerite, and other rocks that included grinding stones, ground-edge hatchet heads (by far the earliest known from anywhere), thinning flakes, and snapped points. Remarkably, sheets of mica were found wrapped around a lump of ground ochre.

This new evidence has huge implications for megafaunal extinctions in Sahul. As observed by Clarkson and colleagues: "Our chronology places people in Australia more than 20 kyr before continent-wide extinction of the megafauna."[13] Moreover, "this evidence sets a new minimum age for the arrival of humans in Australia, the dispersal of modern humans out of Africa, and the subsequent interactions of modern humans with Neanderthals and Denisovans."[14] "It also extends the period of overlap of modern humans and *Homo floresiensis* in eastern Indonesia to at least 15 kyr."[15] However, the claim of human colonization at 65 kya has been disputed by O'Connell and colleagues, who "find that an age estimate of >50 ka for this site is unlikely to be valid," suggesting that younger artifacts had sunk down into older sediments from an overlying deposit. They also argue that a date of ca. 65 kya is anomalous in the context of archaeological dates from southern Asia. However, on the basis of a major stratigraphic and faunal study of Lida Ajer Cave, Sumatra, K. Westaway and colleagues concluded that modern humans arrived in Sumatra between 73 and 63 kya. The island of Sumatra is along a likely route that humans would have taken to reach Sahul. Resolving this important issue requires the discovery of additional early sites in Sahul.[16]

It has been proposed that people were only able to colonize Tasmania from mainland Australia when a land bridge was established between ca. 43 and 40 kya.[17] The likely impact on the Tasmanian megafauna is considered later in this chapter.

8.3 THE LARGEST-EVER MARSUPIAL: *DIPROTODON OPTATUM*

In this section, I discuss several of the best studied and characteristic "vanished giants" of the Australian Ice Age, beginning with the biggest and best known. When considering dates for megafaunal species it should be borne in mind that the OSL method, widely used on Australian sites, dates the enclosing sediment but not the fossils themselves, so that in some cases the date obtained might be inaccurate—for example, if the fossil has been reworked into younger deposits. On the other hand, the ESR (electron spin resonance) method can produces a direct date on the tooth enamel of a megafaunal species, so in theory should be much more reliable. Where bone collagen is well preserved, radiocarbon dating is the most accurate, although the method is limited to samples younger than about 50,000 years.

Diprotodon optatum (sometimes inaccurately called the "giant wombat") was the largest Australian land mammal of all time, and indeed, the world's biggest-ever marsupial.[18] In common with many extinct megafauna from around the world it was the last of its evolutionary line, which in the case of *Diprotodon* extended back to the Oligocene. It was a massive, heavily built quadruped that walked with its feet flat on the ground (plantigrade gait) (fig. 8.3). Evidence from both articulated foot bones and preserved trackways show that it was pigeon-toed with the feet slightly turned in.

At about 3 meters long and 1.8 meters at the shoulder and weighing up to 2,800 kilograms, it was a huge animal, comparable in size to a modern African white rhino, although its nearest living relatives are the much smaller wombats and koala. The term *diprotodont* (Greek: "two forward teeth") refers to the large, forward-pointing incisor tusks in the lower jaw. There were also two large rodent-like curved incisor teeth in the front of the upper jaw, much as in the living wombat, and two smaller upper incisors behind them on each side. As in the more distantly related kangaroos, the molars (four in each jaw half) each have two transverse ridges (lophs), suggesting that the animal was a mixed feeder, on shrubs, herbs, and grasses. Direct evidence of its diet is provided by remains of leaves, stalks, and twigs recovered from the abdominal regions of skeletons excavated by Ruben Stirton in 1900 from Lake Callabonna (fig. 8.4).[19] In 2008 these were identified as saltbush (*Atriplex*)—a salt-tolerant C4 plant adapted to dry environments, abundant in the area today. Additionally, analyses of carbon isotopes of *Diprotodon* tooth enamel from the Last Interglacial site of Katapiri (South Australia) indicate that both C3 and C4 plants were eaten.[20]

Diprotodon optatum was the first of many extinct Australian megafaunal species to be described and named by Sir Richard Owen, initially on the basis of finds from caves in the Wellington Valley, New South Wales.[21] Although several species of *Diprotodon* (such as *D. australis* and *D. minor*) were described and named by Owen and subsequent authors, only *D. optatum* is generally

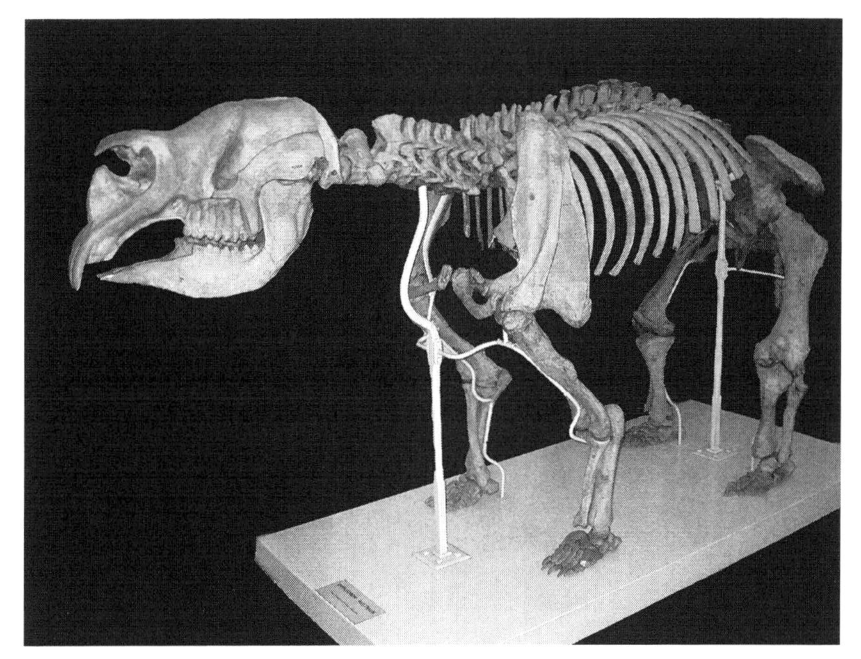

FIGURE 8.3. *Diprotodon optatum* (cast) skeleton from Lake Callabonna, South Australia. (Cat. UMZC. A11. 5/2, Courtesy of the University Museum of Zoology, Cambridge, photo by Nigel Larkin.)

FIGURE 8.4. *Diprotodon optatum* partial skeleton *in situ*, Lake Callabonna, South Australia. (Courtesy of Rod Wells.)

recognized today. The first fossils were shipped to England by Major Thomas Mitchell, a veteran of the Napoleonic Peninsular War, who was then exploring and surveying southeastern Australia. With little to go on at first, Owen erroneously thought he was dealing with the remains of some kind of elephant-like creature; but when more material, became available he realized that this was a giant extinct marsupial entirely new to science. His deep fascination with this extraordinary beast is evident from the following:[22]

> Of no extinct animal of which a passing glimpse, as it were, had thus been caught, did I ever feel more eager to acquire fuller knowledge than of this huge Marsupial. No chase can equal the excitement of that in which, bit by bit, and year after year, one captures the elements for reconstructing the entire creature of which a single tooth or fragment of bone may have initiated the quest; in the course of which one finally realizes, with more or less exactitude, the picture of which the laws of correlation had led one to frame of an animal which may have passed out of existence long ages ago.

However, Owen was destined to be disappointed in that he was never able to study the entire skeleton. His paper in the *Philosophical Transaction of the Royal Society* included a reconstruction of the skeleton, although lacking the feet, which the artist had very conspicuously obscured by vegetation. Owen made strenuous efforts to obtain a set of foot bones, but these were not forthcoming until after his death, when excavations at Lake Callabonna in South Australia began to recover more complete material.

Diprotodon optatum was widespread across Last Glacial Australia; it is not known from the Northern Territory, Tasmania, or New Guinea. Localities include the Darling Downs (Queensland), Wellington Caves and Cuddie Springs (New South Wales), Lancefield Swamp and Bacchus Marsh (Victoria), and Lake Callabonna and Naracoorte (South Australia). The remains of hundreds of individuals have been found at Lake Callabonna.[23] Evidently from time to time—probably over a period of many centuries—animals became mired as they attempted to cross the treacherous sticky mud when the bed of the salt lake had partly dried out. Many skeletons were found lying on their sides but rather gruesomely with their intact articulated feet still upright in the sediment. In the late nineteenth century, at least one near-complete skeleton was restored, molded, and cast, and replicas distributed to museums around the world (fig. 8.3). Tedford reported that one adult *Diprotodon* skeleton was accompanied by the remains of a pouch-young individual situated where the pouch would have been.[24] OSL dating of Lake Callabonna sediments enclosing megafaunal remains gave a figure of ca. 75 ± 9 kya.[25]

For such a large beast, the skull of *Diprotodon* is surprisingly fragile, so that reasonably intact examples are rare (broadly similar adaptations to reduce weight are seen in elephants). In 2011 a fine specimen was discovered eroding out of a creek bank in the Darling Downs, Queensland.[26] Ian Sobbe, a farmer and expert fossil enthusiast, received a phone call from a local who reported, "I've found a skull in the dirt—it looks like a Grand Angus bull." On the basis of photos emailed to him, Ian immediately realized that this was no bull, but a skull of the extinct "mega-marsupial," *Diprotodon*. Subsequently, Ian visited the site at Gowrie Creek, accompanied by the vertebrate paleontologist Gilbert Price (University of Queensland) and the discoverer. Price relates: "Our jaws dropped when we saw it—it was a monster, measuring around 90 cm in length! With the help of the discoverer, we excavated the skull. We put a plaster jacket around the skull to protect it, but it was just so enormous that the three of us were not able to carry it."[27] Eventually it took five people to lift the skull (which, including the surrounding sediment and plaster jacket, weighed around 170 kilograms) out of the gully and across the field, then maneuver it into the back of a vehicle. Ian worked for nearly a year to skillfully prepare the specimen, which is now displayed in the Queensland Museum (fig. 8.5).

Diprotodon optatum exhibited considerable sexual difference in size, with the larger form probably male and the smaller female. At most sites both forms are present, but only the smaller form occurs at Bacchus Marsh. Here the fossil remains appear to have accumulated during a single mass-mortality event in which a segregated group of immature females became trapped in the mud of a drying marsh.[28] The site has produced a beautifully preserved skull about 72 centimeters long.[29]

Diprotodon trackways are known so far from only three localities in Australia. At Lake Callabonna, several trackways 300 to 500 millimeters wide preserved as carbonate-cemented imprints were recorded by the Tedford expedition in 1970, but had disappeared (either eroded or reburied) by the time of Tom Rich's visit in the 1990s.[30] The best-preserved megafaunal trackway site in Australia was described by Carey and colleagues.[31] Known locally as "dinosaur footprints," the tracks occur in marginal lake sediments (derived from volcanic deposits) within the Victorian Volcanic Plains of southeastern Australia. The sediments were dated by OSL to ca. 110 to 60 kya. Two trackways are attributable to *Diprotodon*: the better preserved, 63 meters long (with a 20-meter gap) crosses several sand bars; the second, about 30 meters away, is

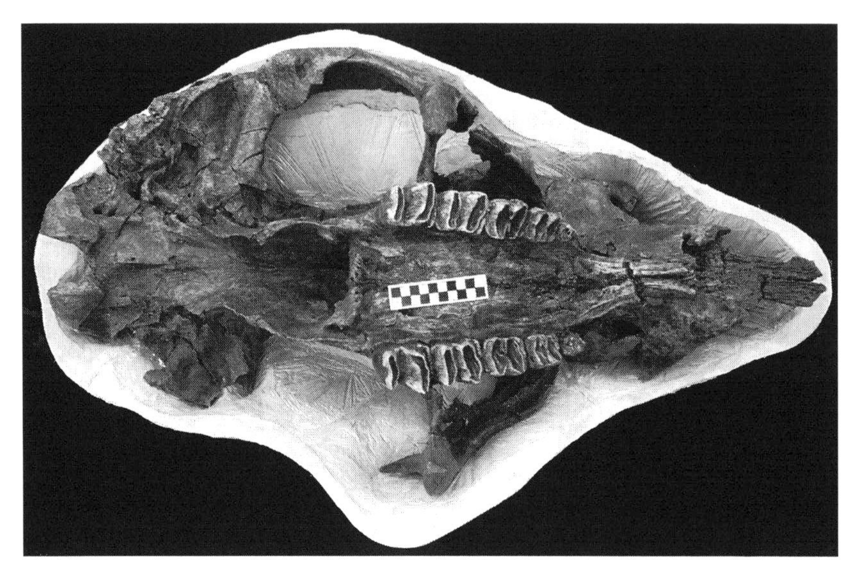

FIGURE 8.5. *Diprotodon optatum* skull, Darling Downs, Queensland. View of palate. (courtesy of Gilbert Price, Ian Sobbe).

short and poorly registered. The site also features tracks of a kangaroo (probably an extinct *Protemnodon* species) and a large wombat (perhaps *Ramsayia* or *Phascolomys*), most of which are at right angles to the main track. There are also a couple of prints that are very likely of a marsupial lion. A portion of *Diprotodon* trackway was replicated for display in the Museums Victoria.

A fascinating new insight into the ecology of *Diprotodon optatum* has been provided by analyses of stable isotopes from the enamel of an upper incisor, dated by U-Th analyses (on dentine from the same tooth) to ca. 300 kya (Middle Pleistocene).[32] This paper presents, for the first time, evidence for seasonal migration in a marsupial, living or extinct. Taking advantage of the fact that the incisor was permanently growing, the enamel was sampled sequentially though the last approximately three years of the individual's life. Variations in strontium isotope ($^{87}Sr/^{86}Sr$) ratios revealed a cyclical pattern, interpreted as annual two-way latitudinal migrations, in which the same locations within the Darling Downs (with different strontium values) were revisited seasonally. The inferred average growth rate of 0.2–0.3 millimeters per day is similar to that of extant mammals. $\delta^{18}O$ values are similar to modern tropical marsupials, indicating a tropical climate, contrasting with the temperate to subtropical climate of the region today. The ^{13}C values suggest that the individual's diet comprised a mixture of C3 and C4 plants. This study opens up some exciting possibilities for further investigations, both on additional *D. optatum* material and on other species of Australian megafauna.

Because of their massive size, adult *Diprotodon* presumably would not have been the easiest prey for contemporary predators, mammalian or reptilian. Evidence of either predation or scavenging by a "marsupial lion" (*Thylacoleo carnifex*) is provided by a *Diprotodon* ulna from near Glen Innes (New South Wales) that has deep, blade-like tooth marks attributable to the shearing premolars of this carnivore, whose remains also occur at the site.[33] Moreover, crocodilian tooth marks on *Diprotodon* bones at the Katapiri Site probably bear witness to predation or scavenging by the extinct crocodile *Pallimnarchus pollens*.[34] Other potential predators include the extinct giant monitor lizard (*Varanus priscus*), which, if it possessed a venomous bite (as seems likely), could well have overcome a *Diprotodon*; as well as the living saltwater crocodile (*Crocodylus porosus*), where their ranges coincided in the north of the continent. Assuming a substantial overlap in time, it is probable that humans would have hunted *Diprotodon* at least occasionally, although no evidence to support this has been discovered so far.

Probably the most reliable age estimates on *Diprotodon* are provided by ESR dates (thirteen in total) on *Diprotodon* tooth enamel. The youngest of these are: 50.67 kya from Cuddie Springs (New South Wales), 48.96 kya from Cooper's Dune (South Australia), 46.3 kya from Lancefield Swamp (Victoria), and 44.25 kya from Hallett Cove (South Australia).

Finally, a brief mention of the "bunyip," a celebrated Australian legendary monster that is said to haunt billabongs and other bodies of water and attack unwary humans. It has been suggested that the legend represents an Aboriginal folk memory from the time *Diprotodon* and other mega-mammals roamed Australia, and the story is supposed to gain credence from instances when Aboriginals, who were shown megafaunal bones, identified them as the remains of "bunyips." However, more prosaically, these legends could well have originated from or been reinforced by finds of fossil bones—as happened with many other beasts in other cultures around the world.

8.4 "MARSUPIAL RHINO": *ZYGOMATURUS TRILOBUS*

Zygomaturus trilobus, a large herbivorous marsupial about 1.2 meters at the shoulder and weighing about 300 to 500 kilograms, was second in size only to *Diprotodon*, to which it was closely related.[35] *Z. trilobus* is recorded from

Western Australia, Queensland, South Australia, Victoria, and Tasmania. In 1920 a well-preserved, substantially complete skeleton of *Z. trilobus* was discovered at Mowbray Swamp near the coast of northwest Tasmania. The mounted skeleton, described at the time as "stained brown by the tannic waters of the swamp" and with a "very perfect skull," is in the Tasmanian Museum and Art Gallery, Hobart.[36] The massive oddly shaped skull, 33 centimeters long and 50 centimeters deep, has an upturned three-lobed snout and strikingly large cheek bones (or zygomatic arches—hence the name). With its squat body, short legs, and huge head, *Z. trilobus* broadly resembled a rhino or hippopotamus. From the distribution of finds, it seems likely that it favored swampy areas and perhaps was semiaquatic, like a hippopotamus. The low-crowned cheek teeth indicate a browsing diet.

Mammoth Cave, a large show cave in the southwest of Western Australia, has produced a number of megafaunal marsupial fossils (but definitely no mammoths!). Visitors can see a lower jaw of *Zygomaturus* in its original position still embedded in deposits in the cave wall.

Reassessed radiocarbon dates on *Z. trilobus* from Mowbray Swamp using XAD-2 resin gave infinite ages.[37] The youngest ESR direct dates on *Z. trilobus* tooth enamel are 127.5 kya and 128.5 kya from Victoria Fossil Cave (South Australia). The youngest OSL dates on sediment enclosing *Z. trilobus* remains include 57 kya from Scotchtown Cave (Tasmania), 63 kya from Mammoth Cave, and 70 kya from Victoria Fossil Cave. A *Zygomaturus* skeleton from the Willandra Lakes, New South Wales, has been dated by two independent techniques—both are outstandingly young in relation to most Australian records. OSL dating on the enclosing sediment gave a maximum age range of 36.7 to 33.3 kya, while U-series of bone from the skeleton gave a minimum range of 32.4 ± 0.5 kya.[38]

8.5 "MARSUPIAL TAPIR": *PALORCHESTES AZAEL*

Working with the very limited material that he had been sent, Owen mistook *P. azael* for a species of giant kangaroo, since its cheek teeth (with their transverse ridges) resemble those of kangaroos.[39] It was only in 1958 that its relationship to *Diprotodon* and *Zygomaturus* was recognized. *Palorchestes azael* is recorded from Queensland, South Australia, Victoria, and Tasmania, but fossils are rare and even today no reasonably complete skeleton is available for study. *P. azael* had shortened nasal bones on its skull, suggesting that it may have had a small trunk rather like a tapir.[40] It had powerful forelimbs and claws, which could have been used to dig up roots or pull down leaves and branches, and no doubt would have been effective for defense. The narrow, elongated skull and jaw suggests that it was a picky eater, possibly using a long prehensile tongue to select particular food plants. These features invite comparison with the medium to large ground sloths from North and South America.

The youngest OSL dates are 56 kya from Scotchtown Cave (Tasmania) and 122 kya from Cave QML 796 (Queensland). No direct dates are available.

8.6 GIANT SHORT-FACED KANGAROO: *PROCOPTODON GOLIAH*

In a continent replete with extraordinary animals past and present, the extinct giant short-faced kangaroo (*Procoptodon goliah*) manages to stand out for sheer size and novelty (fig. 8.6). Like so many other extinct Australian megafauna *Procoptodon goliah* was first described and named by Richard Owen,[41] once again on the basis of fossils shipped to England by his Australian contacts.

P. goliah is the largest and most heavily built kangaroo known; it had long arms, an upright posture, and forward-facing eyes. At about two meters to the top of the head, it was similar in height to the largest existing marsupial the red kangaroo (*Macropus rufus*), but with an estimated weight of 240 kilograms, it was nearly three times as heavy. *P. goliah* was the largest of at least fourteen

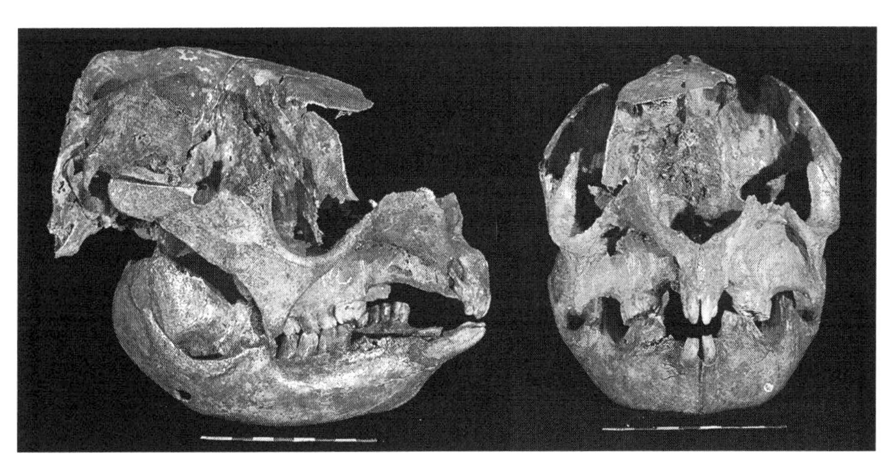

FIGURE 8.6. *Procoptodon goliah* skull, side and front views, Naracoorte, South Australia. (Courtesy of Rod Wells.)

FIGURE 8.7. *Sthenurus occidentalis* (replica) skeleton, Green Waterhole Cave, South Australia. (Courtesy of Rod Wells.)

species of sthenurines (short-faced kangaroos). The next largest, *Sthenurus stirlingi*, would have weighed in at around 180 kilograms. Several, including *Sthenurus* (*Simosthenurus*) *occidentalis* and *Procoptodon rapha*, were rather smaller, while the smallest, *Sthenurus* (*Procoptodon*) *gilli*, weighed only about 60 kilograms (figs. 8.7, 8.8, 8.9).

A study of sthenurine anatomy by Christine Janis and colleagues came to the rather surprising conclusion that, unlike all kangaroos alive today, *Procoptodon goliah* (and at least its larger relatives) would have been incapable of hopping and therefore could not have achieved the high speeds of modern kangaroos. Instead they probably walked bipedally while using the tail as a fifth limb, as do modern kangaroos when moving slowly.[42] They had a single large toe on each foot (whereas modern kangaroo species have three or four), a

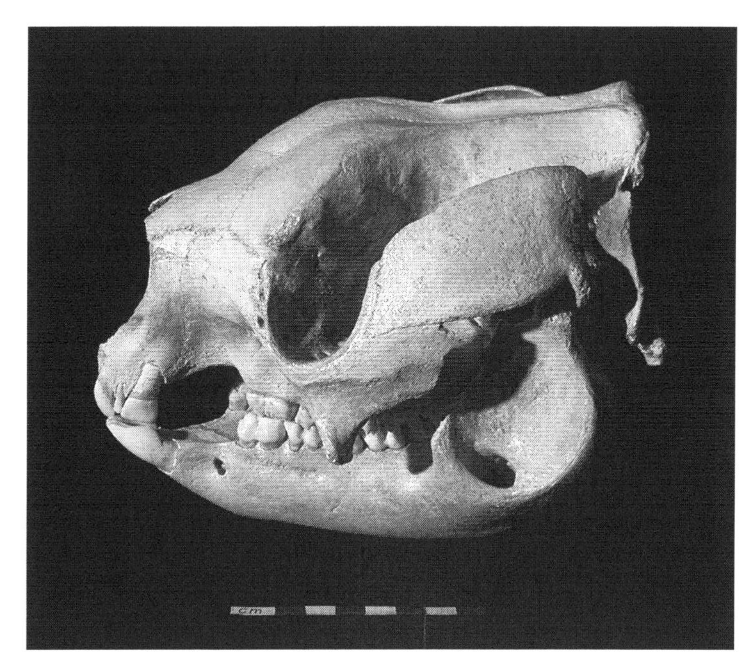

FIGURE 8.8. *Sthenurus occidentalis* skull, Green Waterhole Cave, South Australia. (Courtesy of Rod Wells.)

FIGURE 8.9. *Sthenurus* (*Procoptodon*) *gilli* skull *in situ*, Naracoorte, South Australia. (Courtesy of Steve Bourne.)

conspicuously short face, and forward-facing eyes, which would have helped them to select food plants. Extended forelimbs with two long, clawed fingers on each hand would have enabled them to reach well above their heads and pull down leafy branches on which to feed—in the case of *P. goliah*, up to 3 meters from the ground. However, with this modification it would have been impossible to use the forelimbs for locomotion. *P. goliah* and its relatives had low-crowned, bilophodont cheek teeth, evidently adapted for browsing.[43] Isotope analyses and dental microwear studies indicate that it was a specialist browser on chenopods and may have had a preference for saltbushes (*Atriplex*), which use the C4 photosynthetic pathway.[44] Moreover, oxygen isotope analyses of *P. goliah* tooth enamel show that, like saltbush feeders today, it drank more in low-rainfall areas than contemporary grazers.

OSL dates on sediment associated with *P. goliah* remains include 46 kya and 47 kya from Neds Gully, 52 kya and 54 kya from Lake Victoria, and 164 kya from Victoria Fossil Cave. No direct dates are available.

8.7 OTHER EXTINCT KANGAROOS

Species of the extinct genus *Protemnodon* formerly ranged throughout much of Australia, New Guinea and Tasmania. The largest of these, *Protemnodon anak*, is usually described as a giant wallaby and probably weighed more than 110 kilograms.[45] Especially well-preserved remains (probably of this species) from caves in Tasmania have been radiocarbon dated to 41.8 to 40.9 kya and 42.9 to 42.1 kya (Mount Cripps), and to 44.9 to 43 kya (Titans Shelter).[46] As discussed below, these comparatively recent dates—which are at or close to the limit of radiocarbon dating—have added fuel to the debate about the causes of extinction.

There were also some more conventional-looking kangaroos, such as *Macropus ferragus*, differing from existing species mainly in their larger size

8.8 MARSUPIAL LION: *THYLACOLEO CARNIFEX*

The marsupial lion (*Thylacoleo carnifex*) was perhaps the weirdest of all the weird Australian megafauna (fig. 8.10). It is by far the largest mammalian carnivore known from that continent, and easily holds the world weight record for a flesh-eating marsupial. On the basis of skull and jaw fragments and teeth collected from a cave in the Wellington Valley by Major Thomas Mitchell, *Thylacoleo* was described and named by Richard Owen, who dramatically called it

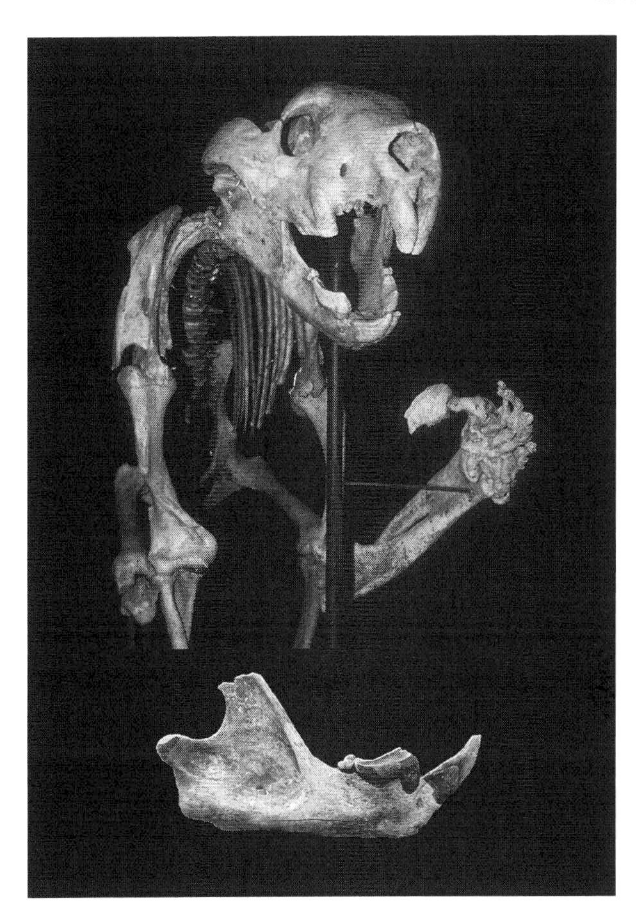

FIGURE 8.10. *Thylacoleo carnifex* (replica) skeleton, Naracoorte (courtesy of Rod Wells), together with mandible, from Neds Gulley, Queensland (courtesy of Gilbert Price).

"the fellest and most destructive of predatory beasts."[47] However, his conviction that it was a fearsome predator was strongly disputed for many years.

The occasion for such contrary opinions stems from the unique anatomy of *Thylacoleo*'s skull, jaws, and teeth, which are quite unlike those of any living animal, and the fact of its relationship to wombats and koalas, which are overwhelmingly herbivorous. Alternative, rather implausible, theories included a diet of fruit, gourds, cycad nuts, or crocodile eggs, and the contentious question of its mode of life has been resolved only in recent years as much more complete material has been discovered, with the result that Owen's original interpretation is now almost universally accepted.[48]

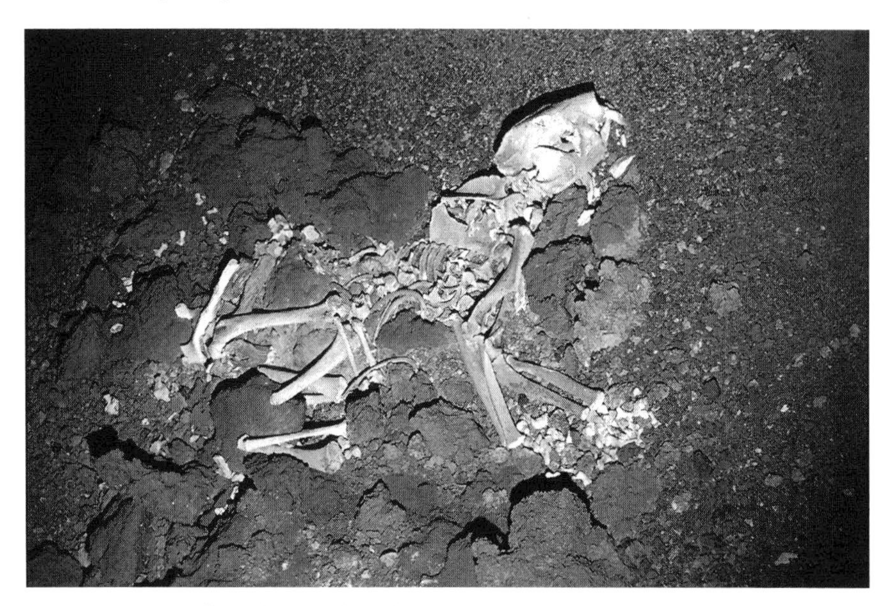

FIGURE 8.11. *Thylacoleo carnifex* skeleton of immature individual *in situ*, Nullarbor Caves, South Australia. (Courtesy of Clay Bryce / Western Australian Museum.)

The marsupial lion is known from many fossil sites across Australia, including partial skeletons of several individuals from Victoria Fossil Chamber, Naracoorte.[49] However, no complete skeleton had been recovered until 2002 when spectacular discoveries were made in remote limestone caves beneath the arid, treeless Nullarbor Plain (South Australia).[50] On first venturing into Flight Star Cave, some 50 meters below ground, John Long, Gavin Prideaux, and their team were greeted by a stunning sight: a beautifully preserved intact skeleton of a *Thylacoleo* still laid out essentially undisturbed from the time of its death many hundreds of thousands of years ago (fig. 8.11).

The unfused ends of limb bones showed that it was an immature individual. The extreme dryness and cool temperature of the cave resulted in extraordinarily good preservation; the bones are white and unmineralized, and look as if they were at most only a few hundred years old. However, the remains proved to be extremely fragile, which necessitated much time treating them with consolidant. Also in the cave were remains of hundreds of other animals, including *Procoptodon goliah*, other kangaroos, and ten incomplete skeletons of marsupial lion.

The carnivorous dental specializations of *Thylacoleo* are unique. Placental carnivores such as lion, tiger, and leopard are equipped with enlarged canines for gripping and killing prey, and carnassials (fourth premolar above, first molar below) for slicing flesh. In marked contrast, the canines in *Thylacoleo* were tiny, whereas the extremely enlarged first incisors possibly fulfilled a similar function to the canines of placental carnivores. The most striking feature of *Thylacoleo* is its enormously enlarged blade-like third premolars, up to 6 centimeters long, which could not only slice flesh but also cut through bone, functioning much like bolt-cutters. Stephen Wroe and colleagues concluded that *Thylacoleo* probably had the most powerful known bite of any mammal, living or extinct.[51] The possession of massive forearms, large retractable thumb claws, and forward-facing eye sockets (for binocular vision) also strongly suggest a predatory mode of life. At around 0.75 meters at the shoulder and with an estimated weight of 87 to 130 kilograms, it was roughly equivalent in size to a modern jaguar.[52] Previous estimates had compared it to a leopard, which weighs only half as much. The greater weight estimate argues against previous suggestions that *Thylacoleo* was primarily arboreal and even carried its prey up into trees (like the leopard) to protect it from other carnivores such as the Tasmanian devil. According to Wroe: "[*Thylacoleo* was] just a lump of muscle and bone, and powerfully built. It had a build that was closer to a bear than a cat. It probably preyed on slow but large prey. This creature was built to wrestle—its arm bones were twice as thick as a leopard's, and its skull was as wide as it was long.' "[53] So, *Thylacoleo* had every indication of having been an ambush predator that probably lay in wait under cover near waterholes or elsewhere until large-mammal prey came within reach. Prey probably included *Zygomaturus*, *Palorchestes*, juvenile *Diprotodon*, and kangaroos up to the size of *Procoptodon goliah*. It seems likely that emu (*Dromaius*) and thunderbird (*Genyornis*) were also on the menu. *Thylacoleo carnifex* fossils from Moree, New South Wales, include an adult female skeleton in association with the lower jaw of a newborn and the skull of an older juvenile; possibly all three were a family group. On the basis of the new finds from Flight Star Cave, Nullarbor (fig. 8.11) and Henschke's Quarry, Naracoorte, Wells and Camens have inferred additional information on the likely behavior of this extraordinary animal.[54] For the first time it proved possible to reconstruct the entire skeleton, including the complete tail and previously unrecognized clavicles. These authors envisage *T. carnifex* as a carnivore/scavenger inflicting massive damage to its prey with its slashing first digit on each hand and by biting the neck region with its paired lower incisors, while the powerful forelimbs and grasping hands were used to restrain a victim or a carcass. They view its feeding behavior as analogous to the much smaller, living Tasmanian devil *Sarcophilus harrisii*. The skeletal anatomy suggests that it was capable of climbing.

Several cave paintings have been claimed to represent *Thylacoleo*, but it seems more likely that they actually depict the marsupial wolf (*Thylacinus*; see below). Tight Entrance Cave (Western Australia) has produced interesting evidence of *Thylacoleo* behavior in the form of numerous claw marks in the cave walls.[55] The widely spaced deep V-shaped scratches points strongly to *Thylacoleo* rather than to any other possible candidate (Tasmanian devils probably also visited the cave). Smaller scratch marks, most of which are attributed to juvenile marsupial lion, suggest that out-of-the-pouch young were left secure in the cave while the mother was out hunting. The lack of tooth marks attributable to *Thylacoleo* on bones at the site is consistent with its inferred flesh-stripping rather than bone-crunching behavior.

There are just two direct ESR dates on *Thylacoleo* tooth enamel: 50 kya from Black Creek Swamp (South Australia) and 53 kya from Titan Shelter (Tasmania). The youngest of 26 (indirect) OSL dates for *Thylacoleo* are 46 and 47 kya from Neds Gully (Queensland).

8.9 OTHER EXTINCT MARSUPIALS

The genuine "giant wombat" *Phascolonus gigas* probably weighed about 200 kilograms, with a wide yet elongated low-slung body less than 1 meter at the shoulder and distinctive, broad, flat incisors.[56] Another wombat, *Ramsayia curvirostris*—known only from partial skeletons—was smaller than *Phascolonus* and had narrow incisors like the three extant species: *Lasiorhinus latifrons*, *L. krefftii*, and *Vombatus ursinus*.

At around 70 kilograms, the poorly known extinct rat-kangaroo *Propleopus oscillans* was much larger than its living relative *Hypsiprymnodon*. Its dentition suggests that it was primarily carnivorous.[57]

Weighing only up to about 30 kilograms, the thylacine *Thylacinus cynocephalus* ("Tasmanian wolf" or "Tasmanian tiger") does not qualify as megafauna, although it was the largest marsupial carnivore other than *Thylacoleo* (and perhaps also *Propleopus*). It was extirpated from mainland Australia sometime after 3.65 kya, very likely due to competition from the introduced dingo, which first appears in the archaeological record ca. 3.5 kya.[58] However, thylacines persisted until the twentieth century CE in dingo-free Tasmania but were then deliberately exterminated. Incredibly, bounty hunters were paid by the Tasmanian government for dead "tigers," and the last wild specimen of this wonderful, unique animal was killed in 1932.[59] The last captive animal—fortunately recorded on film—died in Hobart Zoo in 1936.

8.10 GIANT GOANNA: *VARANUS PRISCUS* (AKA *MEGALANIA PRISCA*)

In addition to giant marsupials, Sahul was and is also home to some giant reptiles. There's an old joke in reference to the wildlife in Australia that "everything is trying to kill you." This would have been only too true for anyone unlucky enough to have encountered *Varanus priscus*, a giant extinct relative of the living Komodo dragon (*Varanus komodoensis*) (fig. 8.12), which is native to Komodo and adjacent Indonesian islands.

The Komodo dragon can attain a length of 3 meters and weigh 70 kilograms, but reliably estimating the size of *Varanus priscus* is very difficult, as fossil material is extremely rare and only fragmentary remains have been discovered so far. Molnar suggested an average weight of 320 kilograms and a maximum of 1,940 kilograms for a 7-meter-long animal, and Fry and colleagues estimated 575 kilograms and 5.5 meters—substantially increased from previous figures of 97 kilograms and 3.5 meters.[60] It is clear that *Varanus priscus* was several times the size of a Komodo dragon, and moreover was probably the largest terrestrial lizard of all time. Judging from isolated fossil finds and in comparison with the Komodo dragon, *Varanus priscus* would have been equipped with a formidable array of sharp, backwardly curved teeth, which were serrated like steak knives (much as in *Tyrannosaurus rex* from the Cretaceous period)—superbly adapted for slicing through flesh and bone.[61] The skin was armored with numerous embedded bony nodules (osteoderms), and it bore an unusual crest on its snout; a similar but smaller crest is seen in the living Australian perentie (*Varanus giganteus*).

The giant goanna was first described by Richard Owen on the basis of three huge vertebrae collected from a tributary of the Condamine River west of Moreton Bay (Queensland), which along with other megafaunal material from this locality had been purchased by the British Museum; additional remains were the subject of a series of subsequent papers.[62] Owen recognized their similarity to the much smaller extant Australian monitor lizards or goannas (he was unaware of the existence of the Komodo dragon). He stated that "the chief peculiarity of the Australian fossil lizard is its great size; the vertebrae rival in bulk those of the largest living Crocodiles." He mused: "Whether among the vast and unexplored wildernesses of the Australian continent any living representative of the more truly gigantic *Megalania* still lingers may be a question worth the attention of travellers. But, most probably, like the gigantic marsupials, *Diprotodon* and *Nototherium*, with whose fossil remains

FIGURE 8.12. Komodo dragon *Varanus komodoensis*. (Photo by the author.)

those of *Megalania* were associated in the tertiary deposits now cut through by the Condamine and its tributaries, the gigantic land-lizard has long been extinct."[63]

Hocknull and colleagues concluded that the extant Komodo dragon *V. komodoensis* originated in Australia by the early Pliocene, and from there dispersed west to Flores (by 900 kya) and Java (by 800 to 700 kya). In Australia, *V. priscus* "reached gigantic proportions by the late Middle Pleistocene," presumably having evolved from *V. komodoensis* or a similar form.[64] Research on the living Komodo dragon revealed that it has a venomous bite—inflicted by numerous slashing curved teeth—which enable it to successfully overcome large-mammal prey.[65] (Previously it was mistakenly believed that bacteria in the dragon's mouth infected bite wounds resulting in blood poisoning.) Death results from a combination of shock, rapid decrease in blood pressure, and blood loss. *Varanus priscus* is thought to have killed in a similar way, and undoubtedly would have been able to tackle much larger prey. However, again by analogy with the living Komodo dragon—it most probably included a substantial amount of carrion in its diet, which according to Wroe is likely to have included leftovers from marsupial lion kills.[66] Unlike the latter, it was probably a highly efficient feeder; a Komodo dragon can consume up to 90% of a carcass. *Varanus priscus* was an animal that you really wouldn't want to meet on a dark night (although actually this might have been rather safer, as lower temperatures would have rendered this cold-blooded animal less active).

Previously the youngest available (OSL) dates for *V. priscus* (*M. prisca*) were about 107 kya and 83 kya, from the Darling Downs (Queensland). However, a paper published by Gilbert Price and colleagues provided the first evidence that in Australia, humans and some kind of giant monitor would have overlapped in time. At Colosseum Chamber, Mount Etna region (Queensland), an excavated sequence with a series of radiocarbon dates (on charcoal) and uranium-thorium dates (on straw stalactites) produced a well-preserved osteoderm of a large monitor from a level dated to around 50 kya. This fossil "represents an extremely large-bodied monitor lizard that is no longer extant in Australia."[67] From its size it could belong either to the living *V. komodoensis* or the extinct *V. priscus*.

Lastly, how do we account for the claim that humans wiped out giant monitor lizards in Australia but somehow failed to do so on Komodo and adjacent islands? As observed by Price: "If humans were directly responsible for the extirpation of large-bodied monitor lizards in Australia, the contrast in survivorship to the similar giant lizards of Southeast Asia is quite remarkable"; and "given the ecological impacts humans famously have had on island ecosystems."[68]

8.11 OTHER EXTINCT REPTILES

The extinct *Pallimnarchus pollens* was a large Pliocene and Pleistocene freshwater crocodile (up to 5 meters long) known from central and northern Australia.[69] An ancestral form is recorded from the famous Miocene fossil site of Riversleigh, Queensland.[70] Fossil finds of *P. pollens* are scarce. The exceptionally broad, flat skull (rather similar to the living saltwater crocodile) and serrated conical stabbing teeth suggest that it was an aquatic ambush predator capable of taking large prey. A series of tooth marks on a probable *Zygomaturus* humerus have been attributed to *Pallimnarchus*.[71] Large osteoderms beneath its skin would have afforded it considerable protection against attack, likely to have come principally from its own kind. There are no direct dates available for this species.

The extraordinary giant snake *Wonambi naracoortensis* is known from the southern half of Australia. First described from Victoria Fossil Cave, Naracoorte (South Australia), it takes its generic name from the legendary rainbow serpent of the Aboriginal Dreamtime.[72] Partial skeletons comprising vertebrae, ribs, and skull elements have been described from Henschke's Cave, Naracoorte, and reconstructed skeletons nearby dramatically depict an imagined mortal struggle between *Wonambi* and the marsupial lion (*Thylacoleo*). *Wonambi*, which was one of the last survivors of an entirely extinct primitive group of snakes known as madtsoiids, is estimated to have grown to more than 5 meters long and would have killed its prey by constriction, like pythons and boas today.[73] Recent research has demonstrated that modern constricting snakes kill (very efficiently) by cutting off their victim's blood flow—and thus oxygen—to the brain, not by suffocation, as long believed. There is insufficient information from the available skull remains to judge how big an animal *Wonambi* could have swallowed, so we don't know what is likely to have been on its menu.

Indirect dates on *Wonambi* include ca. 55.2 kya (uranium-thorium on flowstone) and ca. 63 kya (OSL) from Mammoth Cave (Western Australia). OSL dates on sediment enclosing *Wonambi* remains from Victoria Fossil Cave range from ca. 93 to 76 kya.

8.12 "THUNDERBIRD": *GENYORNIS NEWTONI*

A megafaunal bird, *Genyornis newtoni* was the last representative of the dromornithids: a family of giant flightless birds, variously known as thunderbirds, demon ducks, and mihirungs. Although somewhat smaller than some of its predecessors the largest *Genyornis newtoni* individuals were nearly 2 meters tall and heavily built, probably weighing up to about 515 kilograms—approximately twice as heavy as a male ostrich.[74] *G. newtoni* was not at all closely related to the well-known giant ratites, such as ostrich, emu, and the extinct moa. Its nearest living relatives are ducks and geese, and it is often reconstructed as a huge goose—for example, in the charming 2008 series of postage stamps featuring Australian megafauna by the artist Peter Trusler.

The first *Genyornis newtoni* finds were recovered from Lake Callabonna (South Australia), which has also produced the most abundant and complete remains.[75] Like *Diprotodon*, these huge birds evidently perished by becoming trapped in the glutinous lake sediments. Many bones were also found at Cuddie Springs (New South Wales; see below). Other important sites include Naracoorte Caves (South Australia) and Wellington Caves and Cuddie Springs (New South Wales). An Aboriginal rock art painting at Nawarla Gabarnmang in the Northern Territory depicts two large flightless birds with outstretched necks, which perhaps represent *Genyornis* rather than emu.

G. newtoni had large hoof-like claws on its feet as well as a large beak, although no well-preserved skull has yet been recovered. It was probably herbivorous, with a diet that included fruit and nuts. Groups of rounded and highly polished gizzard stones (gastroliths), used to grind up its food, have been recovered from the gizzard region of *Genyornis* skeletons from Lake Callabonna; significantly, gastroliths are unknown from carnivores.

By far the most intensively researched example of megafaunal extinction in Australia is the extensive study of more than seven hundred dated *Genyornis* eggshell fragments.[76] Amino acid racemization (AAR) of eggshell and optically stimulated luminescence dating (OSL) of the enclosing sediment, gave youngest dates of about 50 kya, the inferred time of extinction. In marked contrast, a series of dates on emu (*Dromaius novaehollandiae*) shell fragments from the same sites continued through to the present day. Furthermore, the authors suggested that this was compelling evidence that megafaunal extinction in Australia was due to human activity, rather than climate change. This view was reinforced by further research on two hundred sites that produced burnt eggshell fragments.[77] Analyses of amino acids revealed a thermal gradient consistent with the eggs having been cooked on ember fires, proving that humans and an extinct megafaunal species coexisted. This association provided compelling evidence for predation of an Australian megafaunal species by humans, with the implication that this sustained practice resulted in *Genyornis* extinction.

However, very soon after the previous paper appeared, another research group suggested that the dated eggshells are not *Genyornis* at all, but pertain to a species of extinct large (but not megafaunal) megapode (genus *Progura*)—a relative of the living brush turkey.[78] In particular, they proposed that the eggshell fragments were too small for *Genyornis*. Inevitably, the story does not end there. Miller and colleagues disputed the criticisms of Grellet-Tinner and colleagues, who responded in turn, reinforcing their original interpretation.[79] The scientific stakes are high. As remarked by Grellet-Tinner and colleagues in their original paper: "The extinction of *Genyornis* became the most well-dated megafaunal extinction in Australia, and one of the best dated in the world."[80] Of course, if Grellet-Tinner and colleagues are correct, it would mean that we have information only on the timing of the demise of a large *Progura*, and unfortunately nothing whatever about the extinction of the megafaunal *Genyornis newtoni*.

8.13 CUDDIE SPRINGS

Cuddie Springs (New South Wales) potentially provides by far the best evidence for the co-occurrence of extinct megafauna and humans in Sahul, and also makes the best case for the survival of several megafaunal species to 30 to

40 kya. However, the site remains highly contentious because of starkly conflicting interpretations of the stratigraphic and dating evidence.[81] Important issues include whether or not the archaeology and megafaunal remains in Unit 6 are *in situ* and contemporaneous, as well as which set of conflicting dating results is closer to the truth.

Fossil bones of extinct animals were first discovered at Cuddie Springs in the late nineteenth century during well-sinking activities.[82] The site is in a treeless pan—the lowest point in an ancient shallow lake bed that occasionally fills with water after rain. Excavations revealed a sequence of vertebrate-bearing layers, two of which also yielded abundant stone artifacts (figs. 8.13, 8.14). Pollen and spores occur at all levels. The literature on this site is extensive, and here I just attempt to summarize what I believe to be the most important points. Dodson and Field give a useful overview of the stratigraphy and paleontology.[83]

Numerous horse and cattle remains occur in the uppermost (post-European contact) levels. Gillespie and Brook argued that the stone horizon, SU5—which effectively separates the underlying Pleistocene deposits from the

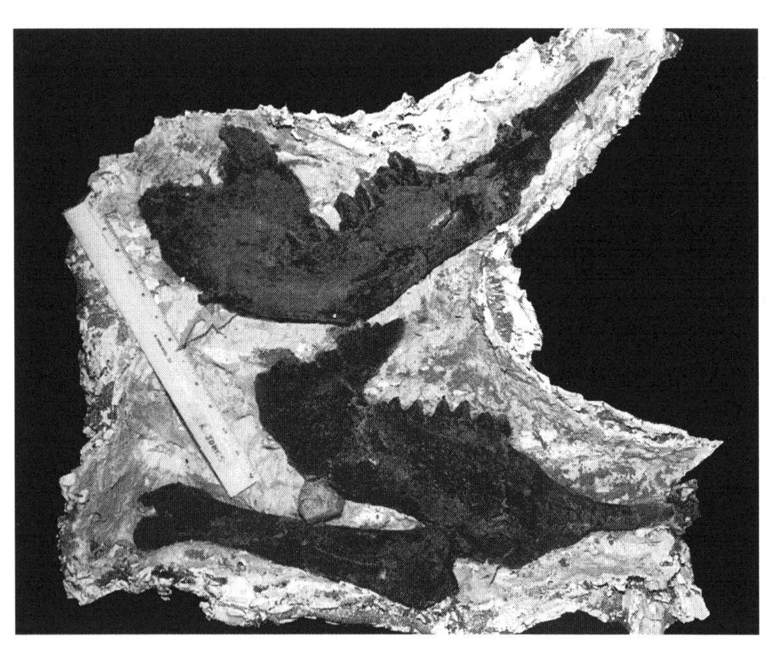

FIGURE 8.13. Associated left and right mandibles of *Diprotodon* and a *Genyornis* tarsometatarsus, with a stone artifact between, in plaster jacket as excavated, Unit SU6B, Cuddie Springs, New South Wales. (Courtesy of Judith Field.)

FIGURE 8.14. *Genyornis*, Unit SU6B, Cuddie Springs, New South Wales. Foreground: left femur and right? tibiotarsus. Next to scale rod: left tibiotarsus. All are probably from a single individual. Stone artifacts are indicated by the white arrows. (Courtesy of Judith Field.)

1 meter of disturbed overburden—was not a natural formation but originated from stones laid down by farmers in the nineteenth century to prevent cattle sinking in the mud.[84] However, Judith Field and colleagues interpret SU5 as a stone deflation pavement, including many artifacts, that caps the underlying Pleistocene sequence, pointing out that this unit contains no post-Pleistocene material.[85] It is bracketed by OSL and radiocarbon dates of ca. 22 and 33 kya.

Bones of extinct megafauna, with associated archaeology (flaked and ground stone tools), were found in two apparently discrete underlying stratigraphic units: SU6A (about 1.05 to 1.35 meters deep) and SU6B (about 1.35 to 1.70 meters deep).[86] Extinct megafauna from SU6 includes *Diprotodon* cf. *optatum*, *Palorchestes* cf. *azael*, *Phascolonus* sp., *Sthenurus* sp., *Protemnodon* cf. *brehus*, *Protemnodon* sp., *Genyornis newtoni*, and possibly *Pallimnarchus* sp.[87] Extant species of kangaroos, wallabies, and others are also represented.

Radiocarbon (charcoal) and OSL (sediment) dating both agree on an age range of ca. 40 to 30 kya for units SU6A and SU6B.[88] However, ESR/U-series analyses of bones and teeth for the same units by Grün and colleagues.[89] produced much older dates, in the range ca. 50 to 40 kya, A third unit, SU9 (about 2 to 2.2 meters deep), containing a concentrated layer of megafaunal remains but lacking associated archaeology, was dated by ESR/U-series to ca. 580 to 379 kya, placing it within the Middle Pleistocene. No artifacts occur in the sequence below 1.7 meters.[90] Grün and colleagues attributed the discrepancies in dating SU6A and SU6B to mixing of the deposits and argue that the megafaunal remains and archaeology were intrusive. However, both units comprise clays and silts, deposited in a low-energy lacustrine environment, so it is very difficult to see how numerous large bones, and artifacts could have been washed in, or otherwise introduced, from the surrounding area. Moreover, there are no indications of abrasion or weathering that would be expected to have occurred during transport, and there are clear instances in both units of associated megafaunal remains from individual animals (fig. 8.13); such associations would not have survived transport. In view of the importance of Cuddie Springs for the Sahul extinctions debate, the dating problems need to be resolved.

The site is believed to have functioned as a waterhole from the mid-Pleistocene onward, with animals probably becoming in essence tethered to a shrinking water source during dry spells. As animals died due to lack of water and/or food, their bones were incorporated in the silty clay sediment. The presence of flaked and ground stone tools attests to the presence of people. The pollen and sediment record indicate marshy conditions, with occasional standing water, during the deposition of SU6B. As the site dried out, people are thought to have camped on the claypan floor.

The reluctance of many researchers to consider a relatively late age for the Cuddie Springs megafauna—and that associated megafaunal remains and artifacts could be in primary context—is arguably related to the widely held view that all Australian megafauna went extinct around 45 to 46 kya—about the same time as humans were believed to have arrived.[91] However, with new evidence of much earlier human arrival, ca. 65 kya, and a date of ca. 34.6 kya on a *Zygomaturus* skeleton from Willandra Lakes, this scenario no longer appears tenable; evidently there was a long temporal overlap between humans and megafauna in Sahul.[92] In the light of these recent developments, Cuddie Springs appears to be less of an anomaly, and one that merits further investigation—notably, targeted dating of megafaunal remains from SU6A and SU6B.

8.14 NEW GUINEA

Today, New Guinea is a large tropical island situated to the north of Australia, whereas during the Last Glacial it was broadly connected to Australia, due to substantially lowered sea level.[93] The large expanse of dry land between the two was submerged by the sea level rise early in the Holocene. New Guinea's varied habitats range from montane in the central mountain chain that forms the spine of the island, to tropical lowland forests and grasslands. The highly varied topography and climate is reflected in the exceptionally diverse flora and fauna. However, only a handful of megafaunal species are known from the late Quaternary.[94] Zygomaturines (that is, *Zygomaturus* relatives) comprise *Maokopia ronaldi* (about 100 kilograms) and *Hulitherium thomasettii* (100 to 200 kilograms). Other recorded megafaunal species are the "giant wallabies": *Protemnodon hopei* (about 50 kilograms), *P. tumbuna* (about 50 kilograms), and *Protemnodon nombe* (around 40 kilograms). Additionally, there is a poorly known tapir-size marsupial, *"Kolopsis" watutense* (estimated at about 300 kilograms). All are distinct from related Australian species, presumably reflecting the very different range of habitats in New Guinea. There is no definite information on likely extinction times for these species.[95]

8.15 EXTINCTION PATTERNS AND THE DATING PROBLEM

Attempting to make sense of extinctions in Australia is to venture into a minefield of controversy.[96] Determining the chronology and geographical patterns of extinction is far more difficult for Australia than for any other region, principally because here most losses occurred beyond the limit of radiocarbon dating, and moreover bone collagen (for direct radiocarbon dating of faunal remains) tends to be less well preserved in warm climates. In addition, the fossil record is not nearly good enough, with taxa such as *Varanus priscus* based only on fragmentary material. We are therefore largely dependent on alternative methods that range much further back in time: OSL, U-series, and ESR. As stated above, the most reliable of these appears to be direct dating of tooth enamel from megafaunal species by electron spin resonance, and in some case U–T dating. Datasets published by Saltré and colleagues very usefully list all the published dates, direct and indirect, currently available.[97] However, these compilations also reveal that the numbers and the quality of these dates are inadequate to reliably determine the extinction chronology for any of the megafaunal species.

The youngest of just thirteen available ESR dates on *Diprotodon optatum* include 44.25 kya (Hallet Cove, South Australia) and 46.3 kya (Lancefield Swamp, Victoria). *Procoptodon gilli* has as its latest ESR dates 41 kya and 45.75 kya (Black Creek Swamp, South Australia), and *Thylacoleo carnifex* 50 kya (Black Creek Swamp) and 53 kya (Titan Shelter, Tasmania). The few direct radiocarbon dates on *Protemnodon anak* from Tasmania—41.8 to 40.9 kya (Mount Cripps) and 42.9 to 42.1 kya (Titan Shelter)—are rather younger than the widely quoted megafaunal extinction date of ca. 46 to 45 kya from mainland Australia.[98] These authors attribute the difference to an inferred, significantly later colonization of Tasmania by humans, who it is proposed were unable to get there until a lower sea level created a land connection. When humans did succeed in reaching Tasmania, it is implied that they rapidly proceeded to wipe out the late-surviving megafauna. However, it is worth mentioning that humans reached Sahul in the first place (from South East Asia) by crossing several water gaps by using boats or rafts—so they may not have had to wait for a land connection before colonizing Tasmania.

If correct, the recently published evidence from the Madjedbebe rock shelter (Northern Territory) that humans colonized Sahul by ca. 65 kya has profound implications for megafaunal extinctions on that continent.[99] Moreover, this new information has starkly highlighted how far we still have to go in reliably determining the pattern of megafaunal extinctions and its relationship to human colonization in Sahul. Nevertheless, even taking the previously widely accepted extinction date of 45 kya, it seems likely that several megafaunal species persisted at least twenty thousand years after human arrival, which would eliminate a "Blitzkrieg" and probably even a modest slow-burn overkill.[100] Moreover, as described above, an articulated *Zygomaturus trilobus* skeleton from the Willandra Lakes, New South Wales, has been dated to ca. 36.7 to 32.4 kya.[101] M. Westaway and colleagues observed: "Regardless of whether one accepts a short (47.5kya) or long (55kya) chronology for Aboriginal occupation of Australia, it would now appear that the second largest marsupial to ever exist was still present for a considerable time after the first arrival of Aboriginal people."[102]

It is not at all clear, given that humans coexisted with megafauna in Sahul for tens of millennia during the Last Glacial, why the record from that continent is so different from the rest of the world. It is remarkable that, with the exception of Cuddie Springs and Warratyi Rock Shelter (see below), there are no sites that provide good evidence of artifacts directly associated with megafaunal remains, whereas in North America, South America, and northern Eurasia, such evidence occurs at many sites. At Warratyi Rock Shelter, in the arid zone

TABLE 8.1A Youngest direct dates on megafauna and *Thylacinus*, Sahul

Species	Method	Lab no.	Date BP	Site
Diprotodon optatum	ESR	ANU-1052A	50,100 ± 9,700	Lancefield Swamp (VA)
Diprotodon optatum	ESR	2084A	48,959 ± 2,449	Cooper's Dune (SA)
Diprotodon optatum	ESR	ANU-1052B	46,300 ± 8,900	Lancefield Swamp (VA)
Diprotodon optatum	ESR	2087	44,250 ± 5,750	Hallet Cove (SA)
Zygomaturus trilobus	ESR	1366A	128,500 ± 5,938	Victoria Fossil Cave (SA)
Zygomaturus trilobus	ESR	1366B	127,500 ± 5,938	Victoria Fossil Cave (SA)
Zygomaturus trilobus	U-series/OSL	n/a	32.4 ± 0.5	Willandra Lakes (NSW)
Procoptodon gilli	ESR	2121	45,750 ± 5,250	Black Creek Swamp (SA)
Procoptodon gilli	ESR	2120A	41,000 ± 5,196	Black Creek Swamp (SA)
Simosthenurus occidentalis	^{14}C	OxA-17143	47,858 ± 1,005	Mt. Cripps (TAS)
Simosthenurus sp.	^{14}C	JF155/Roof	44,700 ± 4,850	Unnamed Cave (TAS)
Sthenurus sp.	ESR	2169A	84,000 ± 6,500	Black Creek Swamp (SA)
Sthenurus sp.	ESR	2166A	69,500 ± 7,750	Black Creek Swamp (SA)
Protemnodon sp.	ESR	E10–15	50,500 ± 5,000	Cuddie Springs (NSW)
Protemnodon anak	^{14}C	OZM-739	41,508 ± 350	Mt Cripps (TAS)
Protemnodon anak	^{14}C	OZM-081	40,943 ± 330	Mt Cripps (TAS)
Protemnodon anak	^{14}C	OxA-16417	40,802 ± 339	Mt Cripps (TAS)
Thylacoleo carnifex	ESR	2739	53,000 ± 5,576	Titan Shelter (TAS)
Thylacoleo sp.	ESR	2165A	50,000 ± 2,367	Black Creek Swamp (SA)
Thylacinus cynocephalus	^{14}C	NSW-28a	**5214 ± 177**	Thylacine Hole (WA)
Thylacinus cynocephalus	^{14}C	?	**4244 ± 128**	Fromm's Landing (SA)
Thylacinus cynocephalus	^{14}C	?	**3469 ± 110**	Murra-el-elevyn Cave (WA)

TABLE 8.1B Youngest OSL (indirect) dates, Sahul

Species	method	Date BP	Site
Genyornis newtoni	OSL/^{14}C	40–30 kya	Cuddie Springs (NSW)
Genyornis newtoni	OSL	60,000 ± 9,000	Warrnambool (VA)
Genyornis newtoni	OSL	55,000 ± 5,000	Wood Point (SA)
Diprotodon cf. *optatum*	OSL/^{14}C	40–30 kya	Cuddie Springs (NSW)
Diprotodon optatum	OSL	47,000 ± 6,000	Neds Gulley (QLD)
Diprotodon optatum	OSL	46,000 ± 6,000	Neds Gulley (QLD)
Metasthenurus newtoni	OSL	56,000 ± 4,000	Scotchtown Cave (TAS)
Palorchestes cf. azael	OSL	56,000 ± 4,000	Scotchtown Cave (TAS)
Palorchestes azael	OSL/^{14}C	40–30 kya	Cuddie Springs (NSW)
Phascolonus sp.	OSL/^{14}C	40–30 kya	Cuddie Springs (NSW)
Phascolonus gigas	OSL	52,000 ± 8,000	Lake Victoria (NSW)
Phascolonus gigas	OSL	47,000 ± 6,000	Neds Gulley (QLD)
Phascolonus gigas	OSL	46,000 ± 6,000	Neds Gulley (QLD)
Procoptodon goliah	OSL	52,000 ± 8,000	Lake Victoria (NSW)
Procoptodon goliah	OSL	47,000 ± 6,000	Neds Gulley (QLD)
Procoptodon goliah	OSL	46,000 ± 6,000	Neds Gulley (QLD)
Procoptodon browneorum	OSL	52,000 ± 3,000	Kudjal Yolgah Cave (WA)
Procoptodon browneorum	OSL	46,000 ± 2,000	Kudjal Yolgah Cave (WA)
Procoptodon browneorum	OSL	40,000 ± 2,000	Kudjal Yolgah Cave (WA)
Protemnodon anak	OSL	56,000 ± 4,000	Scotchtown Cave (TAS)
Protemnodon anak	OSL	54,000 ± 7,000	Lake Victoria (NSW)
Protemnodon anak	OSL	52,000 ± 8,000	Lake Victoria (NSW)
Protemnodon cf. *brehus*	OSL/^{14}C	40–30 kya	Cuddie Springs (NSW)
Protemnodon brehus	OSL	54,000 ± 7,000	Lake Victoria (NSW)
Protemnodon brehus	OSL	52,000 ± 8,000	Lake Victoria (NSW)
Protemnodon roechus	OSL	47,000 ± 6,000	Neds Gulley (QLD)
Protemnodon roechus	OSL	46,000 ± 6.000	Neds Gulley (QLD)
Simosthenurus occidentalis	OSL	56,000 ± 4,000	Scotchtown Cave (TAS)
Simosthenurus occidentalis	OSL	46,000 ± 2,000	Kudjal Yolgah Cave (WA)
Sthenurus atlas	OSL	54,000 ± 7,000	Lake Victoria (NSW)
Sthenurus atlas	OSL	52,000 ± 8,000	Lake Victoria (NSW)
Sthenurus andersoni	OSL	54,000 ± 7,000	Lake Victoria (NSW)
Sthenurus andersoni	OSL	52,000 ± 8,000	Lake Victoria (NSW)
Sthenurus tindalei	OSL	54,000 ± 7,000	Lake Victoria (NSW)
Sthenurus tindalei	OSL	52,000 ± 8,000	Lake Victoria (NSW)
Thylacoleo carnifex	OSL	56,000 ± 4,000	Scotchtown Cave (TAS)
Thylacoleo carnifex	OSL	54,000 ± 7,000	Neds Gulley (QLD)
Zygomaturus trilobus	OSL/ U-series	36.7–33.3kya	Willandra Lakes (NSW)
Zygomaturus trilobus	OSL	63,000 ± 9,000	Mammoth Cave (WA)
Zygomaturus trilobus	OSL	56,000 ± 4,000	Scotchtown Cave (TAS)

Date BP: date before present. Boldface dates: Holocene. VA: Victoria. SA: South Australia. TAS: Tasmania. NSW: New South Wales. QLD: Queensland.
Sources: Willandra *Zygomaturus*, dated by OSL and U-series: Westaway, Olley, et al. 2017. Cuddie Springs OSL/^{14}C dates: Fillios, Field, and Charles 2009. All others: Saltré, Rodríguez-Rey, et al. 2016 and references therein.

of South Australia, a partial juvenile radius of *Diprotodon optatum* and some unburnt eggshell fragments (attributed to a large megapode) were found in association with stone tools and OSL dates of more than ca. 46 kya.[103]

8.16 SURVIVORS

The only surviving Australian mammals that qualify as megafauna (again, 45 kilograms or heavier) are red kangaroo (*Macropus rufus*), at around 85 kilograms, and the rather smaller western and eastern gray kangaroos (*Macropus fuliginosus* and *Macropus giganteus*). There are also the large flightless birds, the ratites: the emu (*Dromaius novaehollandiae*) averages only about 36 kilograms, although some individuals reach 50 kilograms; the somewhat larger southern cassowary (*Casuarius casuarius*, northeastern Australia) and northern cassowary (*Casuarius unappendiculatus*, New Guinea) can reach 58 kilograms (females are larger). However, by far the largest megafaunal survivor is the formidable saltwater or Indo-Pacific crocodile (*Crocodylus porosus*), which ranges from northern Australia to New Guinea, southeast Asia, Sri Lanka, and eastern India. With males weighing up to 1 tonne, this is the biggest extant species of reptile in the world. Although at home in both marine and freshwater habitats, it qualifies for a place among the megafauna, as today it is an important predator on large terrestrial mammals—including the occasional human—and undoubtedly would also have been in the Quaternary.

CHAPTER 9

Madagascar: Giant Lemurs, Elephant Birds, and Dwarf Hippos

In this chapter and the next, I discuss megafaunal extinctions on the islands of Madagascar and New Zealand—which, because of their size, are in effect mini continents. In marked contrast to the continental land masses that we have considered so far, neither was impacted by humans before the Holocene. Until then, their unique animals, including many megafaunal species, had evolved for tens of millions of years in splendid isolation.

The French naturalist and explorer Alfred Grandidier (1836–1921) was one of the first to excavate some of the extraordinary animals that had previously inhabited Madagascar. In "Souvenirs de Voyages d'Alfred Grandidier: 1865–1870," he gave a colorful account of a particularly exciting discovery:[1]

> I had stopped to cook my lunch at Ambohisatrana (that is, the place where satrana, or dwarf palm tree plants grow, later called Ambolisatra, the satrana plantation) where I was visited by the chief of the region, with whom I discussed, as was usual for my conversations with the native people, the local industry and animals (particularly the Song'aomby, a cow-like animal). Because I asked for information regarding the Song'aomby (previously known to me only through a very poor description that Flacourt had provided under the name Mangarsahoc), the chief of the region indicated the location of a nearby marsh and informed me that I could find this animal's bones there. On that advice, I hurried to the location—barefooted and barelegged, with pants cut at the knees, as I am prone to do. I entered the marsh, and lowering myself, tapped the bottom where I sensed a large object, and lifted it. After washing it, I found to my surprise and joy that it was a femur—the thighbone of a bird. The bird must have

been enormous, like the famous Roc of 1001 Nights. Enthusiastically, I returned to the water and, with some of my men, dug into the mud that carpeted the floor of the marsh. I retrieved more bones of the colossal bird, *Aepyornis*, known previously only from its 8-liter eggs and a few indeterminable pieces sent by Mr. Abadi and described in 1850 by Isidore Geoffroy Saint-Hilaire. Alongside these bird bones were numerous other bones belonging to an unknown species of hippopotamus that I named *Hippopotamus Lemerlei* in honor of our odd-job man at Tuléar, as well as bones of other new and interesting animals.

Madagascar is a large tropical island that lies between latitudes 12.14° and 25.72° south in the Indian Ocean, separated from the east coast of Africa by the Mozambique Channel, about 450 kilometers at its narrowest. At the present day the island is about 1,580 kilometers from north to south, with an area of about 587,713 square kilometers. After the breakup of the supercontinent of Gondwana and separation of Madagascar from the African and Indian continents by at least 88 mya (Cretaceous period), its flora and fauna followed their own evolutionary paths, largely, but not entirely, in biological isolation. Some ancestral plants and animals were probably "on board" from the time of the breakup, whereas others, including the iconic lemurs, evolved from African ancestors that arrived through rare chance colonization events. The outcome was a uniquely rich assemblage of wonderful plants and animals that are found nowhere else, making it a "a hotspot of biodiversity."

Endemic animals alive today include some one hundred species of lemur, as well as Malagasy carnivores (Eupleridae). Madagascar also has the most tenrecs (there are also three species in Africa), including the spiny hedgehog-like species; the world's richest and most colorful array of chameleons; the aptly named tomato frog; and numerous endemic insects, such as the bizarre long-necked giraffe weevil, the stunningly beautiful comet moth, and the hissing cockroach. Most of the approximately ten thousand species of native plants are endemic, including 165 palms, such as the traveler's palm; six species of baobab (extraordinary trees with swollen trunks for conserving water in the dry season); and an entire family of spiny plants (Didiereaceae). The latter, which live in the spiny forests or spiny thickets of the arid southwest, look very much like some kinds of cacti, although not closely related, and quite unlike cacti bear numerous small, fleshy leaves that are protected by vicious spines and are shed in times of drought.

The climate and vegetation of Madagascar are very varied and strongly zoned, reflecting differences in topography and precipitation. The Central Highlands, ranging from 800 to 1,800 meters, block most of the moist easterly winds from the Indian Ocean. Consequently, it is very wet in the east, which supports tropical rain forest, and much drier in the west and south. The island can be conveniently divided into five climatic/vegetational units: (1) Humid Forest along most of the east coast (facing the Indian Ocean), (2) Central Highlands along the spine of the island, (3) Dry Deciduous Forest along approximately two thirds of the west coast, (4) Spiny Thicket on the southern end of the island, and (5) a wedge of Succulent Woodland between Spiny Thicket to the south and Dry Deciduous Forest to the north.[2] Before human arrival, most of the island was forested, but perhaps 80 to 90% of these forests has now been cleared, to the inevitable detriment of the wildlife. On the positive side, a number of nature reserves have been established in the different ecological regions, and active conservation initiatives include planting forest corridors to link up surviving populations.

So far there is no fossil record for Madagascar between the Middle Cretaceous and the late Quaternary—a gap of more than 80 million years—so that unfortunately we have no direct information about the origins of the modern and "subfossil" faunas, nor do we know what extinctions might have occurred earlier in the Quaternary (or previously in the Tertiary).

9.1 EXTINCT FAUNA

"Subfossil" remains have been recovered, often in abundance, from caves as well as marsh and river deposits. Artwork by Velizar Simeonovski evocatively depicts scenes of Holocene Madagascar, with its wonderful animals. These faunal assemblages are a strikingly unbalanced mix of seemingly random elements (fig. 9.1).[3] Lemurs, which still dominate the modern mammal fauna, occurred in rich profusion, along with huge flightless birds, giant tortoises, and dwarf hippos. The range of larger carnivorous animals in this lost prehistoric world was remarkably limited; restricted to horned crocodile (see below) and "giant" fossa (in reality, only moderate size). The role of medium-large grazers—elsewhere occupied, for example, by deer, antelope, or kangaroos—was here taken up by pygmy hippos and giant tortoises. But there were no elephants—only "elephant birds."

9.2 ELEPHANT BIRDS: *VOROMBE*, *AEPYORNIS*, AND *MULLERORNIS*

A gigantic bird called a roc (or rukh) is featured in the "Voyages of Sinbad the Sailor" from the *One Thousand and One Nights*, based on stories widespread

FIGURE 9.1. Extinct and extant megafauna from Madagascar. From the top, row 1, left to right: *Vorombe titan*; *Aepyornis maximus*; *Aepyornis hildebrandti*; *Mullerornis modestus*. Row 2: *Archaeoindris fontoynontii*; *Palaeopropithecus maximus*; *Megaladapis grandidieri*; *Hadropithecus stenognathus*; *Megaladapis edwardsi*. Row 3: *Hippopotamus madagascariensis*; *Hippopotamus lemerlei*; *Aldabrachelys grandidieri*; *Indri indri*. Row 4: *Voay robustus*; *Cryptoprocta spelea*. The only living species figured (*Indri*, row 3)—not megafauna—is the largest survivor. The extinct *C. spelaea* (row 4) also does not qualify as megafauna.
Black: selected extinct species. Gray: selected living species. The outlined *Homo sapiens* gives approximate scale.

in the medieval Islamic World and beyond (ca. 1300 CE). This bird, resembling an enormous eagle, was believed to be so strong that it was able to lift an elephant in its talons, let it fall, and then feast on its carcass. The story was related by the famous Venetian traveler Marco Polo in the account of his many adventures, which included twenty years at the court of the Mongol Chinese emperor Kublai Khan. Kublai was said to have been delighted when presented with one of its giant "feathers"—possibly just a large dried palm frond. It seems likely that the legend of the roc originated at least in part from accounts of the huge but flightless Malagasy "elephant bird," later identified as *Aepyornis*, which was probably still in existence in the fourteenth century. It is easy to imagine how finds of its huge eggs would also have lent credence to these tales.

In the 1930s, remains of elephant birds were assigned to two genera, *Aepyornis* and *Mullerornis*, although the number of species was uncertain.[4] However, in a major revision of elephant bird taxonomy based on leg bones, James Hansford and Sam Turvey have distinguished three genera and four species.[5] They assigned the largest material, which is morphologically distinct from *Aepyornis*, to a new genus and species: *Vorombe titan*, the giant elephant bird.

The biggest individuals of *Vorombe titan* probably reached more than 2 meters high at the back and 3 meters to the top of the head (fig. 9.2), making it the largest bird of all time, exceeding both *Aepyornis maximus* (the previous contender) and the South Island moa of New Zealand (chapter 10). Hansford and Turvey estimated the mean weight of *V. titan* at about 643 kilograms (maximum 732 kilograms). They also distinguished two species of *Aepyornis*: *A. maximus* (mean weight about 410 kilograms, total height roughly 2.3 meters) and *A. hildebrandti* (283 kilograms, roughly 1.8 meters). In the following account, it needs to be borne in mind that some material (bone or eggshell) here ascribed to *Aepyornis* may well turn out to be *Vorombe*.

In addition to *Vorombe* and *Aepyornis*, there was another genus of smaller, more gracile elephant bird assigned to the genus *Mullerornis*.[6] The genus was named in honor of the French naturalist and explorer Georges Muller, who was murdered by brigands in 1892 while on a scientific expedition to the interior. Several species have been distinguished, but only one, *Mullerornis modestus*, has been recognized by Hansford and Turvey. The youngest known date (bone), from Ankilibehandry in the west of the island, is 986–1285 BP (665–964 CE).

In common with other ratites, elephant birds lacked a keel on the sternum and so had no flight muscles. They also possessed only rudimentary wings. The fact that in the leg the tibiotarsus (midleg) was longer than the tarsometatarsus (lower leg) suggests that they were "ponderous browsers and grazers. The bill was long and deep with large narial openings, so the birds probably had a well-developed sense of smell."[7]

The eggs of the biggest elephant birds *Aepyornis maximus* (or *Vorombe titan*)—the largest known of any bird, or of any other animal, living or extinct—measured as much as 32 by 24 centimeters and were equivalent in volume to around 160 chicken eggs. Moreover, the shell was thicker than in any other animal (in excess of 4 millimeters), which undoubtedly contributed to the excellent preservation—for example, of the abundant shell fragments that occur so profusely in the dune sands at Faux Cap and elsewhere at the southern end of the island. It is likely that the large size and exceptionally thick shell evolved in part as a defense against predation, as prior to human arrival Madagascar

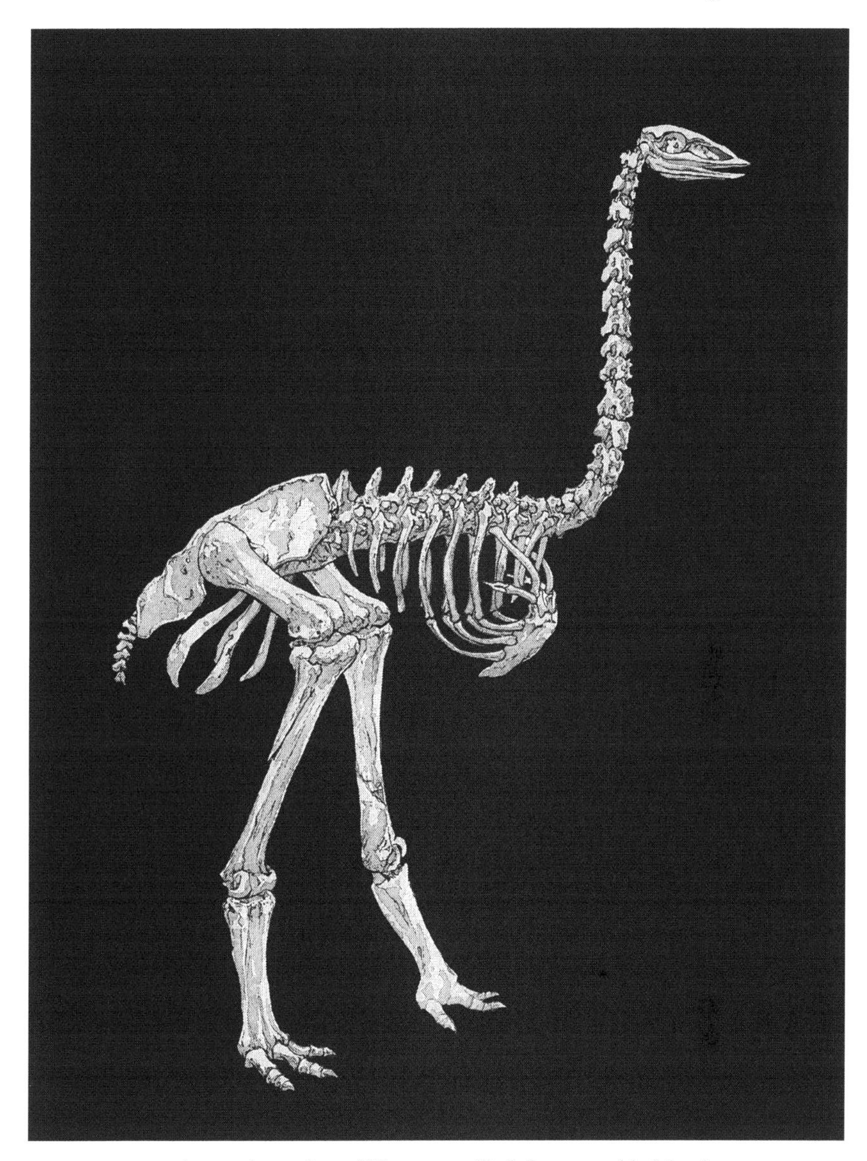

FIGURE 9.2. *Aepyornis maximus* ("Vorompatra") skeleton, total height about 2.3 meters. (Artwork by wildlife artist Alain Rasolo, Madagascar).

possessed no predator (possibly other than crocodiles) that was capable of breaking into these huge eggs. Of course, these adaptations would have been ineffective against predation by humans. Eggshells of *Mullerornis*—thinner than those assigned to *Aepyornis* and with a smoother surface—have been found in several localities, but not in the vast quantities assigned to *Aepyornis*.

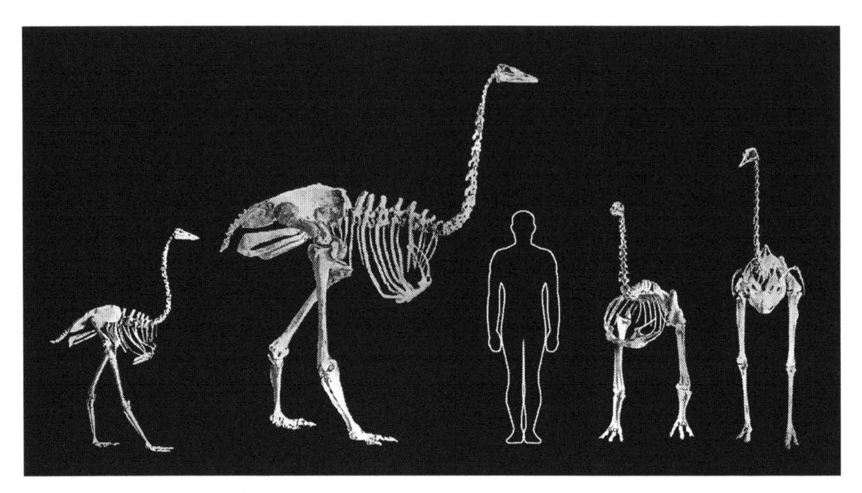

FIGURE 9.3. Mounted skeletons of elephant birds, formerly exhibited in the Queen's Palace in Antananarivo. (Artwork by Rob Stuart, redrawn from Charles Lamberton, 1934; tentative identifications courtesy of James Hansford.)
Left to right: *Mullerornis modestus; Vorombe titan; Aepyornis hildebrandti*. For comparison (extreme right), ostrich *Struthio camelus* from Africa. Note the much longer distal leg in the fast-running ostrich; and the vestigial wings.

The youngest available date for *Aepyornis* species, on eggshell from Talaky in the far south of Madagascar, is 985–1260 BP, 690–965 CE (table 9.1). However, there are reports suggesting that an elephant bird may still have been alive in the seventeenth century. Étienne de Flacourt, a French governor of Madagascar, mentions stories of an ostrich-like bird in remote areas of the island. He describes the "vouropatra—a large bird which haunts the Ampatres and lays eggs like the ostriches; so that the people of these places may not take it, it seeks the most lonely places."[8] According to Hume and Walters, *vouropatra* (or *vorompatra*) is an ancient Malagasy name that translates as "bird of the Ampatres."[9] This, the most southerly region of the island, is now known as Androy. It is in the coastal dunes of this region that large quantities of eggshell are still to be seen.

Analyses of carbon isotope ratios in the organic and calcite fractions of eggshells indicate that *Aepyornis* species primarily browsed C3 vegetation.[10] Since oxygen isotope values are more negative and less variable than those of ostriches also living in semiarid environments, the authors suggest that *Aepyornis* populations relied upon groundwater-fed coastal wetlands for their drinking water.

Midgley and Illing postulated that the bizarre large fruits of *Uncarina*, a genus of Malagasy endemic plants, largely relied on elephant birds for the dispersal of their seeds to new locations—a striking example of evolutionary adaptation.[11] These authors argued that the multiple barbed grappling hooks with which each fruit was equipped were far better adapted to clinging to the feet of elephant birds than to the fur of lemurs. *Uncarina* species mostly occur in dry deciduous forest and arid spiny bush—areas believed to have been favored by *Aepyornis*. After the demise of the elephant birds, cattle, and other introduced hoofed mammals obligingly took over the job of dispersing *Uncarina* fruits.

9.3 GIANT TORTOISES: *ALDABRACHELYS* SPECIES

In the late Holocene, Madagascar was home to two species of giant tortoise: *Aldabrachelys abrupta* and *Aldabrachelys grandidieri*, both of which are now extinct. They were closely related to the Aldabran giant tortoise *Aldabrachelys gigantea*, which happily still thrives on Aldabra Atoll, 400 kilometers to the northwest (chapter 11). Four much smaller endemic tortoises still occur in Madagascar, including the ploughshare tortoise *Geochelone yniphora* and radiated tortoise *Astrochelys radiata*. Unfortunately, both are critically endangered and under intense threat from illegal poaching, as these highly attractive animals are in great demand for the pet trade.

The two species of giant tortoise are readily distinguished by the shape of the carapace (upper shell). *Aldabrachelys grandidieri* had a flattened, very thick carapace up to 125 centimeters long, with bulging flanks; the closely related *Aldabrachelys abrupta* was rather smaller, up to 115 centimeters long, with a domed, relatively thinner carapace and less bulging flanks.[12] Isotope evidence suggest that *A. grandidieri* consumed a high proportion of C4 plants and was predominantly a grazer, preferring open, herb-rich habitats, whereas *A. abrupta* was mainly a browser on C3 plants in woodland.[13] The slow rate of reproduction in these giant tortoises would have rendered them especially vulnerable to human hunting, even if they were present at high densities. The youngest available date on *Aldabrachelys* (listed as *Geochelone*) is 1305–1510 BP (440–645 CE).[14]

There is a current proposal to introduce three hundred juvenile tortoises to Madagascar from the captive-bred stock of Aldabran giant tortoises on the island of Mauritius. This rewilding scheme aims to replace the recently extinct tortoises with this closely related species that has a presumed similar ecology.[15]

9.4 HORNED CROCODILE: *VOAY ROBUSTUS*

The extinct Malagasy horned crocodile *Voay robustus* was more closely related to the living African dwarf crocodiles (*Osteolaemus* species) than to the living Nile crocodile (*Crocodylus niloticus*) that occurs in Madagascar at the present day.[16] The ancestor of *V. robustus* probably arrived from Africa several millions of years ago by "sweepstakes route," either on rafted vegetation or by swimming across the Mozambique Channel. *V. robustus* had two prominent triangular "horns" near to the back of the skull. Its snout was deep relative to its length, and it had robust limb bones and limb girdles—all characteristics that indicate adaptation to a more terrestrial mode of life. With an estimated length of up to 5 meters and weight of 170 kilograms, it would have been a formidable predator, and by far the largest to have inhabited late Quaternary Madagascar. Prey could have included various lemurs, elephant birds, and the young of dwarf hippos.

Recently, a calibrated radiocarbon date of 1813 ± 78 BP (CAMS 167399) for *Voay robustus* has been obtained from Ankilibehandry on the coast of western Madagascar by Samonds and colleagues showing that, along with other Malagasy megafauna, it survived into the late Holocene.[17]

The Nile crocodile *Crocodylus niloticus* is the only megafaunal species that occurs in Madagascar today, so it might therefore be claimed as the sole megafaunal survivor. However, the single available radiocarbon date of 385 ± 75 BP (CAMS 150524) from Anjohibe in the northwest of the island tells us nothing about a possible temporal overlap with *V. robustus* and whether *V. robustus* disappeared because it was outcompeted by *C. niloticus*. Alternatively, possibly *V. robustus* was extirpated before the arrival of *C. niloticus*.

The absence of "subfossil" finds of Nile crocodiles in Madagascar and the close genetic similarity of living animals to Nile crocodiles from the African mainland suggest that it was a very late arrival from Africa.

9.5 DWARF HIPPOS: *HIPPOPOTAMUS LEMERLEI* AND *HIPPOPOTAMUS MADAGASCARIENSIS*

Hippos arrived relatively late in Madagascar, very likely derived from an ancestral common hippo, *Hippopotamus amphibius* (a minimum of a pregnant female or a male and female) that swam across the Mozambique Channel from Africa at some time in the Quaternary. The oldest radiocarbon dated record of hippo on Madagascar (*H. lemerlei*) is fairly early in the Holocene, about 7150 BP. When in Madagascar, the ancestral stock underwent island dwarfing and diverged into two species. *Hippopotamus lemerlei* and *H. madagascariensis* were rather larger than the living pygmy hippo *Choeropsis liberiensis* of West Africa, which is up to 1 meter high at the shoulder and weighs around 228 kilograms. Weston and Lister estimated *H. lemerlei* at 374 kilograms and *H. madagascariensis* at 393 kilograms.[18] There are a number of distinctions between the two species: *H. madagascariensis* has a broader, more robust skull than *H. lemerlei*; there are differences in the dentition; and the mandible is also more compact and robust.[19] Two morphs of *H. lemerlei* have been recognized—females have a narrower muzzle than males. The inferred female skulls often bear healed wounds, as seen in common with hippos, in which the male bites the head or neck of the female during mating. *H. lemerlei* exhibits a number of features which indicate that it retained the aquatic habits of its presumed *H. amphibius* ancestor, including more elevated orbits than in *H. madagascariensis*, enabling the animal to see even when mostly submerged.

On the other hand, *H. madagascariensis* is thought to have led a more terrestrial existence, much as in the extant *Choeropsis liberiensis*. With only a few equivocal exceptions, *H. lemerlei* remains have been recovered from the Coastal Lowlands, whereas *H. madagascariensis* records are confined to the Central Highlands. The youngest radiocarbon date for *H. lemerlei* is 1270–1350 BP (600–680 CE) and for *H. madagascariensis* 982–1179 BP (771–968 CE). A third Madagascan species, *H. laloumena*, has been claimed on the basis of sparse material from Mananjary on the east coast. These remains are very similar to those of common African hippo.

Hippos and some other now-extinct Madagascan fauna may have survived substantially later than suggested by the radiocarbon dates. As previously mentioned in relation to elephant birds, in the seventeenth century the French colonial governor Flacourt reported eyewitness accounts from remote areas of what could have been giant lemurs, elephant birds, and hippos. Hippos in particular have also featured in other colonial accounts from as late as the nineteenth and early twentieth centuries.[20]

9.6 GIANT LEMURS

Lemurs are a remarkable, extremely diverse group of primates unique to Madagascar that have evolved in the absence of competition from monkeys, as well as having had to contend with few predators. A curious and unique anatomical feature of most lemurs, living and extinct, is the modification of the lower incisors and canines into a "comb" for grooming the fur. How and when lemurs colonized Madagascar remains uncertain, but it most likely occurred by the Middle Eocene (ca. 43 mya) through the chance rafting of individuals of a single ancestral species from Africa, probably on a floating mat of vegetation swept down a river into the sea by a tropical storm—a prime instance of George Gaylord Simpson's "sweepstakes dispersal," whereby success or failure of a species to cross a sea barrier is primarily a matter of chance.[21] Subsequently these pioneers diversified into many species, adapting to the wide range of topography, climate, and vegetation of this large island. The one hundred or so currently recognized species of extant lemurs (some authorities consider this figure too high) include the well-known ring-tailed lemur (*Lemur catta*), twenty-six sportive lemurs (*Lepilemur* spp.), twenty-two tiny mouse lemurs (mainly *Microcebus* species), and nine sifakas (*Propithecus* species). The latter are remarkable for their unique mode of locomotion, leaping gracefully on their hind legs like furry ballet dancers. In size, lemurs range from Madame Berthe's mouse lemur (*Microcebus berthae*), weighing only 30 grams and the world's smallest primate, up to the indri (*Indri indri*), the largest living lemur, at 6 to 9.5 kilograms. Confined today to the forests of the northeast, the indri is listed as "critically endangered" by IUCN. Of a total of ninety-five lemur species listed by IUCN, it is alarming to see that nineteen are classified as critically endangered, forty-seven are endangered, and that all are decreasing. These potential extinctions threaten to add to the approximately seventeen larger species of "subfossil" lemurs exceeding 10 kilograms that disappeared within the last thousand years or so, including six weighing 45 kilograms or more that qualify as megafauna.

9.7 SLOTH LEMURS: *ARCHAEOINDRIS* AND *PALAEOPROPITHECUS*

The extinct large lemurs belonging to the genera *Archaeoindris* and *Palaeopropithecus* (palaeopropithecids) are known as "sloth lemurs" because of the similarities in their postcranial skeletons to the living tree sloths of Central and South America. *Archaeoindris fontoynontii*, the largest known species of lemur, is recorded only by sparse skeletal remains from a single site: Ampasambazimba, a marsh deposit in the western highlands of central Madagascar. Its extreme rarity might imply a restricted geographical distribution. The species is known from a single complete skull, 269 millimeters long (fig. 9.4), together with fragmentary jaws, a fragmentary adult humerus and femur, and four incomplete long bones of an immature individual.[22] With an estimated weight of about 160 kilograms, *Archaeoindris* was comparable in size to a medium-size male gorilla, and the resemblance is enhanced by the fact that, as in the

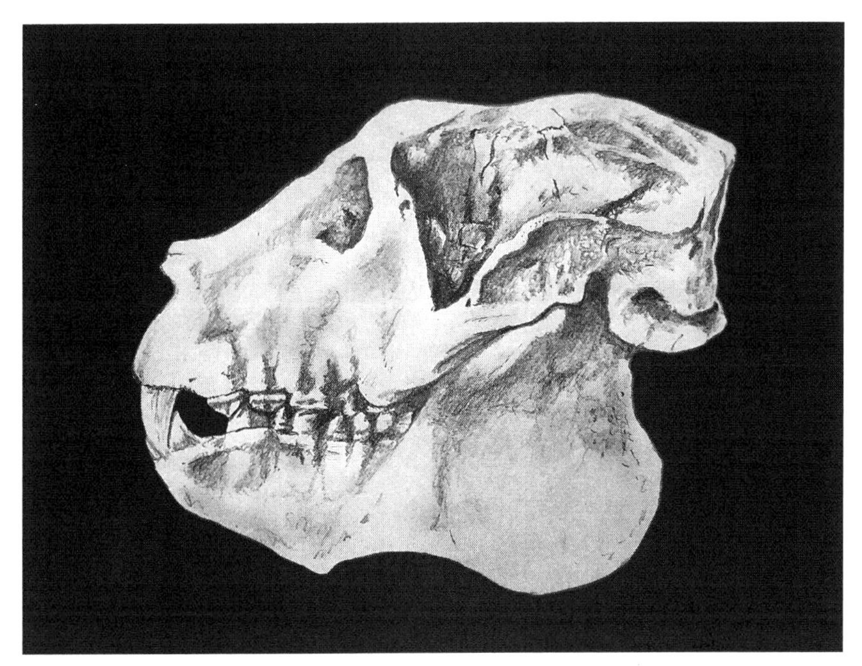

FIGURE 9.4. Giant lemur: *Archaeoindris fontoynontii* skull and mandible (length about 25 cm), Ampasambazimba. (Artwork by Rob Stuart, redrawn from Charles Lamberton, 1934b.)

gorilla, the arms are longer than the back legs suggesting that it was equally at home in the trees as on the ground. *Archaeoindris* has also been compared with the smallest of the New World extinct ground sloths (*Nothrotheriops* and *Nothrotherium*, chapters 6, 7). The low-crowned, multicusped cheek teeth were adapted to crushing leaves, fruits, and other plant food. It lacked a tooth comb, having "re-evolved" stout front teeth for cropping vegetation.[23] In the smaller sloth lemurs, genus *Palaeopropithecus*, the metapodial and phalangeal bones of the hands and feet are strongly curved, with deep flexor grooves, allowing them to hang from branches; and the arms are strikingly longer than the hind limbs. These suspensory adaptations imply lives spent almost entirely in the trees, in both dense forest and areas that pollen evidence shows were a mosaic of woodland and grassland environments.[24] The lower-front dentition was modified from a true tooth comb into four short and stubby procumbent (forward-pointing) teeth. *Palaeopropithecus maximus*, with an estimated weight of 46 kilograms, is recorded from central and possibly northern Madagascar. The youngest available date is 4415–4144 BP (before present) (2464–2205 BCE). Slightly smaller at about 41 kilograms, *Palaeopropithecus*

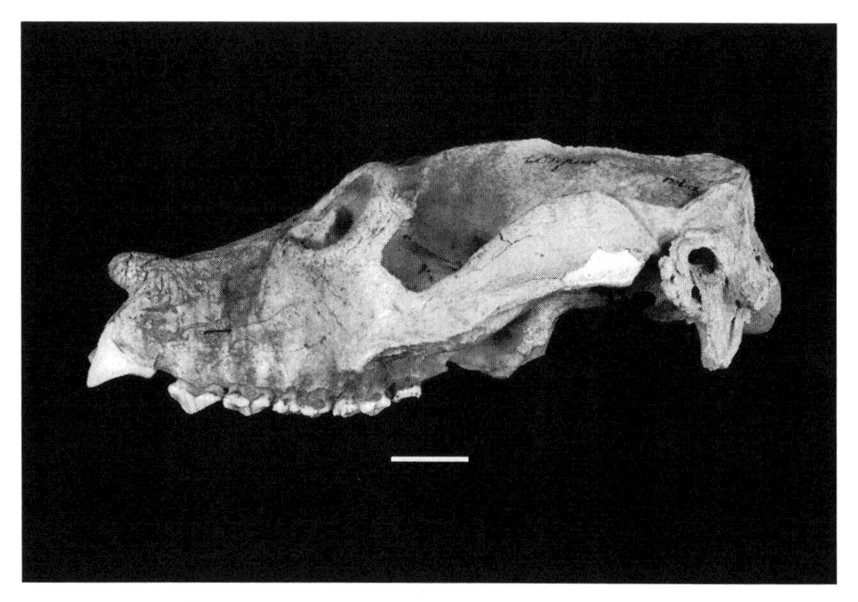

FIGURE 9.5. *Palaeopropithecus ingens* skull (scale bar equals 2 cm). Photo by Laurie Godfrey.

ingens (fig. 9.5) is known from southern and western Madagascar. According to Godfrey and colleagues, "Further analysis may make it difficult to maintain the specific distinction between *P. maximus* and *P. ingens*."[25]

The younger of only two radiocarbon dates on *Archaeoindris* is 2355–2104 BP (405–154 BCE). There are far too few dates to estimate its likely time of extinction. The latest date for *P. ingens*, from Ankilitelo Cave, is 322–635 BP (1316–1628 CE)—which qualifies as the youngest known record of a giant lemur (table 9.1). Remarkably, according to Godfrey and Jungers, "it is likely that *Palaeopropithecus* is the animal described by Etienne de Flacourt in 1658 as the 'tretretretre' and represented in Malagasy folklore as an ogre incapable of moving on smooth rocky surfaces."[26] This account suggests that it could still have been alive in Flacourt's time. An additional—so far undescribed—new species of *Palaeopropithecus*, the smallest and most gracile of the genus, is known from two fossil localities in the northwest. Weighing about 21 kilograms, *Babakotia radofilai* is notable as the first new genus of extinct lemur to be discovered in Madagascar since 1909. All the fossil remains of this genus recovered to date were found in Antsiroandoha Cave in northern Madagascar. Completing the roster of sloth lemurs: *Mesopropithecus dolichobrachion* (about 14 kilograms) is recorded from northern Madagascar; *Mesopropithecus globiceps* (about 11 kilograms) is known from the south, southwest, and southeast; and

Mesopropithecus pithecoides from the center of the island. The latter has a youngest date of 1185–1350 BP (600–765 CE).

9.8 KOALA LEMURS: *MEGALADAPIS* SPECIES

With an estimated weight of about 85 kilograms, *Megaladapis edwardsi* was the largest of three species of koala lemurs (Megaladapidae). It lacked upper incisors and had projecting upper canines, extremely large molars, and robust limb bones. The second largest, *Megaladapis grandidieri*, weighed about 74 kilograms, while the smallest, *Megaladapis madagascariensis*, was about 46.5 kilograms. *M. edwardsi* and *M. madagascariensis* are known from southern and southwestern Madagascar, while *M. grandidieri* is recorded from the central region. The youngest available dates on *Megaladapis* spp. are 1065–1268 BP (682–885 CE) for *M. edwardsi*, 1280–1353 BP (597–670 CE) for *M. grandidieri*, and 524–655 BP (1206–1427 CE) for *M. madagascariensis* (table 9.1).

9.9 MONKEY LEMURS: *HADROPITHECUS* AND *ARCHAEOLEMUR*

The term *monkey lemur* refers to a superficial resemblance to monkeys, to which these animals are not closely related. The largest *Hadropithecus stenognathus*, at about 35 kilograms, is recorded from southern, western, and central Madagascar. The youngest date is 1074–1415 BP (535–876 CE). *Archaeolemur majori*, known from southern and western Madagascar and possibly central and northern Madagascar, weighed about 18 kilograms. *Archaeolemur edwardsi*, from central Madagascar and possibly western, northern, and southeastern Madagascar, weighed about 26.5 kilograms. It gave a relatively young date: 774–961 BP (989–1176 CE).

Other extinct lemurs include (Lemuridae) *Pachylemur insignis*, at about 11.5 kilograms; *Pachylemur jullyi*, at about 13 kilograms; and (Daubentonidae) *Daubentonia robusta*, at about 14 kilograms. *P. insignis* has a youngest date of 960–1060 BP (890–990 CE), and *D. robusta* 800–1050 BP (900–1150 CE).

9.10 THE UNDERWATER LEMUR GRAVEYARD

Following the initial discovery of subfossil material in Tsimanampetsotsa (aka Tsimanampesotse) National Park in the arid southwest of Madagascar, in 2014 a team of scuba divers led by Phillip Lehman of the Dominican Republic Speleological Society investigated three extensive flooded caves: Aven, Mitoho, and Malaza Manga. These caves (especially Aven Cave) have produced a spectacularly large and exquisitely preserved assemblage of lemur remains, together with those of other mammals, reptiles, and birds. The fauna identified so far includes these extinct animals: lemurs *Megaladapis edwardsi*; *Mesopropithecus globiceps* and *Pachylemur insignis*; "giant" fossa *Cryptoprocta spelaea*; dwarf hippo *Hippopotamus lemerlei*; giant tortoise *Aldabrachelys* sp.; horned crocodile *Voay robustus*; and elephant bird *Mullerornis* sp., as well as living species such as ring-tailed lemur *Lemur catta* and Malagasy giant rat *Hypogeomys antimena*.[27]

9.11 "GIANT" FOSSA: *CRYPTOPROCTA SPELAEA*

The "giant" fossa was an enlarged version of the endemic fossa *Cryptoprocta ferox*, which today is the largest native mammalian carnivore in Madagascar. Some authorities have questioned whether it qualifies as a distinct species. All the Malagasy carnivores (family Eupleridae) are thought to have been derived from a common ancestor that was rafted across from Africa ca. 20 mya. Although cat-like in appearance, fossas are most closely related to the mongoose family (Herpestidae). The living fossa is a ferocious carnivore, generally feared by the indigenous human population. With a maximum estimated weight of 20 kilograms—twice that of its living relative—the extinct *C. spelaea* must have been formidable indeed.[28] Its prey probably included even the larger extinct lemurs and smaller elephant birds. The youngest record is 1620–1860 BP (90–330 CE). *Cryptoprocta spelaea* is thought to have opportunistically preyed on lemurs weighing up to around 85 kilograms.[29]

9.12 "MALAGASY AARDVARK": *PLESIORYCTEROPUS*

This poorly known animal, with digging adaptations and probably toothless, was originally described on the basis of sparse material (limb bones, pelvic bones, and partial skulls) as a relative of the African aardvark, but protein sequencing has now revealed that is has close affinities to tenrecs.[30] Estimates of body weight fall between 6 and 18 kilograms.

9.13 HUMAN COLONIZATION

The story of the human colonization of Madagascar is much less straightforward than for New Zealand. Both events occurred within the Holocene, but happened considerably earlier in Madagascar. On the basis of pollen of the introduced *Cannabis* plant and dated signs of butchery (cut marks) on bones of extinct animals, it was thought that humans first arrived in Madagascar (either from Indonesia or Africa) before ca. 2300 BP.[31] However, exciting new evidence published by Hansford et al. revises previous archaeological and genetic evidence, pushing back human arrival by more than 6,000 years.[32] Bones of *Aepyornis maximus* and *Mullerornis* sp. from a 2009 excavation at Christmas River (Ilaka), in the interior of the southern part of the island, have been directly radiocarbon-dated to more than 10,500 years BP. The bones exhibit "perimortem chop marks, cut marks and depression fractures consistent with immobilization and dismemberment of carcasses."[33] Hansford and colleagues comment that "this revision of Madagascar's prehistory suggests prolonged human-faunal coexistence with limited biodiversity loss."[34]

The present-day population, which has now reached more than 22 million, principally comprises people of African and southeast Asian descent. The ancestors of the latter, much like the Polynesians in the Pacific (chapter 10), crossed thousands of kilometers of ocean, probably in outrigger canoes, bringing with them domestic animals and crops—especially rice, which is now extensively cultivated on this tropical island. Other important introduced crops include bananas, yams, and cassava. Humped zebu cattle, which originated in southern Asia and are tolerant of heat, were introduced to Madagascar ca. 1000 CE; as shown by pollen in sediments, at about the same time there was a marked increase in grassland, reflecting major forest clearance. The subsequent history of Madagascar has been rich and varied, involving, for example, Arab traders, pirates, native chiefs and kings, and the colonial aspirations of Britain and France. It was incorporated in the French colonial empire from 1897 until 1958, when the country gained its independence.

9.14 EXTINCTION

Madagascar is "the only primate-dominated assemblage among the world's extinct late Quaternary megafaunas," and "extinction losses here were more severe than on any of the continents and most other large islands."[35] As might have been expected, all species of megafauna went extinct, but more surprising, all animals weighing 10 kilograms or more disappeared as well. As elsewhere, the key issue is, what or who was responsible for these severe losses? Several hypotheses have been proposed to explain "this last of the great megafaunal extinctions":[36] unsustainable hunting ("overkill") by colonizing humans; large-scale burning of the forests; and, most innovatively, "hypervirulent diseases."[37] Paul Martin perceived what he believed was compelling evidence in the Madagascan record for his "Blitzkrieg" hypothesis, an extreme variant of "overkill" (chapter 4).[38] Both hyperdisease and Blitzkrieg predict very rapid extinction; but, as discussed below, such scenarios are not supported by the dating evidence, which clearly indicates a substantial interval between human arrival and megafaunal extinctions.

Although all agree that humans played a significant role in Madagascar's Holocene extinctions, there is disagreement as to the relative contributions of humans and climate, mirroring the debates in other parts of the world. A sharp and widespread increase in grass pollen and decline in tree and shrub pollen ca. 1000 CE has been variously attributed to an island-wide drought, to human activities, or to a combination of both.[39] Crowley and colleagues analyzed stable nitrogen isotope ($\delta^{15}N$) values from radiocarbon-dated subfossil vertebrates.[40] If increasing aridity were responsible for megafaunal decline, they reasoned that there should have been an island-wide rise in $\delta^{15}N$ values, peaking at the time of proposed maximum drought, ca. 1000 BP (ca. 950 CE). As no such rise was found, they inferred that human activities were responsible for the vegetational changes, and that "increases in grasses at around that time may signal a transition in human land use to a more dedicated agro-pastoralist lifestyle, when megafaunal populations were already in decline. Land use changes ca. 1,000 years ago would have simply accelerated the inevitable loss of Madagascar's megafauna."[41] Taking another approach to this question and coming to a similar conclusion, Burns and colleagues analyzed two stalagmites from Anjohibe Cave in northwestern Madagascar that provided a high-resolution record of ecological and climatic change in this region over the past 1,800 years.[42] Stable carbon isotope data showed a rapid transformation ca. 890 CE from an open woodland dominated by C3 plants to a C4 grassland similar to the modern landscape. The authors inferred that this change—which occurred within a century—resulted from a dramatic increase in the use of fire to promote grazing for cattle, while the lack of significant variation in oxygen isotopes at this time indicates that the vegetational turnover was not caused by increased aridity. They concluded that the rapid loss of forest habitat "very likely increased environmental pressures on Madagascar's megafauna and accelerated their disappearance."[43]

Although there is a strong case for humanly induced habitat destruction substantially contributing to extinctions, direct evidence for human predation on the extinct fauna is generally scarce. Two sites that were excavated by Alfred Grandidier, Lamboharana and Ambolisatra, produced four hippo femora with unequivocal cut marks made by sharp metal objects.[44] Further evidence of butchery of extinct Madagascan animals is provided by cut marks on skeletal remains of extinct lemurs *Palaeopropithecus ingens* (three humeri, two ulnae, one radius, two tibiae) and *Pachylemur insignis* (two femora) from Taolambiby and Tsirave in the southwest.[45] The bones exhibit sharp cuts and chop marks near joints, oblique cuts along the shafts, spiral fractures, and percussion marks consistent with skinning, disarticulation, and filleting of the carcasses. The *P. ingens* radius with cut marks, from Taolambiby, has been dated to 2350–2150 BP.[46] Previously, there was little indication that elephant birds were butchered and eaten.[47] However, as described above, important new evidence in the form of cut-marked bones of *Aepyornis maximus* from Christmas River dated to more than 10,500 BP has been published recently by Hansford and colleagues.[48] In addition, clear evidence of butchery is seen on leg bones of *A. maximus* from Ambolisatra on the southwest coast, radiocarbon-dated to 1182–1057 BP (OxA-33535). Undated cut-marked leg bones of *Aepyornis hildebrandti* are recorded from Antsirabe (central Madagascar), while similar evidence is seen on *Mullerornis* sp. from Lamboharana (southwest coast): 6415–6282 BP (UBA-29726), and an unknown locality (table 9.1).

No evidence of human predation on elephant bird eggs has been reported, even though this seems very likely to have happened—a single egg would have fed many people. Although much excellent work on the chronology of extinctions and possible causes has been done, there is considerable scope for future research, including dating. Nevertheless, it is already clear that even on the basis of limited numbers of available radiocarbon dates on faunal material such as bone or eggshell, fifteen out of nineteen extinct species have youngest dates that fall within the second half of the first millennium CE (roughly 1000–1500 BP) or a little later (table 9.1). Therefore, we can safely say that most of the extinct species were still alive more than 9,000 years after human arrival, which excludes a Blitzkrieg. Moreover, written and oral accounts suggest that at least some species of elephant bird, giant lemurs, and especially dwarf hippos might have survived in remote areas as late as the seventeenth, nineteenth, or even twentieth centuries CE.[49] So, the available evidence is consistent with a drawn-out period of extinctions resulting from habitat destruction (forest clearance) and "overkill" by people.

The fungus *Sporormiella*, which we have encountered previously for other parts of the world, grows on dung, and the abundance of its spores in sediments has been widely taken as a proxy for the abundance of megafauna in the surrounding area (see chapter 6). In Madagascar, *Sporormiella* levels suggest that "megafaunal biomass in the high-elevation ericoid bushlands and humid lowland forests was relatively lower than in wooded grasslands of the semiarid southwest and central highlands."[50] Hippos, giant tortoises, and elephant birds

TABLE 9.1 Youngest radiocarbon dates for extinct Malagasy genera/species and extant Nile crocodile *Crocodylus niloticus*

Species	Lab. No.	Cal BP	Cal CE/BCE	Site	Source
Voay robustus	CAMS 167399	1735–1891	60–216 CE	Ankilibehandry	10
Crocodylus niloticus	CAMS 150524	310–460	1491–1641 CE	Anjohibe	10
Aldabrachelys sp.*	CAMS 143057	1305–1510	440–645 CE	Taolambiby	3
*Vorombe titan**	OxA-33531	2352–2699	749–402 BCE	Ankazoabo Grotte	9
*Vorombe titan**	OxA-34776	3478–3680	1730–1528 BCE	Amposa	9
*Aepyornis maximus**	OxA-33535	1057–1182	768–893 CE	Ambolisatra	8
*Aepyornis hildebrandti**	OxA-34758	1315–1420	530–635 CE	Masinandreina	9
Aepyornis sp.*	OxA-8270	985–1260	690–965 CE	Talaky	1
Mullerornis sp.*	B-103349	986–1285	665–964 CE	Ankilibehandry	2
Mullerornis sp.*	UBA-19725	1074–1270	680–876 CE	Unknown	8
Cryptoprocta spelea	CAMS 143077	1620–1860	90–330 CE	Ankazoabo Grotte	6
Plesiorycteropus sp.	NZA-16994 R-28139/3	2000–2303	353–50 BCE	Masinandreina	4
*Hippopotamus lemerlei**	CAMS 142734	1270–1350	600–680 CE	Taolambiby	6
*Hippopotamus madagascariensis** (aka *Hexaprotodon guldbergi*)	CAMS 143065	982–1179	771–968 CE	Antsirabe	6
*Archaeoindris fontoynontii**	NZA-18519 R-28331/1	2104–2355	405–154 BCE	Ampasambazimba	4
*Palaeopropithecus ingens**	NZA-10059 R-24649/1	971–1285	665–979 CE	Ankazoabo Grotte	4
*Palaeopropithecus ingens**	Beta	635–322	1316–1628 CE	Ankilitelo Cave	7
*Palaeopropithecus maximus**	CAMS 142607	4415–4144	2464–2205 BCE	Ampasambazimba	3

TABLE 9.1 (*continued*)

Species	Lab. No.	Cal BP	Cal CE/BCE	Site	Source
Mesopropithecus pithecoides	NZA-18523 R-28331/6	1180–1340	610–770 CE	Ampasambazimba	6
Mesopropithecus globiceps	CAMS 142801	1310–1510	440–640 CE	Ankazoabo Grotte	6
*Megaladapis grandidieri**	CAMS 142606	1280–1353	597–670 CE	Ampasambazimba	6
*Megaladapis madagascariensis**	CAMS 142908	1381–1541	409–569 CE	Andolonomby	3
*Megaladapis madagascariensis**	Beta	655–524	1296–1427 CE	Ankilitelo Cave	7
*Megaladapis edwardsi**	NZA-18999 R-28421/10	1065–1268	682–885 CE	Anavoha	4
Hadropithecus stenognathus	NZA-12582 R-26341/1	1074–1415	535–876 CE	Ankilibehandry	2
Archaeolemur cf. *edwardsi*	B-60797 CAMS-5484	774–961	989–1176 CE	Antsiroandoha	5
Pachylemur insignis	UCIAMS 159131	960–1060	890–990 CE	Tsirave	6
Daubentonia robusta	NZA-18524 R-28331/7	800–1050	900–1150 CE	Anavoha	6

*megafaunal species (>45 kg). Cal BP: calibrated date before present. Cal CE/BCE: calibrated date, Common Era or Before Common Era. All dates fall within the Late Holocene.

Sources: 1. Bronk Ramsey et al. 2002. 2. Burney 1999. 3. Crowley 2010. 4. Burney et al. 2004. 5. Simons et al. 1995. 6. Crowley et al. 2017. 7. Simons 1997, quoted in Burney, Burney, et al. 2004. 8. Hansford, Wright, et al. 2018. 9. Hansford and Turvey 2018. 10. Samonds et al. 2019.

that grazed in more open habitats probably attained the greatest biomasses and were the main producers of dung, whereas the forest-dwelling large lemurs contributed much less. Sediment cores from sites in the southwest record a drastic decrease in fungal spores ca. 1720–1540 BP (230–410 CE), interpreted as reflecting a substantial decrease in megafaunal biomass. This is significantly earlier than the latest dates that we have for most of the megafauna (table 9.1), so may record a decline but not an extinction. The subsequent increased levels of charcoal particles suggest that both megafaunal reduction and increased fire regimes were due to human activities. Interestingly, other sites record a resurgence of *Sporormiella* ca. 1000 CE, attributed to the widespread introduction of zebu cattle.

As eloquently summarized by Godfrey, Jungers, and Burney: "Although the fingerprints of humans are surely present at this Holocene crime scene, a recently compiled ^{14}C chronology for late prehistoric Madagascar is incompatible with the extreme versions of these extinction scenarios. The anthropogenic 'smoking gun' smoldered for a very long time, much too long, in fact, to validate the predictions of any model of overnight eradication of the subfossil lemurs and other megafauna."[51]

CHAPTER 10

New Zealand: Land of the Moa

New Zealand (Maori: Aotearoa) comprises two large islands, North and South Island, separated by the Cook Strait, and many smaller ones. Situated between about 34° and 47° south, it ranges from warm subtropical in the far north to cool temperate in the south. In addition, alpine conditions prevail in the north-south mountain range (Southern Alps) that extends the entire length of the South Island. The mountains pose a major barrier to westerly winds so that the west coast is especially wet. New Zealand is separated from Australia, the nearest continent, by approximately 1,900 kilometers of sea and by similar distances from major Pacific islands. The contrast between New Zealand and Australia in topography, climate, fauna and flora, and history of human settlement could hardly be greater. New Zealand was the last major land mass to be settled by humans, and also witnessed the last great prehistoric megafaunal extinctions.[1]

New Zealand is overwhelmingly a land of birds. Its diverse modern bird fauna includes flightless endemic species such as the iconic kakapo (a parrot), takahe (a rail), weka (a rail), Auckland Island teal, Campbell Island teal, and five species of kiwi. The kakapo and Okarito kiwi are listed by the IUCN as critically endangered, and the takahe and North Island brown kiwi as endangered, while many other native birds have become recently extinct. The native terrestrial vertebrates include geckos, skinks, tuataras (*Sphenodon punctatus*, *S. guntheri*—unique lizard-like reptilian survivors from the Jurassic period[2]), and four surviving species of frog. In the last few million years New Zealand has been situated too far from Australia for land vertebrates to have swum across or been rafted on floating vegetation, and it has neither snakes (in marked contrast to Australia) nor native mammals other than three species of bat (one is extinct). Remarkably, however, fossil finds from St Bathans, South Island, show that in the early Miocene period (19 to 16 mya), New Zealand did support at least one terrestrial small mammal (of unknown affinities).[3] Other vertebrates from the same site include several bats, a crocodilian, tuatara, lizards, and a possible moa as well as other birds.[4]

The first humans to arrive in New Zealand—the Polynesian ancestors of the Maori—brought several animals with them, including the Pacific rat (*Rattus exulans*) and a domestic dog (called a *kuri*). Then from the 1820s onward, European (mostly British) settlement was accompanied by the introduction of a "witless menagerie" of alien mammals from Europe: red deer, hedgehogs, rabbits, stoats, weasels, and ferrets (to control previously introduced rabbits); the Australian brushtail possum; together with the feral domestic cat and the ubiquitous brown rat, black rat, and house mouse. Fortunately, however, the European red fox—a big problem in Australia—was not introduced. These alien intruders have severely impacted on a native fauna that had evolved in the blissful absence of ground-dwelling predators.

The endearing ground-dwelling kakapo (*Strigops habroptilus*), the world's only flightless parrot, barely escaped extinction at its lowest point in the mid-1990s, when only about fifty individuals remained. As the result of intensive conservation efforts—including hand-rearing chicks and translocation of the survivors to predator-free islands—the total population of this extremely slow-breeding bird has so far more than doubled. Alas, it is far too late to save the nine species of moa, which had disappeared well before Captain Cook's pioneer circumnavigation and mapping of New Zealand in 1769/70 and some 400 years before European colonization in the nineteenth century. As in other parts of the world, a key question is, to what extent were humans to blame for the loss of megafauna?

10.1 THE FIRST NEW ZEALAND HUMANS: POLYNESIAN COLONIZATION

The Maori are a Polynesian people whose ancestors originated from southeast Asia and voyaged eastward via Indonesia, coastal New Guinea, and the islands of Melanesia, eventually reaching the immensity of the Pacific Ocean with its thousands of widely scattered islands. The Polynesians were evidently highly accomplished seafarers, although in the absence of direct evidence we can only speculate on the technologies they used based on historical observations from approximately the last 300 years. Lacking magnetic compasses or other navigational instruments, they probably navigated by the stars in combination with observations of natural phenomena such as ocean currents, clouds, winds, and the flights of birds. Canoes could have been constructed by hollowing out tree trunks or alternatively from sewn-together planks, powered by sails when the wind was favorable and paddles when it was not. Their craft were evidently capable of transporting several families—together with cultivated plants, dogs, and Pacific rats (which were also eaten)—across as much as 3,000 kilometers of open ocean.

Hundreds of years before Europeans first ventured into the region (in the sixteenth century), Polynesians had settled virtually all the habitable Pacific islands and were able to voyage back and forth between islands. Traveling eastward from Tonga and Samoa, they reached the Society Islands about 1025–1120 CE and subsequently other islands in the "Polynesian Triangle," north to Hawaii and east to Rapa Nui (Easter Island) ca. 1190–1290 CE.[5] New Zealand—the southernmost, largest, and last island to be settled—was not colonized until the early fourteenth century CE.

Until recently, an age of ca. 1280 CE, based on radiocarbon-dated moa eggshell from earth ovens at Wairau Bar (northeastern South Island), had been widely quoted for the Polynesian settlement of New Zealand[6]; but a recent reassessment estimates that it occurred approximately 65 years later (see below). The age of the tephra from the North Island Kaharoa eruption, which has now been fixed very precisely at 1314 ± 6 CE, is highly relevant to this important question.[7] Crucially, so far no archaeological deposit has been found underlying this tephra, implying that the first settlers arrived after this date. Revised analyses of radiocarbon dates, now calibrated using the Southern Hemisphere curve, are compatible with first Polynesian occupation early in the fourteenth century, soon after the Kaharoa eruption.[8]

10.2 PRODIGIOUS BIRDS

In 1912, Urupeni Puhara, an 88-year-old Maori chief, said that "the name 'moa' was not the name that the great bird that lived in this country was known to my ancestors. The name was 'Te Kura,' or the red bird; and it was only known as 'moa' after the pakehas [Europeans] said so. Te Kura was known to all ancient people, and was handed down from father to son, who spoke about the big bird that was as high as the top of the door (pointing to the door of the room in which we were sitting). Its legs were thick as those of a bullock. Neither his father nor grandfather had seen Te Kura, but it had been told to them, and tales were told of what it did. He did not know how it was caught or snared, nor had he seen or heard of moa's eggs, and knew nothing of its footmarks. The moa loved, he had heard, all over the North Island, but they disappeared after the coming of Tamatea,' who set fire to the land."[9]

Moa, which were entirely confined to New Zealand, have been extinct for several hundred years. Fossils include bones, mummified soft tissues, feathers, coprolites, and eggshell recovered from swamps, caves, and rock shelters, as well as food remains recovered from Maori archaeological sites. Moa were ratites (from the Latin for "raft")—flightless birds lacking a keel on the sternum. (Unlike all other ratites, moa had no vestige of wings.) Extant ratites include the ostrich (Africa), two species of rhea (South America), emu (Australia), two cassowaries (Australia, New Guinea), and five kiwis (New Zealand). In addition, the extinct ratite elephant birds of Madagascar probably disappeared only in the last few hundred years (chapter 9). Unlike other ratites, moa possessed a smaller rear toe in addition to three front-facing toes. It should be stressed that all but three moa skeletons displayed in museums are incorrectly mounted (as also depicted in most artistic renderings), showing the neck held upright like an ostrich or emu, whereas the position of the articulations on the back of the moa skull show that the neck and head were carried forward, much as in a kiwi.

For more than a century and a half, an inflated number of species was distinguished, to a large extent because the considerable difference in size between the sexes was not appreciated—females were larger. Today nine moa species are recognized, on the basis of both morphology and ancient DNA (fig. 10.1).[10] Of these, two species were endemic to the North Island (North Island giant, Mantell's), five to the South Island (South Island giant, heavy-footed, crested, eastern, upland), and two common to both (stout-legged, little bush). In the absence of the mammalian herbivores present on other continents—such as deer, antelope, and kangaroos—in New Zealand, moa radiated into a range of different sizes, beak shapes, and leg forms, allowing them to adapt to a range of environments.

Moa first came to scientific attention in 1839 when Richard Owen received a letter from a retired naval surgeon, John Rule, offering to sell for ten guineas "a portion or fragment of a bone recovered from the mud of a river that disembogues into one of the bays in New Zealand."[11] Rule brought the bone to London and showed the 15-centimeter-long fragment to the eminent anatomist, stating that, according to Maori tradition, it came from a large flighted bird "now considered to be wholly extinct."[12] Initially Owen rather

FIGURE 10.1. Megafauna (all extinct) from New Zealand. From the top, row 1, left to right: *Dinornis robustus*; *Dinornis novaezealandiae*; *Pachyornis elephantopus*. Row 2: *Pachyornis australis*; *Megalapteryx didinus*; *Anomalopteryx didiformis*; *Pachyornis geranoides*; *Emeus crassus*; *Eurapteryx curtus*.
The outlined *Homo sapiens* gives approximate scale.

contemptuously dismissed it as a "tavern delicacy"—"merely the marrow-bone of an ox." However, after more careful examination he realized that the bone was hollow, with internal struts characteristic of a bird, although he recognized that it was much too large and heavily built to have been capable of flight. Drastically revising his first hasty opinion Owen then proclaimed it to be the femur of a very large hitherto unknown bird "nearly, if not quite, equal in size to an ostrich."[13] In fact, the North Island giant moa (*Dinornis novaezealandiae*) subsequently proved to have been more than twice as heavy as its largest modern relative. Owen sent out requests for further material to his contacts in New Zealand and within two to three years received very many more bones, enabling him to reconstruct entire composite moa skeletons. The confirmation of his diagnosis was widely hailed as a triumph for Owen and the science of comparative anatomy. On the basis of the new material, Owen went on to describe and name many new species for which he coined the generic name *Dinornis*, meaning "prodigious (or terrible) bird." A much-reproduced nineteenth-century photograph (fig. 10.2) shows Owen standing next to an impressively tall reconstructed skeleton of a North Island giant moa (see below) and holding the historic bone fragment (both currently on display at the London Natural History Museum). However, it should be noted that the skeleton is incorrectly mounted with an upright neck.

As in chapter 9, here I include radiocarbon-based LADs (last appearance dates), both as BP (calibrated years before present) and as historical dates CE.[14] By convention, *present* is taken as 1950 CE, the year when the first radiocarbon dates were published.

10.3 NORTH ISLAND GIANT MOA: *DINORNIS NOVAEZEALANDIAE*

Females of *D. novaezealandiae*—the slightly smaller of the two giant moa species—weighed about 76 to 242 kilograms, with a maximum body height of 2 meters, whereas males were significantly smaller. The size also varied according to habitat, with the lowland populations in mosaic shrubland larger than upland populations in closed forest. Like its larger relative, *D. novaezealandiae* had a tall and rather slender body with a relatively small, broad head and a robust, slightly downward-curved bill. The North Island giant moa occupied a wide range of habitat, from coastal dunes to inland shrublands and wet forests.

The partial skeletons of an adult female giant moa and a chick were recovered from a natural pumice pitfall trap at Turangi, North Island.[15] Based on skull and bill morphology and the frequent presence of large masses of gizzard stones, it is likely that North Island giant moa had a fibrous diet of twigs and leaves. A large white egg measuring 190 by 150 millimeters, found in a rock shelter near Waitomo, has been attributed to this species.

The youngest of the few available dates on this species are 1295–1392 CE and 1276–1389 CE, both on a single individual from Aorangi Awarua, Ruahine Range, North Island;[16] and 1229–1402 CE from the Turangi adult.

10.4 SOUTH ISLAND GIANT MOA: *DINORNIS ROBUSTUS*

The South Island giant moa, with a height at the back (females) of 1.8 to 2 meters (the total height of a tall man) and estimated weight of around 76 to 249 kilograms, was the largest species of moa. It ranked second in weight only to the elephant birds (*Vorombe titan* and *Aepyornis maximus*) of Madagascar (chapter 9). In all species of moa, females were larger than males, but the size difference was especially marked in *D. robustus*, in which females could be as much as three times heavier (fig. 10.3).[17] Dark annual lines in the leg bones, recording ceased or slowed winter growth, reveal the surprising fact that these giant birds grew to their full height in only 3 years from hatching. In marked contrast, similar studies on the smaller moa species *E. crassus*, *E. curtus*, and *P. elephantopus* (see below) show that these species, which are placed in a separate family, took at least ten years to reach full size.

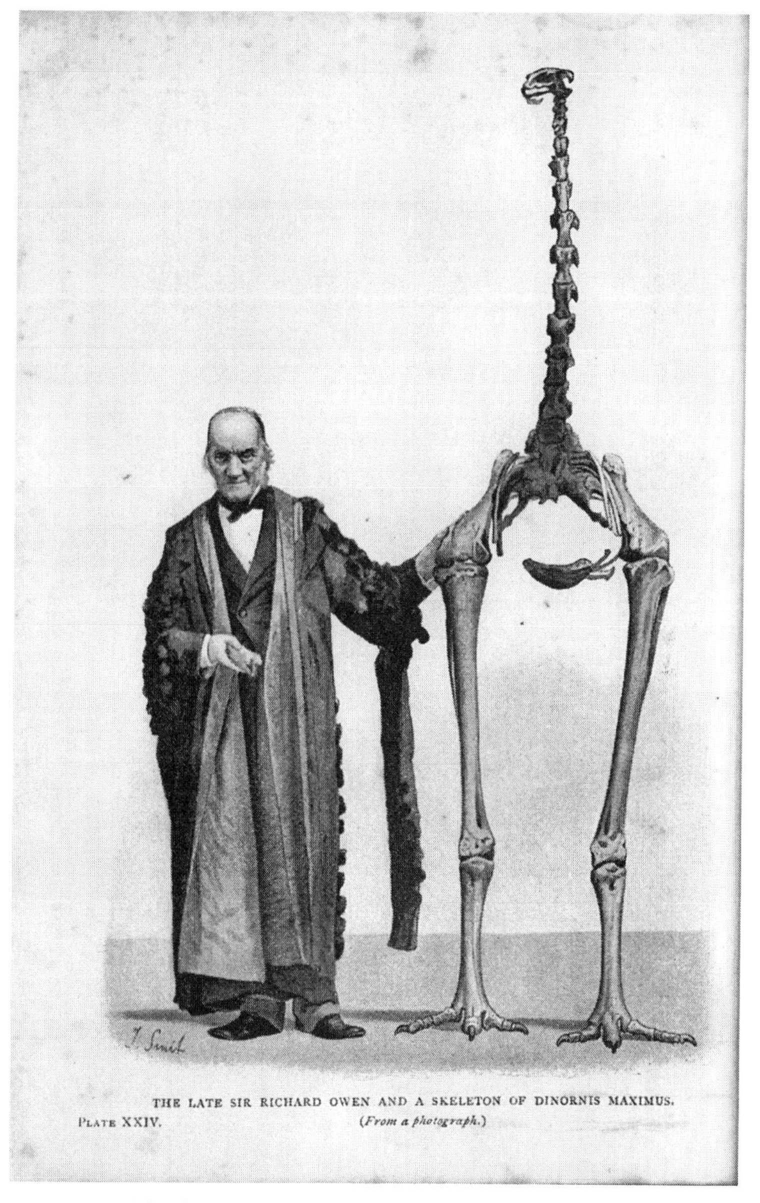

FIGURE 10.2. Richard Owen with (incorrectly) mounted skeleton of North Island giant moa *Dinornis novaezealandiae* (nineteenth-century engraving from a photograph).

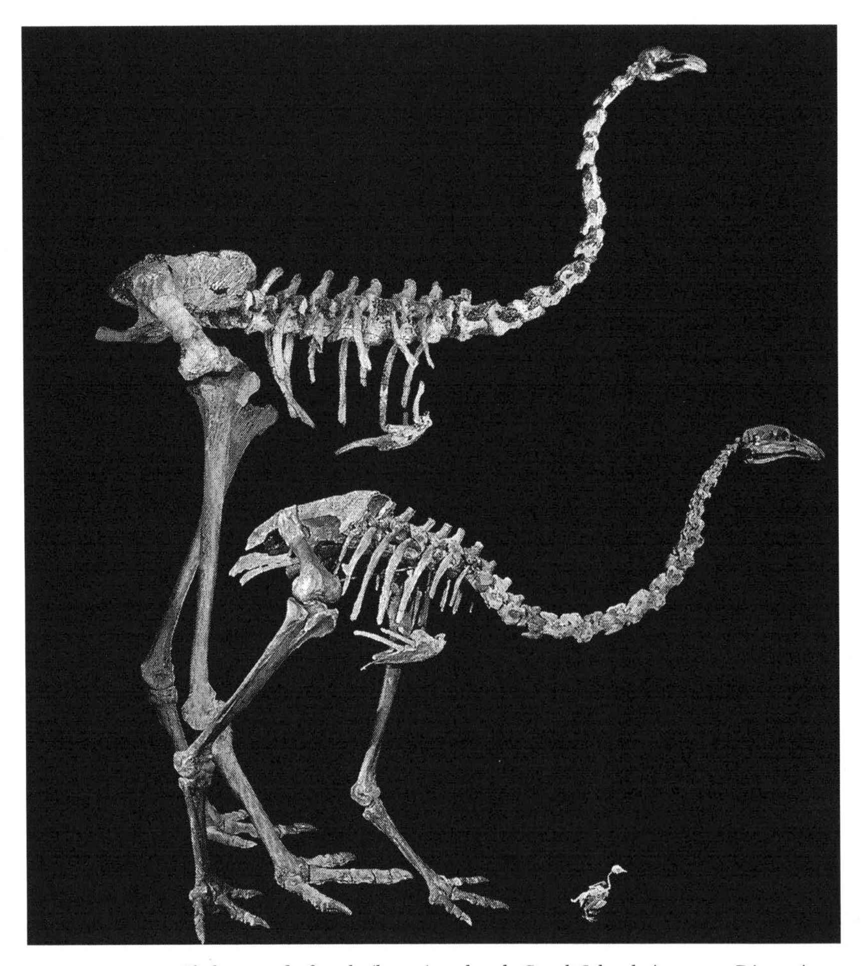

FIGURE 10.3. Skeletons of a female (larger) and male South Island giant moa *Dinornis robustus* compared with a pigeon. The male skeleton is mounted correctly, with the neck extended forward. (Courtesy of Michael Bunce.)

D. robustus was a relatively slender moa with a broad, flattened head, and robust flattened, slightly downward-curved bill. Both *D. robustus* and *D. novaezealandiae* had longer necks than other moa species, with three additional vertebrae. The large olfactory lobe indicates a well-developed sense of smell.

D. robustus is recorded from most areas of the South Island, with the exception of the high alpine areas of the Southern Alps. Its habitat ranged from coastal dunes to shrublands, forests, and subalpine herb fields and grasslands;

it was also present on D'Urville Island (north coast) and Stewart Island (south coast). It browsed a variety of plants in upland, lowland, and open forest habitats. Larger animals were present in areas of lower rainfall. Fossil finds suggest that *D. robustus* nested in rock shelters. Nests comprised a shallow bed of twigs clipped to 20 to 60 millimeters long, together with shrubs, stripped bark, and other plant material. The largest known moa egg, white and measuring 240 by 178 millimeters, was reputedly found in association with a Maori burial site in Kaikoura (northeast South Island). Probably a single egg was

FIGURE 10.4. Skull and gastroliths from an individual *Dinornis robustus*. (Photo copyright Rod Morris/www.rodmorris.co.nz.)

laid. Masses of up to several hundred rounded gizzard stones (gastroliths), weighing as much as 5 kilograms, have been found in association with adult giant moa remains (fig. 10.4). In the absence of teeth these stones were used to grind up fibrous plant food.

The youngest available dates on *D. robustus* are 521–649 BP (NZA-33540), 1301–1429 CE, from Pyramid Valley, near Christchurch, and 509–542 BP (NZA-52929), 1408–1441 CE, on eggshell from an archaeological site at Pounawea, South Island.

10.5 HEAVY-FOOTED MOA: *PACHYORNIS ELEPHANTOPUS*

The heavy-footed moa weighed about 73 to 163 kilograms and was up to 120 centimeters high at the back. It was a large, thick-set bird with short massive legs, as its name suggests (fig. 10.5). The limb structure of this and the other species with thick-set legs indicates that they would have waddled rather than run. The heavy-footed moa had a long, sturdy, downcurved bill with a pointed tip, which implies that it could carefully select food items. It inhabited lowland forest edges in eastern South Island and was relatively rare.

The youngest known date for *P. elephantopus* is 665–531 BP (NZA-9069), 1285–1419 CE, from Mason Bay, Stewart Island, off South Island.

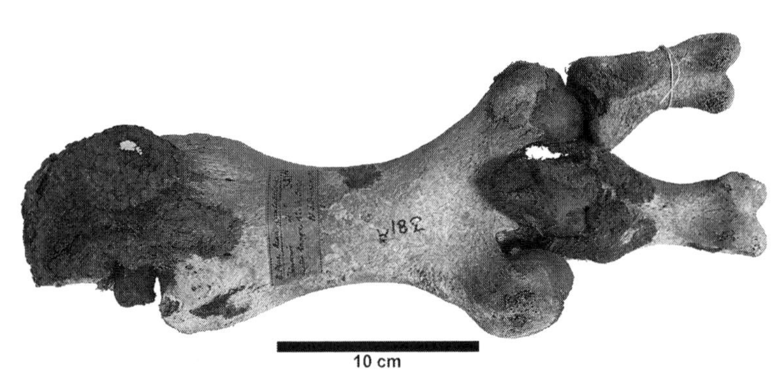

FIGURE 10.5. Tarsometatarsus (lower leg), proximal phalanges, and adhering mummified tissue of *Pachyornis elephantopus*, probably from a rock shelter, Hector range, Nevis, Otago. (Cat. UMZC.381.A, courtesy of the University Museum of Zoology, Cambridge. Photo by the author.)

10.6 STOUT-LEGGED MOA: *EURAPTERYX CURTUS*

The short and stocky stout-legged moa (aka broad-billed moa) weighed around 12 to 34 kilograms (males) or 49 to 109 kilograms (females) and was 51 to 103 centimeters high. It had a relatively small head, a blunt downcurved bill, thick legs, and very wide feet. Present over most of the North Island and the eastern strip of the South Island, it inhabited drier areas that supported a vegetational mosaic of grassland and shrubland, where it browsed a diverse range of shrubs and herbs. The presence of an extraordinarily elongated trachaea in both this species and eastern moa suggests the ability to make loud, resonating calls that could carry long distances.

Stout-legged moa were abundantly represented at Polynesian archaeological sites on both the North and South Islands. It was the one of the most abundant species in Polynesian middens at Wairau Bar (Marlborough) and other moa-hunter archaeological sites along the Canterbury and Otago coasts.

The youngest available date on *E. curtus* is 1395–1427 CE from Monck's Cave, South Island.

10.7 CRESTED MOA: *PACHYORNIS AUSTRALIS*

The crested moa was a large, thick-set bird, weighing about 44 to 90 kilograms, with a body height of up to 150 centimeters. It had a short, robust pointed bill and stout legs. The common name refers to the possible presence of a feather crest, inferred from pits in the front of the skull (no feathers have been found in association with this species). Unlike most other moa species, which were described by Richard Owen in the nineteenth century, *P. australis* was named (by W. R. B. Oliver) as late as 1949.

Crested moa inhabited subalpine shrublands and fellfields on the South Island, together with coastal dunelands in Southland (the far south region of South Island). Its remains are commonly found in caves above 800 meters, but it is unknown from archaeological sites, so there is no direct evidence that it was hunted. Until fairly recently it was thought to have gone extinct around the Pleistocene–Holocene transition (ca. 11.7 kya), but several records now have been dated to the fourteenth and fifteenth centuries CE.

The youngest available date for *P. australis* is 1396–1442 CE, from Bulmer Cave at 1,500 meters above sea level in a remote area of northwestern South Island.[18] So far, this is also the youngest known date for any moa species.

10.8 EASTERN MOA: *EMEUS CRASSUS*

The eastern moa weighed about 36 to 79 kilograms, measured 73 to 99 centimeters high, and was stocky, with short, stout legs. It was confined to a strip along most of the east of South Island and mainly inhabited lowland forests, usually below 200 meters.

The youngest date on eastern moa is 1314–1419 CE, from Wairau Bar in the north of South Island.

10.9 UPLAND MOA: *MEGALAPTERYX DIDINUS*

Weighing about 28–80 kilograms and measuring 65 to 95 centimeters in body height, *M. didinus* was a small, slender moa with a small head and downcurved bill. It was restricted to the upland regions of South Island, mainly above 900 meters, in areas of subalpine scrub, grassland, and forests. Its legs were covered with feathers—presumably an adaptation to the cold of its subalpine habitat.

The youngest available date on upland moa is 1300–1422 CE, from Takahe Valley Rockshelter, Murchison Mountains, South Island.

10.10 LITTLE BUSH MOA: *ANOMALOPTERYX DIDIFORMIS*

The little bush moa, the second smallest moa species, weighed only around 24 to 64 kilograms and was some 50 to 90 centimeters high at the back. It occupied lowland and montane forests with continuous canopy over most of the North Island. Both the distribution of its remains and isotope analyses show that it fed around the edges of clearings in wet forests.

The youngest available date for little bush moa is: 1310–1420 CE, from Echo Valley, Fiordland, in the far south of South Island.

10.11 MANTELL'S MOA: *PACHYORNIS GERANOIDES*

At only 17 to 36 kilograms and 54 centimeters in height, Mantell's moa was the smallest moa. It was present over most of the North Island in lowland forests.

The youngest available date on Mantell's moa is 1072–1301 CE, from remains found at Riverlands on the south coast of North Island.

10.12 CHRONOLOGY OF EXTINCTION

The abundance of moa remains throughout the South Island, combined with more than six hundred radiocarbon dates made directly on moa remains (bones, soft tissues, coprolites, eggshell) and the recent ages of the available material, allows their extinction chronology to be followed in much greater detail and precision than is possible with megafauna in other parts of the world.[19] Moreover, the quality of preservation of ancient DNA is outstanding.

All nine moa species known from the late Holocene were also present during the Last Glacial, and there is no record of any species going extinct during the Last Glacial. All survived until the late Holocene, and most were still in existence until a century after Polynesian settlement. It is almost universally accepted that the extinction of moa resulted from hunting and habitat destruction by Polynesians.

There is abundant evidence that Polynesians in New Zealand actively hunted moa for food, very likely using snares and possibly clubs. A moa's forward-directed neck would have made it easy to snare. In addition, remains from archaeological sites show that many eggs were collected, cooked, and eaten. Moa bones were carved into fishhooks and pendants, and the skins and feathers made into clothing and used for decoration. Hundreds of archaeological sites have been discovered, especially on the South Island, ranging from single finds to fields of earth ovens full of moa bones and extensive middens—some containing the butchered remains of thousands of individuals. Anderson estimated that, on the east coast of South Island, the Waitaki River Mouth Site alone contained the remains of five thousand to eight thousand birds.[20]

Two recent studies using different methods give estimates for the chronology of moa extinction. The first, by Perry and colleagues, employed probabilistic sightings (a statistical technique) to estimate the timing of human settlement from radiocarbon-dated records of introduced Pacific rat, in the form of coprolites and gnawed seeds, plus a chronology of moa extinction from 111 radiocarbon dates made directly on moa remains that met their quality criteria. The broader of their two estimates for human settlement was 1096–1291 CE, and for moa extinction 1418–1473 CE. By comparing national and local extinction dates, they estimated that total extinction occurred about 200 years after settlement. They also suggested that extinction occurred contemporaneously at sites several hundred kilometers apart and that there was little difference between the extinction times of the smallest and largest moa species. They concluded: "Our results demonstrate how rapidly megafauna were exterminated from even large, topographically and ecologically diverse islands such as New Zealand and highlight the fragility of such ecosystems in the face of human impacts."[21]

Richard Holdaway and colleagues used a Bayesian statistical approach to evaluate 270 radiocarbon dates on moa remains from natural contexts, and ninety-six radiocarbon dates on moa eggshell from archaeological sites; the two independently derived chronologies were in agreement. The timing of Polynesian colonization is constrained to the early fourteenth century CE, with the best estimate about 1345 CE, about 65 years younger than estimates based on the Pacific rat evidence (see above).[22] According to Holdaway and colleagues, "New Zealand offers the best opportunity to estimate the number of people involved in a megafaunal extinction event because, uniquely, both the Polynesian settlement of New Zealand and moa extinction are recent enough to be dated with a high degree of precision. In addition, the founding human population can be estimated from genetic evidence."[23] They concluded that the human population of New Zealand would not have exceeded two thousand individuals before extinction of moa populations in the habitable areas of the eastern South Island, and proposed that "an extremely low-density human population exterminated New Zealand moa."[24] Determining the duration of Polynesian–moa interactions is complicated by a large "wiggle" in the terrestrial radiocarbon calibration curve during the fourteenth century CE. Using Bayesian statistics to overcome this problem, Holdaway and colleagues concluded that it took less than 150 years to exterminate moa throughout the South Island, similar to the estimate of Perry and colleagues.

When summer temperatures were 30°C or more and fanned by northwesterly gales, the drought-prone, drier eastern South Island forests would have been destroyed by firestorms, started deliberately or accidentally by Polynesian settlers.[25] The forests were replaced by tussock grasslands, resulting in the loss of the most productive habitats that had supported the densest and most diverse moa populations. A find of butchered remains of an upland moa (*Megalapteryx didinus*) from the Takahe Valley (table 10.1) demonstrates that Polynesian hunters were able to reach even this remote alpine valley and potentially kill off even the small populations in the mountainous and wet western areas.

Although the early Polynesian settlers had a long history of horticulture on other Pacific islands, the South Island climate was largely inimical to this activity, as shown by the sparse evidence for early gardening in the northeastern South Island and none south of Banks Peninsula.[26] However, there is abundant evidence for hunting and gathering of wild terrestrial and marine sources of food. During the first century after colonization, Polynesians in the South Island evidently reverted largely to a hunter-gatherer economy, in which moa

TABLE 10.1 Youngest radiocarbon dates for moa species

Species	Lab. No.	Cal BP	Cal CE	Site	Source
Dinornis robustus [E]	NZA-52929	542–509	1408–1441	Pounawea (SI)	1
Dinornis robustus	NZA-33540	649–521	1301–1429	Pyramid Valley (SI)	1
Dinornis novaezealandiae	Wk-33997	656–558	1295–1392	Aorangi Awarua (NI)	2
	Wk-33998	561–674	1276–1389	(same individual)	
Dinornis novaezealandiae	NZA-11600	548–721	1229–1402	Turangi (NI)	6
Pachyornis elephantopus	NZA-9069	665–531	1285–1419	Mason Bay (SI)	3
Eurapteryx curtus	NZA-52715	555-523	1395–1427	Monck's Cave (SI)	1
Pachyornis australis	OxA-20287	554–508	1396–1442	Bulmer Cave (SI)	4
Emeus crassus	NZA-34414	635–531	1314–1419	Wairau Bar (SI)	3
Megalapteryx didinus	NZA-2227	650–528	1300–1422	Takahe Valley (SI)	3
Anomalopteryx didiformis	OxA-12729	640–530	1310–1420	Echo Valley (SI)	5
Pachyornis geranoides	NZA-15319	878–649	1072–1301	Riverlands (NI)	3

E: eggshell from archaeological site. Cal BP: calibrated date before present. Cal CE: calibrated date, Common Era. NI: North Island. SI: South Island.
Sources: 1. Holdaway, Allentoft, et al. 2014. 2. Wood and Wilmshurst 2013. 3. Perry et al. 2014. 4. Rawlence and Cooper 2013. 5. Bunce, Worthy, Phillips, et al. 2009. 6. Worthy 2002.

hunting was a major component. It is an interesting question to what extent the Polynesian settlers relied on this ready source of high-quality protein. They practiced a wide-ranging economy, in which coastal/marine sources were prominent (as throughout Polynesia); hunting fur seals, sea lions and elephant seals, other marine mammals, and a range of birds[27], as well as moa. They also fished, gathered shellfish, and grew crops (mainly in the north). It seems likely that this mixed economy could have allowed the continued hunting of moa even as they were becoming scarce. In addition to hunting pressure, moa populations might have been susceptible to removal of eggs and predation of chicks by dogs. However, Holdaway and Jacomb found that moa populations were not damaged by loss of eggs or young; predation of adults was the key factor, whereas a later study, using aDNA, found human predation occurred at all life stages.[28] At the same time, burning of forests would have severely reduced suitable habitat.[29]

10.13 HAAST'S EAGLE

Before humans arrived in New Zealand, the major predator of moa was the extinct giant Haast's eagle (*Harpagornis moorei*), described and named in 1871 by the naturalist Julius von Haast. During the Last Glacial it was widespread in the South Island, but by the Holocene it was present only in the eastern part. Females weighed as much as 14 kilograms with a wingspan up to 3 meters, making this the world's largest known eagle—30 to 40% heavier than the largest extant eagle, the harpy eagle (*Harpia harpyja*).[30] It preyed on moa, including even female giant moa (weighing 200 kilograms or more), as shown by damage inflicted by 75 millimeter-long eagle talons to moa skulls and pelvic bones excavated from Pyramid Valley and Bell Hill Vineyard.[31] The eagle's powerful claws were capable of penetrating skin and flesh to crush and pierce bone at least 6 millimeters thick. The fossil evidence shows that attacks were not made from the rear, which would have meant encountering solid muscle and bone. Instead they were made from the side, killing by targeting vulnerable vital organs in the pelvic region, including major blood vessels and the kidneys, while an attack on the head—penetrating the brain—would have killed instantly. It is interesting to speculate whether or not Haast's eagle, which over millions of years had evolved its large size and power to prey on moa, would have been sufficiently discriminating to avoid attacking the newly arrived human biped.

Haast's eagle almost certainly disappeared as a consequence of the loss of its moa prey.

CHAPTER 11

Island Megafauna

In this chapter I do not attempt to cover all island megafauna, but instead focus on selected species, both living and extinct, including comparison with the fate of their mainland relatives (11.1). As is well known, there are many instances of initially small mainland animals evolving to become larger on islands, while conversely large animals became dwarfed in relation to their mainland ancestors.[1] Nevertheless, many of the latter remained large enough to qualify as megafauna. Islands provide excellent opportunities to test between the overkill and Blitzkrieg hypotheses by estimating the time elapsed between human arrival and LADs (last appearance dates) for each extinct species. Human activities—including hunting, habitat destruction, and especially deforestation—and the introduction of alien species such as rats, cats, dogs, pigs, and goats appear to have been responsible for wiping out many of the numerous species, large and small, that have disappeared from islands.[2]

11.1 GIANT TORTOISES

Two giant tortoise populations survive to the present day, on opposite sides of the globe.[3] The remote Galápagos Islands in the Pacific some 970 kilometers from the coast of Ecuador, were not permanently settled by people from the South American mainland until 1832.[4] On his visit in 1835, Charles Darwin made his groundbreaking observations on their unique fauna, including the famous giant tortoises after which the islands are named (Spanish, *galápago*: "tortoise"). Until recently the various island tortoise populations were treated as subspecies of *Chelonoidis nigra*, but now each is regarded as a species of

FIGURE 11.1. Examples of extinct and extant island megafauna. From the top, row 1, left to right: *Meiolania platyceps* (Lord Howe Island, southwest Pacific); *Stegodon florensis* (Flores, Indonesia); *Praemegaceros cazioti* (Sardinia). Row 2: *Mammuthus exilis* (Channel Islands, California); *Megalocnus rodens* (Cuba); *Myotragus balearicus* (Balearic Islands, western Mediterranean). Row 3: *Chelonoidis* spp. (Galápagos Islands, equatorial east Pacific); *Aldabrachelys gigantea* (Aldabra, Indian Ocean). *Varanus komodoensis* (Komodo and adjacent islands, Indonesia). Black: extinct species. Gray: living species. The outlined *Homo sapiens* gives approximate scale.

Chelonoidis in its own right—nine species in all, of which two are entirely extinct and one is extinct in the wild.[5] Males average about 175 kilograms. All ultimately originated from mainland South America although the mainland *Chelonoidis* species still found there are much smaller. The unique biological importance of the Galápagos Islands is recognized by their designated status as a UNESCO World Heritage Site and the Galápagos National Park.

The other extant native population of giant tortoises, *Aldabrachelys gigantea*, is found on the Aldabra Atoll in the Seychelles Islands in the Indian Ocean. On average, males weigh about 205 kilograms. The ancestors of the Aldabra tortoises originated in Madagascar, from where they disappeared less than 1,500 years ago (chapter 9). Declared a protected nature reserve in 1981, Aldabra is now uninhabited except for a small research station. On both the Galápagos and Aldabra, the survival of giant tortoises can be attributed to comparative lack of human interference, highlighted by the fact that within the last few hundred years similar populations have disappeared from several other islands in the Indian Ocean, hunted for food by visiting sailors—a clear instance of recent Blitzkrieg.

The bizarre meiolaniid tortoises, characterized by horns on the rear of the skull and a heavily ossified tail club (comparable to certain glyptodonts and ankylosaurs; chapter 7), are known from the Quaternary and Cainozoic of Australia and several islands in the southwest Pacific; all are extinct. Superbly preserved subfossil material of *Meiolania platyceps* (for example, the mounted skeleton in the American Museum of Natural History, New York) have been recovered from Pleistocene deposits on Lord Howe Island, off the east coast of New South Wales, Australia, but there is no information on when it went extinct. Dates have been obtained on some of the abundant remains of a giant tortoise "*Meiolania*" *damelipi*. which were excavated from the Teouma Lapita archaeological site on the island of Efate, Vanuatu, in New Caledonia.[6] The presence of horns and referral to the genus *Meiolania* is uncertain.[7] Unequivocal evidence that the remains were butchered clearly shows contemporaneity of humans and the extinct tortoises. Two of the tortoise remains were directly radiocarbon-dated to 2.89 and 2.76 kya (WK-25601, WK-25602) demonstrating survival into the late Holocene, in contrast to the much earlier disappearance of *Meiolania* in Australia, although that date is unknown. People of the Lapita culture were the first to colonize Vanuatu, about 3.1 to 3 kya. White and colleagues estimated that the giant tortoises on Efate were wiped out within 300 years of human arrival, probably not only as a result of predation by humans, but also because introduced pigs would have eaten both young and eggs.[8]

11.2 KOMODO DRAGON

The Komodo dragon *Varanus komodoensis*, a giant monitor lizard, is happily not extinct, although today found only on the island of Komodo in Indonesia and some adjacent islands.[9] Weighing up to 80 kilograms, it is the world's largest living lizard (see fig. 8.12). *V. komodoensis*, or a similar species, was considerably more widespread during the Pliocene and Pleistocene, including Australia, where a much larger form, *Varanus priscus*, was also present (chapter 8).[10] The Komodo dragon is entirely carnivorous, mainly feeding on carrion but also actively hunting large mammals such as deer and water buffalo, which it overpowers with the aid of its venomous bite. The survival of *V. komodoensis* to modern times is remarkable in view of the fact that it shares these islands with people, and occasionally attacks and kills humans.

11.3 HOBBIT ISLAND

The 2003 discovery of a previously unknown small-bodied species of hominin (*Homo floresiensis*) in Liang Bua Cave on the Indonesian island of Flores was a worldwide sensation.[11] Inevitably, because of its strikingly small size (estimated at 1.1 meters tall and 16 to 36 kilograms in weight, thus not megafauna), it was immediately dubbed the "Hobbit." The hominin remains were accompanied by those of the truly megafaunal extinct endemic dwarf elephantid *Stegodon florensis insularis*, as well as the extant Komodo dragon, which still lives elsewhere on Flores. Previously the hominin remains had been dated ca. 38 to 18 kya.[12] However, subsequent work has found that the stratigraphic sequence was originally misinterpreted. The revised stratigraphy, accompanied by infrared stimulated luminescence dating of sediment and uranium-series dating on *in situ* stalagmite, indicates that the *H. floresiensis* skeletal remains date to between about 100 and 60 kya; stone artifacts attributable to this species range from about 190 to 50 kya.[13] A molar of *Stegodon florensis insularis*, closely associated with hominin remains, was directly dated by coupled electron spin resonance/uranium series to ca. 74 kya. This date is younger than any mainland or other island record of *Stegodon*.[14]

11.4 MEDITERRANEAN ISLANDS

The numerous Mediterranean islands were colonized at various times by a range of megafauna, including various deer, elephants and hippos. The dwarfed endemic deer *Praemegaceros cazioti* of Sardinia and Corsica very probably evolved from the early Middle Pleistocene *Praemegaceros verticornis* that swam across from mainland Europe. With a LAD of 7.64 kya (radiocarbon and U-series dating) on a near-complete skeleton from Grotta Juntu on Sardinia, *P. cazioti* is known to have outlived its mainland relatives by some 400,000 years.[15] In this case, the island descendant was safely isolated from competition from other herbivores and also from predators with which its mainland relatives had to contend. As observed by Benzi and colleagues: "*P. cazioti* survived on Sardinia for about 1 Ma and therefore it is a clear example of a species well adapted to the ecological conditions of the island: in fact, it successfully overcomes the numerous climatic-environmental crises of Middle and Late Pleistocene, which on the other hand provoked many extinction events in the mainland."[16] The final disappearance of *P. cazioti* might be linked to the arrival of humans on Sardinia, thought to have occurred around 9 kya. However, there was clearly a substantial gap of at least two millennia between the two events, even in the extremely unlikely event that the Grotta Juntu individual was the last of its species.

The highly specialized extinct dwarf "cave goat" *Myotragus balearicus* was present on several of the Balearic Islands in the western Mediterranean. Uniquely among ungulates, it had large forward-directed incisors in its lower jaw.[17] It had short legs, having evolved in the absence of predators; a single pair of short horns; and cheek teeth adapted to browsing. Measuring only about 50 centimeters at the shoulder and weighing some 50 to 70 kilograms, it was small in comparison with mainland goats and sheep. Remarkably, a study of the bone microstructure indicates that *Myotragus* was ectothermic like a reptile, not endothermic, as are nearly all mammals; and grew at slow and variable rates—an adaptation to the severely limited resources available on these islands.[18] On the basis of associated radiocarbon dates, the earliest estimates of human presence on the Balearic Islands are Mallorca ca. 3.98 kya, Menorca ca. 3.88 kya, and Eivissa ca. 3.83 kya.[19] As this curious animal had flourished for millennia in the absence of ground-dwelling predators, it seems likely that the arrival of humans was responsible for its extinction. However, the radiocarbon-dating evidence is inadequate to reliably estimate extinction times for *Myotragus*, which allows the possibility that it had disappeared before the arrival of people.[20]

11.5 WEST INDIAN GROUND SLOTHS AND GIANT RODENTS

Ground sloths were present on some West Indian islands in the Holocene, having survived considerably later than their relatives in mainland South and Central America (chapter 7).[21] The smallest, Haiti's *Neocnus comes*, was similar in size to extant tree sloths. The youngest of several dates on this species is 5060–4840 BP (AA-58439). The other four or so species all qualify as megafauna. The largest, *Megalocnus rodens* from Cuba, was broad-bodied and barrel-chested, had short legs, and weighed probably around 150 kilograms.[22] (A mounted skeleton of *M. rodens* is in the American Museum of Natural History, New York, and another in the National Museum of Natural History, Havana.) *Megalocnus* had rodent-like front teeth, whereas the front teeth of the rather smaller *Parocnus browni*, also from Cuba, resembled canines. A cheek tooth of *M. rodens* from Solapa de Silex, Lomas de Cacahual, Cuba, gave a calibrated radiocarbon date of 4.84 to 4.58 kya (Beta-206173).[23] The youngest available (calibrated) date for *P. browni*, on a humerus, is 6.35 to 4.95 kya (AA-35290).[24]

Archaeological evidence indicates that humans arrived on Cuba by ca. 6.28 to 5.59 kya, which indicates a lag of at least 1,300 years between human arrival and the disappearance of *M. rodens*. As observed by MacPhee and colleagues, this pattern rules out a Blitzkrieg.[25]

Two related species, *Megalocnus zile* and *Parocnus serus*, are known from the island of Hispaniola (Haiti plus Dominican Republic), but currently no dates are available for these. As pointed out by MacPhee and colleagues, bone collagen is often degraded by the warm moist conditions in the Caribbean, so radiocarbon dating is especially difficult in this part of the world.

Estimates of body size for a large hystricomorph rodent *Amblyrhiza inundata* from cave deposits on Anguilla and Saint Martin, in the Lesser Antilles, range from just under 50 to more than 200 kilograms—well in excess of the megafaunal limit of 45 kilograms.[26] *A. inundata* remains are absent from the archaeological record and there are no available dates, so we don't know when it went extinct, or whether or not humans are likely to have been involved in the demise of this species.

11.6 ISLAND MAMMOTHS

A dwarf species of mammoth, *Mammuthus exilis*, evolved in isolation on the Channel Islands that are now about 30 kilometers from the California coast. At around 1.7 to 2 meters at the shoulder and weighing 750 to 1300 kilograms, it was descended from Columbian mammoths that managed to swim across from the mainland.[27] A near-complete skeleton was discovered in 1994 on Santa Rosa Island.[28] The youngest available date for *M. exilis* is 12.895 kya (CAMS-71697).

In the case of woolly mammoth, we can directly compare the fates of identical species on islands with those on the mainland. Far from going extinct at the end of the Last Glacial, as had been believed prior to 1993, woolly mammoths survived until ca. 4.02 kya on Wrangel Island off northeast Siberia—more than 6,000 years later than the youngest known from the Siberian mainland (ca. 10.71 and ca. 11.07 kya; chapter 5). There is a gap of some 400 years between the last known mammoths and the first evidence of humans on Wrangel Island, and the archaeological evidence indicates that they hunted marine mammals, not mammoths. Moreover, as described in chapter 6, woolly mammoths also survived on Saint Paul Island (in the Bering Sea) to within about a century of 5.6 kya, more than seven thousand years longer than on mainland Alaska (ca. 13.34 kya). However, in this case their eventual demise evidently did not result from human activities, but from a drastic reduction of available freshwater and reduced grazing as the island shrank in area due to rising sea level.[29] Rising sea level probably also caused the extirpation of mammoths on Saint Lawrence Island (also in the Bering Sea) soon after ca. 15.92 kya, which was significantly earlier than on mainland Alaska.[30]

11.7 DISCUSSION

Instances of later survivals of particular species on islands might seem counterintuitive; small populations on islands, with nowhere else to go, would appear to be more vulnerable to environmental changes. The fact that in many cases this did not happen argues for the opposite scenario. Isolated for many millennia from the climatic, vegetational and faunal changes that prevailed on the continents, these island populations may have succumbed only after the arrival of humans, which would be consistent with the overkill hypothesis. However, in several cases where we have data, there was a substantial lag between human arrival and megafaunal extinction—ca. 300 years (Vanuatu); ca. 2,000 years (Sardinia); ca. 1,300 years (Cuba)—ruling out Blitzkrieg. Moreover, the last known population of woolly mammoths on Wrangel Island apparently died out some 400 years prior to any archaeological evidence from the island, so humans were probably not responsible for their final disappearance.

CHAPTER 12

Megafaunal Survival: Sub-Saharan Africa and Southern Asia

Unlike the major extinctions that occurred during the late Quaternary in the rest of the world, sub-Saharan Africa experienced very few losses. Although Southern Asia lost a greater percentage of megafaunal species than Africa, most are also extant, so that for both regions most of their wonderful megafauna have survived to the present day. The reasons for their survival when so many megafauna perished in other regions are unclear. However, the fact that many of these species are now threatened with extinction is a potentially irremediable disaster that needs to be recognized and addressed as a global responsibility before it is too late. Quite apart from the impact on the ecosystem and biodiversity, it would be an appalling indictment of the greed and thoughtlessness of our present civilization if future generations could see elephants, rhinos, giraffes, lions, tigers, and many others only in zoos—or maybe not at all.

12.1 SUB-SAHARAN AFRICA (AFROTROPIC ECOREGION)

Here I consider just sub-Saharan Africa, separated from North Africa by the vast Sahara Desert. North Africa is generally regarded as part of the Palearctic Ecozone, although also transitional between the two, as it shares some sub-Saharan elements. This is an important distinction; including North Africa considerably inflates the number of extinct taxa, which for sub-Saharan Africa alone is very modest. It is striking that in this region most of the exceptionally rich Late Pleistocene megafauna, including all the largest species (such as elephants, rhinos, hippos, buffalo and giraffe), survive to the present day.

Papers by Klein, Steele and Faith[1] have provided up to date information on estimated last occurrences for several extinct species in the Late Pleistocene or Holocene. Faith gives a useful table of associated dates (including records from North Africa; table 2). However, most of these dates have wide ranges, and none were made directly on megafaunal remains. Moreover, several of the taxa are known from only a few fragmentary finds, so it is not possible to construct a meaningful extinction chronology for the region. Given the uneven geographical coverage and the relative lack of research on late Quaternary sites in sub-Saharan Africa, it seems very probable that additional extinct taxa will be discovered in the future, although new data are most unlikely to change the view that the impact of extinctions in sub-Saharan Africa was small.

Extinct species include: *Syncerus antiquus* (the spectacular giant longhorn buffalo), *Megalotragus priscus* (giant wildebeest/ hartebeest), *Hippotragus leucophaeus* (blue antelope), and *Equus capensis* (cape zebra). *M. priscus* may have persisted into the Holocene and *E. capensis* to the Late Pleistocene, while *H. leucophaeus* disappeared as recently as 1800 CE.[2] *H. leucophaeus* can claim the dubious distinction of being the first African large mammal to be hunted to extinction by European settlers.[3] *S. antiquus* is recorded from several Late Pleistocene sites in sub-Saharan Africa and from Holocene archaeological sites in North Africa.[4]

According to Tryon and colleagues, the fossil-bearing Late Pleistocene Wasiriya Beds of Rusinga Island, Kenya, range from ca. 33 to 45 kya (AMS radiocarbon dates on gastropod shells) near the top of the sequence to ca. 100 kya on volcanic deposits at the base (by comparison with dated volcanic rocks (phonolites) elsewhere in Kenya).[5] In addition to a range of living species, the fauna includes an extinct giant wildebeest *Megalotragus* sp.; giant longhorn buffalo *Syncerus antiquus*; and, of particular interest, the bizarre "trumpet-nosed wildebeest" *Rusingoryx atopocranion*. Several associated skulls of the latter, probably resulting from a mass-mortality event, were recovered by recent excavations in the Wasiriya Beds on Rusinga Island.[6] The skulls have hollow nasal crests—a feature previously unknown in mammals, but remarkably similar to the nasal crests of certain Cretaceous hadrosaur dinosaurs. The inferred probable function in both was to produce low-frequency sounds for intraspecific communication. Paleoenvironmental evidence from the Wasiriya Beds indicates that *Rusingoryx* inhabited semiarid grasslands, which is consistent with its very high-crowned molars, made for coping with abrasive grasses and accidentally ingested grit.[7]

12.2 SURVIVORS: SUB-SAHARAN AFRICA

The extensive inventory of sixty-four extant megafaunal species includes sixty mammals, of which no fewer than thirty-two are antelope; one bird, ostrich; and three reptiles (fig. 12.1). The mammals include: *Loxodonta africana* (bush elephant), *Loxodonta cyclotis* (forest elephant), *Diceros bicornis* (black rhinoceros), *Ceratotherium simum* (white rhinoceros), *Hippopotamus amphibius* (hippopotamus), *Choeropsis liberiensis* (pygmy hippopotamus), *Giraffa camelopardalis* (giraffe), *Okapia johnstoni* (okapi), *Syncerus caffer* (cape buffalo), *Hylochoerus meinertzhageni* (giant forest hog), *Phacochoerus africanus* (common warthog), *Taurotragus durbianus* (giant eland), *Alcelaphus buselaphus* (hartebeest), *Tragelaphus strepciseros* (greater kudu), *Tragelaphus angasi* (nyala), *Hippotragus equinus* (roan antelope), *Connochaetes taurinus* (common wildebeest), *Equus grevyi* (Grevy's zebra), *Equus quagga* (plains zebra), *Gorilla gorilla* (western lowland gorilla), *Pan troglodytes* (chimpanzee), *Pan paniscus* (bonobo), *Panthera leo* (lion), *Panthera pardus* (leopard), *Crocuta crocuta* (spotted hyena), *Orycteropus afer* (aardvark). Non-mammalian megafauna comprise *Struthio camelus* (ostrich); *Crocodylus niloticus* (Nile crocodile); *Mecistops cataphractus* (slender-snouted crocodile), and *Python sebae* (African rock python). The 2017 IUCN Red List categorizes black rhinoceros, western lowland gorilla, and slender-snouted crocodile as critically endangered, and pygmy hippopotamus, okapi, Grevy's zebra, chimpanzee, and bonobo as endangered.

12.3 SOUTHERN ASIA (INDO-MALAY ECOREGION)

This complex and varied zoogeographical region includes the Indian subcontinent, South China, Southeast Asia, and the numerous islands, both large

FIGURE 12.1. Extinct and extant megafauna from Sub-Saharan Africa (Afrotropic Ecoregion). From the top, row 1, left to right: *Syncerus antiquus*; *Rusingoryx atopocranion*; *Megalotragus priscus*; *Hippotragus leucophaeus*;[E] *Equus capensis*.[E] Row 2: *Loxodonta africana*; *Loxodonta cyclotis*; *Ceratotherium simum*. Row 3: *Diceros bicornis*; *Equus grevyi*; *Equus quagga*; *Giraffa camelopardalis*. Row 4: *Hippopotamus amphibius*; *Choeropsis liberiensis*; *Okapia johnstoni*. Row 5: *Phacochoerus africanus*; *Hylochoerus meinertzhageni*; *Syncerus caffer*; *Connochaetes taurinus*; *Alcelaphus buselaphus*. Row 6: *Oryx gazelle*; *Tragelaphus angasi*; *Hippotragus equinus*; *Tragelaphus strepciseros*; *Taurotragus durbianus*. Row 7: *Panthera leo*; *Acinonyx jubatus*; *Panthera pardus*; *Crocuta crocuta*. Row 8: *Orycteropus afer*; *Gorilla gorilla*; *Pan troglodytes*; *Pan paniscus*. Row 9: *Struthio camelus*; *Crocodylus niloticus*; *Mecistops cataphractus*; *Python sebae*. Extinct megafaunal species (black), selected living species (gray). [E]Extinct following European settlement. The outlined *Homo sapiens* gives approximate scale.

FIGURE 12.2. Extinct and extant megafauna from Southern Asia (Indo-Malay Ecoregion). From the top, row 1, left to right: *Stegodon orientalis*; *Megatapirus augustus*; *Bubalus mephistopheles*. Row 2: *Crocuta [crocuta] ultima*; *Ailuropoda baconi*; *Palaeoloxodon namadicus*. Row 3: *Elephas maximus*; *Dicerorhinus sumatrensis*; *Rhinoceros unicornis*; *Rhinoceros sondaicus*. Row 4: *Tapirus indicus*; *Bos javanicus*; *Bubalus arnee*; *Bos gaurus*; *Boselaphus tragocamelus*. Row 5: *Sus scrofa*; *Rusa unicolor*; *Axis axis*; *Ursus thibetanus*; *Ailuropoda melanoleuca*; *Melursus ursinus*. Row 6: *Panthera leo*; *Panthera tigris*; *Acinonyx jubatus*; *Panthera pardus*; *Pongo pygmaeus*. Row 7: *Crocodylus palustris*; *Crocodylus porosus*.
Black: extinct megafaunal species. Gray: selected living species. The outlined *Homo sapiens* gives approximate scale.

and small, of Indonesia and the Philippines. To the east a faunal boundary—Wallace's Line—runs between Borneo and Sulawesi and through the Lombok Strait, delineating the subregion of Wallacea. Although there are abundant records of late Quaternary megafauna from across Southern Asia, stratigraphic control is generally so poor that in many cases not only is it unknown when each species went extinct, but also which of them were present during the late Quaternary.[8] Moreover, there are additionally many taxonomic uncertainties. Extinct species that may have existed during the Last Glacial in South China, include *Stegodon orientalis* (a large elephant relative), *Palaeoloxodon namadicus* (an elephant), *Ailuropoda baconi* (a giant panda), *Megatapirus augustus* (a large tapir), and *Crocuta* [*crocuta*] *ultima* (a hyena).[9] *Stegodon orientalis* and *Ailuropoda baconi* are also recorded from Vietnam and Java.[10] A *Stegodon* tusk from the island of Timor, Indonesia has been dated by the uranium–thorium method to ca. 130 kya.[11] At most, probably about seven species out of a total of forty-one were lost—that is, about 17%. The total of extinct plus extant megafaunal species breaks down to thirty-eight mammals and three reptiles, which, although impressive, is considerably less than for sub-Saharan Africa (which, however, is much larger in area).

Other notable megafaunal taxa—such as the extraordinary, but imperfectly known, giant ape *Gigantopithecus*—almost certainly disappeared much earlier, in the Middle Pleistocene. Claimed Holocene records of an extinct straight-tusked elephant *Palaeoloxodon* sp. from China are based on misidentified remains and misinterpreted artistic representations of the extant Asian elephant *Elephas maximus*.[12] Claims that several megafaunal species survived into the Holocene in China do not stand up to scrutiny. A critical reassessment of the evidence by Turvey and colleagues failed to confirm such survival, except in the case of the short-horned water buffalo, imaginatively named *Bubalus mephistopheles* (presumably alluding to its Faustian demonic horns).[13] Although no direct radiocarbon dates are available, *B. mephistopheles* is present in a series of early-middle Holocene (Neolithic–Bronze Age) zooarchaeological deposits across southern, central, and eastern China in association with representatives of the modern Chinese large-mammal fauna. Ancient DNA analysis demonstrates that it is phylogenetically distinct from domesticated water buffalo *Bubalus bubalis*.

12.4 SURVIVORS: SOUTHERN ASIA

Surviving megafaunal species include (fig. 12.2): *Elephas maximus* (Asian elephant), *Rhinoceros unicornis* (Indian rhinoceros), *Rhinoceros sondaicus*

(Javan rhinoceros), *Dicerorhinus sumatrensis* (Sumatran rhinoceros), *Tapirus indicus* (Malayan tapir), *Bubalus arnee* (Indian water buffalo), *Bos javanicus* (banteng), *Bos gaurus* (gaur), *Boselaphus tragocamelus* (nilgai), *Sus scrofa* (wild boar), *Rusa unicolor* (sambar), *Axis axis* (chital or axis deer), *Pongo pygmaeus* (Bornean orangutan), *Pongo abelii* (Sumatran orangutan), *Panthera pardus* (leopard), *Panthera tigris* (tiger), *Panthera leo* (lion), *Panthera pardus* (leopard), *Acinonyx jubatus* (cheetah), *Ailuropoda melanoleuca* (giant panda), *Ursus thibetanus* (Asian black bear), *Melursus ursinus* (sloth bear), *Crocodylus porosus* (saltwater crocodile), *Crocodylus palustris* (mugger crocodile), *Gavialis gangeticus* (gharial), *Tomistoma schlegelii* (false gharial), *and Python reticulatus* (reticulated python). Of these, the 2017 IUCN Red List categorizes Javan rhino, Sumatran rhino, Sumatran orangutan and gharial as critically endangered, and Asian elephant, Indian water buffalo, banteng, Malayan tapir, Bornean orangutan, tiger and giant panda as endangered. Asiatic lion, subspecies *Panthera leo persica*, survives today only as a small population in a reserve in the Gir Forest of northwest India, although within the past 200 years it ranged from North Africa, through Turkey and Iraq, to Iran, Pakistan, and northwest India.[14]

CHAPTER 13

Summary and Conclusions: The Global Pattern of Megafaunal Extinctions

It is clear that patterns of megafaunal extinction differed enormously according to zoogeographical region (ecoregion), both in extent and timing. North America, South America, and Sahul suffered severe losses, whereas in northern Eurasia the losses were more modest but still substantial. Moreover, all megafauna as well as many smaller species disappeared from the "island continents" of Madagascar and New Zealand, and almost all megafaunal species were lost from islands in the Mediterranean and Caribbean Seas and the Pacific and Indian Oceans. In marked contrast, few extinctions occurred in sub-Saharan Africa and southern Asia, where most of the megafauna has managed to survive to the present day—although, for many of these, the future prospects are precarious.

The timings of extinctions varied very widely from one region to another (figs. 13.1, 13.2); from mostly before ca. 40 kya in Sahul, to mostly between ca. 14.4 and 11.6 kya in North America, and between ca. 0.62 and 0.53 kya in New Zealand. On the basis of a number of radiocarbon dates made directly on megafaunal material, it currently appears that now-extinct megafauna persisted into the Holocene in South America as well as in northern Eurasia, Madagascar and New Zealand. It is very apparent that in all regions other than New Zealand, many megafaunal species survived thousands of years after the arrival of modern humans. The key features for each region are summarized below.

13.1 SAHUL: AUSTRALASIAN ECOREGION

Mainly because of considerable dating uncertainties, it is very difficult to accurately estimate late Quaternary megafaunal losses in Sahul (Australia plus New

Guinea and Tasmania), but they probably exceeded 90% of species. Although data are limited, it is clear that extinctions in Sahul occurred significantly earlier than elsewhere, with direct LADs ranging from ca. 129 kya to ca. 32.4 kya, mostly beyond radiocarbon-dating range (chapter 8). Species affected included a wide range of giant marsupials, such as *Diprotodon optatum*, *Zygomaturus trilobus*, and *Thylacoleo carnifex*; the giant goanna *Varanus priscus* (aka *Megalania prisca*); and a huge flightless bird, *Genyornis newtoni*. So far, there is no evidence for further megafaunal losses in the Late Glacial or the Holocene.

As described in chapter 8, the recently published evidence from the Madjedbebe rock shelter that humans had colonized Sahul by ca. 65 kya has profound implications for megafaunal extinctions on that continent.[1] With available dates indicating that many megafaunal species persisted for at least another 20,000 years after human arrival, it appears that Blitzkrieg can be discounted. Perhaps megafaunal extinctions in Sahul were staggered over many millennia, much as in northern Eurasia. However, until a lot more data become available, we won't be able to reliably reconstruct the pattern of extinctions and consequently will not be in a position to properly evaluate probable causes.

13.2 NORTHERN EURASIA: PALEARCTIC ECOREGION

Extinctions in northern Eurasia were spread over a much longer time span than in North and South America. The number of megafaunal species lost, although substantial (about 41% of total), is modest compared with, say, North America. However, the dating evidence is much better known, so that the history of extinctions can be followed in more detail, both temporally and geographically. The first wave of extinctions (*Palaeoloxodon antiquus*, *Stephanorhinus kirchbergensis*, *S. hemitoechus*) is poorly dated but probably occurred before ca. 50 kya—essentially, beyond radiocarbon range and before modern humans arrived in the region. However, Neanderthals were present at the time and might conceivably have contributed to the demise of these beasts (chapter 5). The earliest known modern humans (*Homo sapiens*), dating to ca. 300 kya, are recorded from North Africa. They entered Europe ca. 45 kya, and within about 5,000 years the Neanderthals who preceded them had disappeared.

Radiocarbon-dated LADs in northern Eurasia range from before the Last Glacial Maximum to the Holocene, from about 36.8 kya (*Elasmotherium sibiricum*) to ca. 11.07 kya (*Mammuthus primigenius*), ca. 7.66 kya (*Megaloceros giganteus*) and ca. 3.23 kya (*Ovibos moschatus*). Several phases of extinction and marked geographical differences in timing are apparent, corresponding well with climatic phases (fig. 5.15). A marked fragmentation or reduction of range prior to extinction is seen in *Mammuthus primigenius*, *Panthera spelaea*, *Crocuta crocuta*, *Coelodonta antiquitatis*, *Megaloceros giganteus*, and the *Ursus spelaeus* group. *P. spelaea* and *M. primigenius* also declined in genetic diversity throughout northern Eurasia and North America.[2] Severe range shrinkage and reduced genetic diversity are also seen in *Saiga tatarica* and *Ovibos moschatus*; although both experienced similar bottlenecks, they nevertheless managed to avoid global extinction.[3] The histories of these two species suggest that in some instances the starkly contrasting fates of extinction or survival could be largely a matter of chance.

In the case of woolly mammoth in northern Eurasia, evidence for hunting by humans is sparse, and is almost nonexistent for other extinct megafauna. Paradoxically, however, the abundance of remains at archaeological sites strongly suggests that extant species such as red deer, ibex, horse, and reindeer were definitely on the menu, while there is direct evidence—in the form of embedded spear points and impact marks on bones—that Eurasian elk/moose (*Alces alces*), reindeer, and horse were hunted. Moreover, losses in the Holocene, including *Megaloceros giganteus*, *Ovibos moschatus*, and *Equus hydruntinus*, do not obviously correlate with environmental changes, and by default an anthropogenic cause seems likely. The extinction of *Bos primigenius*, including the last remaining population in Poland in the seventeenth century, was very probably due to a combination of hunting and forest destruction.

13.3 NORTH AMERICA: NEARCTIC ECOREGION

Approximately 73% of megafaunal species disappeared from North America. Available LADs suggest that most extinctions occurred in a relatively short period, within the Late Glacial, ranging from ca. 13.67 kya (*Mylohyus nasutus*) to ca. 12.48 kya (*Mammuthus columbi*). As outlined in chapter 6, Faith and Surovell interpreted these data as consistent with "a synchronous event," such as Blitzkrieg or extraterrestrial impact.

Interestingly, a recently published date on *Mammut americanum* is significantly younger than the rest—with a median value of ca. 11.58 kya, it marginally falls within the earliest Holocene. Some geographical differences in extinction chronology are apparent when comparing Alaska/Yukon with the contiguous United States, but this aspect needs to be investigated much more thoroughly. To resolve these critical issues, a comprehensive dating program for North

American megafauna is needed, especially focused on obtaining many more dates for the commoner species and assessing geographical variation in LADs. As Don Grayson has observed, "The continuing debate over the causes of North American losses is not likely to be resolved unless the history of each species is analyzed individually."

13.4 SOUTH AMERICA: NEOTROPICAL ECOREGION

In South America, approximately 80% of megafaunal species—including eleven or more ground sloths, about a dozen glyptodonts, and many more wonderfully extraordinary animals—were lost. As in North America, there are many records for the Late Glacial, ranging from ca. 13.88 (*Cuvieronius hyodon*) to ca. 11.90 kya (*Toxodon* sp.). However, in marked contrast to North America, nine extinct species appear to have persisted into the Holocene, with median dates ranging from ca. 11.28 kya (*Macrauchenia patachonica*) to ca. 7.84 kya (*Doedicurus clavicaudatus*). Modern humans (*Homo sapiens*) had arrived in the Southern Cone by 14.8 kya, and there are several records from Patagonia ca. 13 kya.

This region offers exciting prospects for future research on megafaunal extinctions, especially obtaining many more dates on a wider range of species and in investigating geographical patterns in relation to the vast range of climate and topography on this continent.

13.5 MADAGASCAR

The island continent of Madagascar had a rich, unique extinct megafauna, including elephant birds, giant lemurs, pygmy hippos, and giant tortoises. Moreover, all other species exceeding 10 kilograms also disappeared. There is no information on faunas before the end of the Last Glacial, and all the known extinct species were present late in the Holocene. Most have youngest dates of ca. 1 to 1.5 kya, and it has also been claimed that some of these persisted in remote areas as late as the seventeenth or even nineteenth centuries CE. As humans had arrived by ca. 10.5 kya and most of the extinct species were still alive at least 8,000 years later, a Blitzkrieg can be emphatically ruled out.

13.6 NEW ZEALAND

The late Quaternary megafauna of New Zealand consisted entirely of moa (nine species)—large to huge flightless ratite birds unique to this island continent. All became extinct in the latest Holocene, with tightly grouped LADs ranging from 526 to 615 BP. The time of human arrival is even more tightly constrained: to the 1340s CE, with the best estimate about 1345 CE (605 kya). There is abundant and unequivocal evidence that people killed and ate moa, and fires resulting directly or indirectly from human activities destroyed much of the moa's forest habitat. The interval from human arrival to the total extermination of moa is estimated at 150 years or less—the clearest instance of Blitzkrieg in the fossil record. Moreover, Holdaway and colleagues concluded that "an extremely low-density human population exterminated New Zealand moa." The fact that moa had evolved over millions of years in the complete absence of large ground-dwelling predators would no doubt have resulted in naive prey, highly vulnerable to human hunters.

13.7 ISLANDS

Most Holocene island extinctions, of both large and small animals, across the globe were probably due to humans; but evidence for Blitzkrieg of megafauna is surprisingly sparse. Indeed, in several cases where we have good data, there was a substantial lag between human arrival and megafaunal extinction: ca. 300 years for giant tortoises on Vanuatu; at least 2,000 years for the endemic deer *Praemegaceros cazioti* on Sardinia; and ca. 1,300 years for the ground sloth *Megalocnus rodens* on Cuba. The last known population of woolly mammoths on Wrangel Island apparently died out some 400 years prior to any indication of human presence (their disappearance on Saint Paul, long before human arrival, has been convincingly attributed to local environmental changes).

13.8 AFRICA AND SOUTHERN ASIA: AFROTROPIC AND INDO-MALAY ECOREGIONS

Present knowledge of megafaunal extinctions in sub-Saharan Africa is seriously restricted by very poor dating coverage. However, the region boasted by far the greatest number of megafaunal species in the late Quaternary, and happily the vast majority of these—including elephants, rhinos, hippo, and giraffe—survive today, although in much reduced numbers. The region also experienced by far the smallest percentage losses of all the continents; approximately three out of a total of seventy-one megafaunal species—that is, about 4%, not including two species that only disappeared within the last 200 years, following the arrival of Europeans with firearms.

The late Quaternary megafauna of southern Asia—including Asian elephant, three species of rhino, tiger, and much else—also largely survives to the present day; of course, there are serious conservation issues with each of these species. It is even more difficult to estimate the extent and timing of extinctions in the late Quaternary of this region because of the generally poor quality of the data, including, as for Africa, a lack of radiocarbon dates. Taking an extreme maximum estimate, about seven species out of a total of forty-one went extinct.

13.9 OVERVIEW

I believe that we still have a long way to go in determining the cause or causes of megafaunal extinctions; much more research is needed. The available evidence differs widely from one region to another: (1) each region shows a different pattern in timing of extinctions; (2) the severity of extinctions varied widely from one region to another; (3) in each region there was a marked surge in extinctions only after human arrival; (4) in all regions except New Zealand, there was a substantial delay between human arrival and the majority of extinctions; (5) there are several instances of megafaunal species surviving significantly longer on islands than on the mainland; (6) extinctions in northern Eurasia—the best-studied region—convincingly correlate with environmental changes rather than the archaeological record.

In each region, the majority of extinctions demonstrably occurred after the arrival of *Homo sapiens*—a pattern suggesting that either the activities of modern humans (hunting and/or habitat destruction) caused the extinctions, or alternatively, that a combination of human activities and environmental change was responsible. In either case, the entry of modern humans appears to have been a significant factor (although perhaps other hominins were responsible for some earlier extinctions in northern Eurasia). As megafaunal remains in archaeological sites may result from either hunting or scavenging, it is notoriously difficult to definitively prove the former. Unequivocal evidence for hunting is confined to the rare instances where hunting weapons have been found in intimate association with remains of extinct megafauna—namely, projectile points embedded in the bones of the victims.

The fact that in many cases megafaunal species survived significantly longer on islands than on the mainland is consistent with the overkill hypothesis. However, the evidence for Blitzkrieg is generally very thin; again, in nearly all regions there was a substantial gap between the earliest record of *Homo sapiens* and the disappearance of megafauna. Only in New Zealand does the extreme shortness of the interval between human arrival and moa extinction provide compelling evidence for a Blitzkrieg. However, the claim of rapid extinction in North America requires to be carefully investigated on the basis of many more radiocarbon dates.

In northern Eurasia there is clear evidence that extinctions of a range of megafaunal species during the Last Glacial in this region can be linked to major climatic/environmental changes that occurred during this time. Extinctions in the last thousand years or so were almost certainly due to human activities, and—in the absence of marked climatic shifts—those earlier in the Holocene may possibly be attributed to the same cause.

In my view, we need to address the question of what caused past megafaunal extinctions by means of additional and better data, not more debate

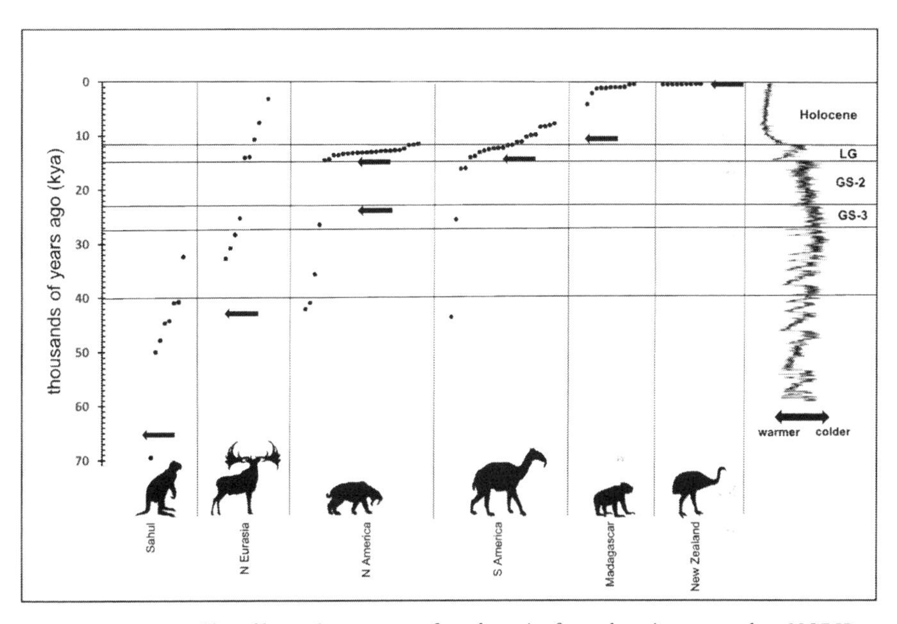

FIGURE 13.1A. Plot of latest dates on megafaunal species for each region, 70 to 0 kya. NGRIP Greenland ice core curve to the right provides a proxy for Northern Hemisphere climate change (see also fig. 5.17). LG (Late Glacial – comprising warm and cold phases); GS-3 and GS-2 (Last Glacial Maximum) are major cold phases. Silhouettes indicate representative species for each region. Arrows indicate inferred arrival times of *Homo sapiens*. North America: lower arrow indicates the earliest record for Yukon; upper arrow indicates the earliest record south of the ice sheets. The very wide range in the timing of extinctions and survivals between regions is evident. Note: many megafauna are recorded long after human arrival in Sahul, northern Eurasia, North America, South America, and Madagascar.

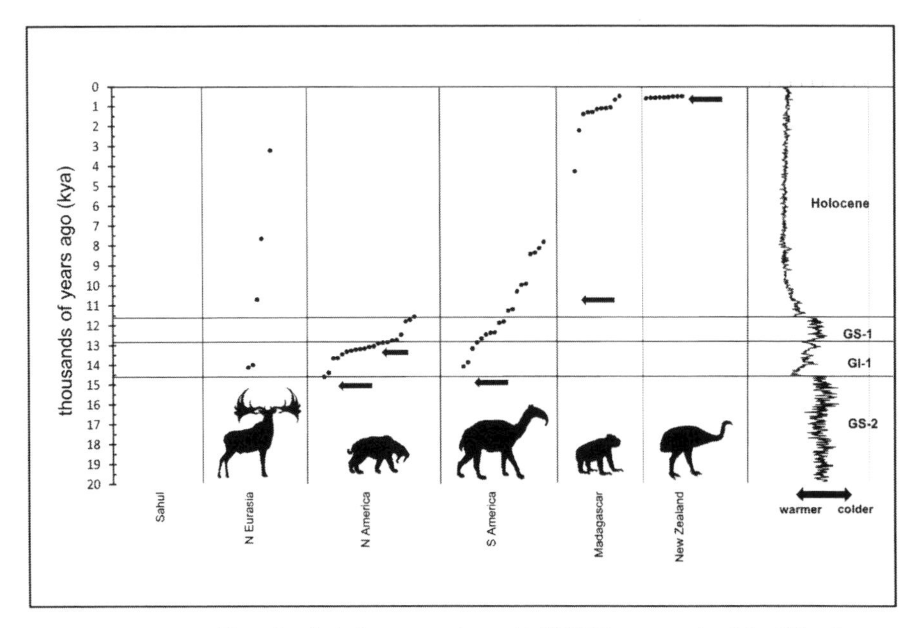

FIGURE 13.1B. More detailed plot, 20 to 0 kya, with NGRIP curve to the right. GS-2: Last Glacial Maximum. GI-1: Late Glacial Interstadial. GS-1: Younger Dryas Stadial. Note: extensive megafaunal survival into the Holocene in northern Eurasia and South America, and many megafaunal records posthuman arrival (arrows) from northern Eurasia, North America, South America, and Madagascar. North America: upper arrow indicates first appearance of Clovis. Only New Zealand shows rapid disappearance of megafauna (moa) after human arrival.

or all-encompassing "solutions" based on inadequate evidence. We need to know first and foremost when each species became extinct, both temporally and geographically, on the basis of many more high-quality dates for each region, made directly on securely identified megafaunal remains. Only when we possess detailed chronologies for the faunal histories and good estimates of extinction times for a wide range of species can we confidently proceed to seek robust correlations with the climatic and archaeological records and properly elucidate the cause or causes of the late Quaternary megafaunal extinctions. Given the considerable differences that we perceive between regions, it seems unlikely that we shall find a one-size-fits-all explanation for late Quaternary megafaunal extinctions that would apply across the globe. Instead, a more complex picture emerges, in which the relative contributions of environmental changes and overkill were different for each region. In many ways, this is excellent news for present and future researchers, in that there is still so much to discover about this fascinating subject—the ultimate "cold case."

The "vanished giants" and other rich wildlife of the Quaternary Ice Age offer huge opportunities for continuing investigations, including the further development and application of a range of dating techniques, ancient DNA, stable isotopes, and morphological and tooth-wear studies. The link between the present day and deeper geological time provides a vital perspective on the modern world and gives us valuable insights into wildlife conservation and the effects of climate change. Such research has enormous implications for how we value and conserve the megafauna (and indeed, all animals) that we have left. As is all too evident, very many present-day species are in serious trouble, predominantly due to human activities, including habitat destruction, unregulated hunting, and global warming. Whatever the cause or causes of prehistoric megafaunal extinctions, it is safe to say that unless immense, unprecedented, and effective international efforts are made, this state of affairs may well prove to have been just the beginning of a major environmental catastrophe, including many more extinctions of wild megafauna in the near future. At best, many or most of these marvelous animals may survive only in zoos. It would be a sorry legacy indeed to bequeath to future generations if the tiger, Asian and African elephants, and many others were to go the same way as sabertooth cats, mammoths, and mastodons.

ACKNOWLEDGMENTS

I am most grateful to the following for kindly commenting on and correcting text within their particular areas of expertise: Luis Borrero (South America), Judith Field (Cuddie Springs), Laurie Godfrey (Madagascar), Don Grayson (North America), John Harris (Rancho La Brea), Richard Holdaway (New Zealand), Adrian Lister (Northern Eurasia), and Gilbert Price (Australia). My special thanks to Henry Gee for his advice and encouragement in writing this book.

I am indebted to Adrian Lister, with whom I have collaborated fruitfully for many years on megafaunal extinctions and other research on Quaternary vertebrates; to my colleagues at Durham University: Judy Allen, Brian Huntley, and Yvonne Collingham; and to Tom Higham and the staff of the Oxford Radiocarbon Accelerator Unit. The Natural Environment Research Council (UK) supported our research on extinctions. I am also most grateful to the late Paul Martin, who encouraged my early interest in megafaunal extinctions.

I warmly thank Kate Scott for creating the evocative paintings of *Elasmotherium* and *Macrauchenia*. Alain Rasolo kindly supplied superb artwork of an *Aepyornis maximus* skeleton, and my sons Rob and Doug provided very able input on the illustrations. The following generously supplied photographs: Pavel Kosintsev; Leonid Petrov; Alexei Tikhonov; Igor Doronin; Gernot Rabeder; Matt Lowe; Nigel Larkin; Sandra Swift; Jim Mead; the Mammoth Site, South Dakota; Thomas Jorstad; Robert Young; Grant Zazula; Brian Switek; John Harris; Luis Borrero; Margarita Belinchón García; Patricio López Mendoza and Ismael Martinez; Rod Wells; Gilbert Price; Steve Bourne; Clay Bryce; Judith Field; Richard Holdaway; Rod Morris; Mike Bunce.

I am grateful to many colleagues for stimulating discussion and/or help in various ways, including: Brian Cooper; Jennifer Crees; Andy Currant; James Hansford; Nigel Larkin; Mark Lewis; Kate Scott; Sam Turvey; John Stewart; Chris Stringer; Ian Barnes; Ross Barnett; Phil Gibbard; the late Roger Jacobi; Tom Lord; the late Alan Turner; Andrew Kitchener; Don Grayson; Russ Graham; Dale Guthrie; Beth Shapiro; Gary Haynes; Jim Mead; Kate Lyons; John Harris; Brooke Crowley; Dick Harington; Grant Zazula; Robert Sommer; Nicholas Conard; Martin Street; Elaine Turner; Dorothée Drucker; Hervé Bocherens; Michi Hofreiter; Ralf Kahlke; Lutz Maul; Wighart von Koenigswald; Susanne Münzel; Martina Pacher; Doris Nagel; Gernot Rabeder; Thijs van Kolfschoten; Dick Mol; Hans van der Plicht; Mietje Germonpré; Adam Nadachowski; Piotr Wojtal; Kim Aaris-Sørensen; Eline Lorenzen; Eske Willerslev; Ole Bennike; Love Dalen; Gennady Baryshnikov; Irina Foronova; Pavel Kosintsev; Yaroslav Kuzmin; Tatiana Kuznetsova; Evgeny Maschenko; Anastasia Markova; Pavel Nikolskiy; the late Andrei Sher; Marina Sotnikova; the late Leopold Sulerzhitsky; Alexei Tikhonov; Sergey Vartanyan; Paula Campos; Ana Pinto; Mauricio Anton; Marzia Breda; Benedetto Sala; Maria Rita Palombo; Vesna Dimitrijevic; Marius Robu; Nikolai Spassov; Richard Holdaway; Alex Baynes; Alan Cooper; Judith Field; Chris Johnson; Gilbert Price; Gavin Prideaux; Tom Rich; Pat Vickers-Rich; Rod Wells.

Lastly, I would like to warmly thank my editors Christie Henry and Scott Gast as well as the rest of the book team at the University of Chicago Press for their support and encouragement.

APPENDIX: DATING THE PAST

Accurate dating is crucial to determining what happened to the megafauna, and we are very fortunate that exciting advances in dating techniques have been made over the past few decades. Some of the principal methods used to date megafaunal remains are outlined below.[1]

A.1 RADIOCARBON

Radiocarbon dating is by far the most useful and accurate method for dating vertebrate remains, in favorable circumstances back to around 50 kya—a time range that fortuitously covers most Quaternary megafaunal extinctions in most parts of the world. In the 1950s, Willard Libby of the University of Chicago successfully developed his idea that the radioactive isotope carbon fourteen (^{14}C, radiocarbon) could be used as a dating tool, and the technique has been much improved since. Today there are over 130 radiocarbon-dating laboratories worldwide. It is difficult to overstate the importance of this enormous step forward, which has transformed our understanding of events over this recent period of Earth's history wherein so much has happened.

^{14}C is constantly being produced in the upper atmosphere by cosmic rays bombarding nitrogen atoms and is absorbed in tiny amounts into plant tissues through photosynthesis in the form of carbon dioxide, along with the much more plentiful stable isotope carbon twelve (^{12}C). In turn animals incorporate both isotopes of carbon by directly eating plants (herbivores), indirectly by eating other animals (carnivores), or by both feeding modes (omnivores). The result is that, when alive, all plants and animals living on land or in freshwater have the same ^{12}C to ^{14}C ratio as was present in the atmosphere at that time. The unstable ^{14}C undergoes radioactive decay at a constant known rate, but this is continually replenished while the organism is alive. However, the "clock" starts at the time of death, as new ^{14}C is no longer taken in and consequently its percentage compared with ^{12}C decreases, so that by accurately measuring the ratio of ^{12}C to ^{14}C, the age of the remains of a plant or animal can be calculated. ^{14}C decays exponentially; after 5,730 years (the "half-life"), half the original amount will have gone; after a further 5,730 years, half of the remainder, and so on. For samples dating from around 50 kya, the residual amount of ^{14}C is so small that it is very difficult to measure accurately, which sets a practical age limit on dating a sample.

A wide range of organic materials can be dated by radiocarbon, including charcoal and wood, as well as—of especial importance here—vertebrate bones, teeth, antlers, horn, skin, other soft tissues, bird eggshell, and dung. Preserved mainly in arid caves and in permafrost, the latter is an especially valuable source of dates. However, the protein collagen extracted from bones, teeth, and antlers is the primary source of radiocarbon dates for megafaunal and other vertebrate remains. Plant material, including charcoal, can give us only associated (indirect) dates, and as we usually don't know how closely the dated sample and the megafaunal remains corresponded in time, the estimated age for the latter can be wildly in error. In order to reliably date megafauna we need samples of material from the animal itself. The increasing use of AMS (accelerator mass spectrometry)—for example, at the Oxford Laboratory—allows very small samples (0.25 g or less of tooth or bone) to be dated accurately.[2] Of course, it is essential that the sample submitted for dating should be securely identified to species, by means of morphological characters and/or ancient DNA analysis. However, sad to say, this principle is not always adhered to.

There are further complicating factors to consider. Levels of radiocarbon in the atmosphere have varied significantly in the past, which means that measured radiocarbon dates need to be calibrated in order to convert them to a close estimate of calendar years, and alas, there is no simple relationship between the two. Wiggly calibration curves have been constructed by comparing radiocarbon dates with known ages obtained by other methods, mainly based on counted tree rings (dendrochronology). The radiocarbon dates quoted in this book are all calibrated using the OxCal program (available on the ORAU website), based on IntCal13 for the Northern Hemisphere and SHCal13 for the Southern Hemisphere. (Another curve, "Marine13," is used for marine samples.)

Another important consideration is possible contamination of the sample with organic material. For example, humic acids that have soaked into the sample from overlying soil will give dates that are spuriously young. Various chemical treatments intended to consolidate and preserve the specimen can also give a false apparent age, according to the substance applied: older—for example, petroleum-based chemicals such as PVA); or younger—for example, shellac or fish glue). Chromatographic techniques are used to remove such contaminants. The isolation and analysis of amino acids specific to bone collagen, such as hydroxyproline, is proving a powerful tool for tackling the problem of contamination, resulting in more accurate and reliable dates.[3]

The rigorous application of objective criteria in both submitting material for radiocarbon dating and assessing published dates is vitally important for establishing reliable chronologies for each species and each geographical region.[4] It would seem obvious that dates should be confined to securely identified material of the species in question. Our experience has shown that dates for faunal material made on associated material such as charcoal (context dates) are often incorrect, sometimes by a wide margin (usually too young). Published dates should be rejected if they lack laboratory identification; if the material dated is not specified; or if it comprises more than one skeletal element or comes from more than one individual. Dates on apatite or bone carbonate should also be rejected, as these have been shown to be unreliable. Lister and Stuart also advocate disregarding dates that were done pre-1980 (an arbitrary but convenient cutoff point), as laboratory methods have improved considerably in recent decades. Important or unexpected results, for example, indicating previously unknown Holocene survivals should be subject to independent repeat dating by another laboratory. Similarly, if a date is an outlier (that is, if in comparison with a reasonably large set of dates, it significantly extends a species' range in time and/or space), it also should be corroborated by independent dating in one or more other laboratories before it is accepted.

A.2 OPTICALLY STIMULATED LUMINESCENCE (OSL)

This method has the considerable advantage that it can date samples as old as 500,000 years—well beyond the range of radiocarbon. However, because OSL gives a date for the sediments in which the fossil is buried and not the fossil itself, it can only be used to date vertebrate remains indirectly. Moreover, the margins of error are much greater than are typical for radiocarbon. The principle of the method relies on the fact that electrons become trapped in the crystalline structure of minerals when subject to radiation from the surrounding sediments, which is measured in the field by a dosimeter. Exposure to sunlight resets the clock, so it is essential that samples are taken in circumstances that exclude light. In the laboratory, stimulated by a pulse of light, trapped electrons in the sample are freed, resulting in the emission of light, which can be measured. Thus, the radiation is absorbed, and hence the time since the sample was last exposed to sunlight can be calculated.

A.3 ELECTRON SPIN RESONANCE (ESR)

This technique also involves measuring the trapped electrons resulting from radiation absorbed from the surrounding sediment during burial. Significantly, it can be used to date tooth enamel, allowing direct dating of megafaunal remains well beyond the range of radiocarbon, although with less precision. As with OSL, the radiation dose from the surrounding sediment needs to be determined. In the laboratory, the sample is placed in a strong magnetic field and subjected to high-frequency electromagnetic radiation. At a certain frequency, the electrons resonate and the electromagnetic radiation is absorbed in proportion to the number of electrons present, which is a function of the elapsed time since burial.

A.4 AMINO ACID RACEMIZATION (AAR)

This technique produces relative ages, not absolute dates, but is very useful for the period beyond the range of radiocarbon, especially in conjunction with other dating methods. It entails analyzing amino acids from protein that is preserved within the crystalline calcium carbonate of shells, including snail shells and the eggshells of birds; but apart from the latter, this method cannot be used to date vertebrate material directly. The basis of AAR is that, on the death of the animal, optically left-handed "L" amino acid molecules spontaneously start to convert to the right-handed "D" form, and this process—called racemization—progresses with time. Amino acid ratios in fossil material are affected by the type of material analyzed and the temperature; it is therefore necessary to understand the thermal history of the sample.[5]

A.5 URANIUM-THORIUM DATING (U-TH)

The decay of the uranium radioisotope ^{234}U to thorium radioisotope ^{230}Th is part of a much longer complex decay series, beginning with ^{238}U via a series of short-lived isotopes and ultimately ending with the stable lead isotope ^{206}Pb. For uranium–thorium dating, the initial ratio of $^{230}Th/^{234}U$ at the time of sample formation must be known or calculated. The principle of U–Th dating of teeth or bone is based on the process whereby uranium (essentially absent in the tissues of a living animal) is taken up from the enclosing sediments, whereas thorium, which is normally insoluble in groundwater, is not. Over time, ^{230}Th accumulates in the sample through radiometric decay. The age is calculated from the difference between the initial ratio of $^{230}Th/^{234}U$ and the ratio in the sample being dated, assuming that the sample has not exchanged ^{230}Th or ^{234}U with the environment (that is, has remained a closed system). The method is used for direct dating of bones and teeth, up to around 600 kya to 500 kya, and also for indirectly obtaining ages of vertebrate fossils by dating speleothems (such as stalagmites) in deposits of limestone caves.

NOTES

CHAPTER ONE

1. Regarding Quaternary time divisions: The Holocene, which began about 11,700 years ago (11.7 kya), is the most recent period of geological time extending to the present day. The Pleistocene covers the period from about 2.58 million years ago (2.58 mya) to 11.7 kya, while the Quaternary comprises the Pleistocene plus the Holocene. Confusingly, the term *Ice Age* has been employed in various ways for different parts of the Pleistocene/Quaternary and has no formal definition. For convenience I use it here as equivalent to the late Quaternary, covering the period from the beginning of the Last Interglacial warm episode (ca. 130 kya) through the Last Glacial (ca. 71 to 11.7 kya) to the present day (see chapter 3).

2. Wallace 1876, 150.

3. Darwin 1845, 173.

4. Johnson 2002.

5. Except those extinctions occurring within the last few hundred years: Dulvy, Pinnegar, and Reynolds 2009; Turvey 2009a, 2009b.

6. Martin 2005.

7. Donlan et al., 2005.

8. Zimov 2005.

9. Oliveira-Santos and Fernandez 2010, 4

10. Shapiro 2015.

11. Lister 2014.

CHAPTER TWO

1. Newell 1963.

2. Lyell 1830–33.

3. Raup and Sepkoski 1982. Other useful references: A. Hallam and Wignall 1997; A. Hallam 2005; MacLeod 2013; Archibald and MacLeod 2013.

4. Erwin 2006; Benton 2015.
5. Benton 2015.
6. Ruppel and Kessler 2017.
7. Benton 2015.
8. MacLeod 2013.
9. Whiteside et al. 2010.
10. Alvarez et al. 1980.
11. Archibald 2012.
12. See, for example, Keller, Adatte, et al. 2008; Keller, Sahni, and Bajpai 2009; Keller, Punekar, and Mateo 2015; MacLeod 2013.
13. Archibald 2014; Brusatte et al. 2012.
14. F. A. Smith et al. 2010.
15. F. A. Smith et al. 2010.
16. MacLeod 2013.
17. MacLeod 2013.
18. MacLeod 2013.
19. Turvey 2009a.

CHAPTER THREE

1. Agassiz 1840, in Woodward 2014, 57. Original in French.
2. Ehlers and Gibbard 2004a, 2004b, 2004c.
3. Darwin 1859, 366
4. Penck and Brückner, 1901/1909.
5. Woodward 2014
6. Hays, Imbrie, and Shackleton, 1976.
7. See Woodward 2014; NGRIP Dating Group 2008.
8. EPICA Community Members 2004.
9. De Beaulieu and Reille 1992.
10. Allen, Watts, and Huntley 2000; Allen, Watts, et al. 2002; Allen and Huntley 2009.
11. Prokopenko et al. 2006.
12. See Woodward 2014.
13. Gibbard and Lewin 2016.
14. Stuart 1982.
15. Woodward 2014; Lambeck and Chappell 2001.
16. Walker et al. 2012.
17. Zalasiewicz et al. 2017.
18. Jefferson 1799, 72.
19. Cuvier 1812.
20. Grayson 1984b, 6.
21. See, for example, Lorenzen et al. 2011; Ersmark et al. 2015.
22. Campos, Willerslev, et al. 2011; Campos, Kristensen, et al. 2010.

CHAPTER FOUR

1. See, for example, Lyons, Smith, and Brown 2004; Koch and Barnosky 2006.

2. For example, compare Grayson and Meltzer 2003, 2004, with Fiedel and Haynes 2004, Sandom et al. 2014.

3. Stringer and Andrews 2012.

4. Hublin et al. 2017.

5. Hershkovitz et al. 2018; Stringer and Galway-Witham 2018.

6. K. E. Westaway et al. 2017; Clarkson, Jacobs, et al. 2017; Higham, Douka, et al. 2014.

7. Stringer 2016; Pinhasi et al. 2011.

8. Goebel, Waters, and O'Rourke 2008.

9. Bourgeon, Burke, and Higham 2017.

10. Tamm et al. 2007.

11. Dillehay 1997.

12. Alcover, Sans, and Palmer 1998; Martin and Steadman 1999; Turvey 2009a.

13. See Martin 1967, 1973, 1984, 2005.

14. Martin 1973, 1984.

15. Martin 1984, 2005.

16. Alroy 2001.

17. Grayson 1984a; Grayson and Meltzer 2002, 2003, 2004.

18. Berger, Swenson, and Persson 2001.

19. Serangeli et al. 2015.

20. Schoch et al. 2015.

21. Stringer 2012.

22. K. Scott 1980; B. Scott et al. 2014.

23. M. White, Pettitt, and Schreve 2016.

24. M. White, Pettitt, and Schreve 2016, 1.

25. See, for example, Guthrie, 1984, 2001; Allen, Hickler, et al. 2010; Huntley et al. 2013.

26. See, for example, Graham and Lundelius 1984; Guthrie 1984.

27. Cooper et al. 2015.

28. See, for example, Barnosky, Koch, et al. 2004; Koch and Barnosky 2006; Nikolskiy, Sulerzhitsky, and Pitulko 2011; Stuart 1991; Haynes 2018.

29. MacPhee and Marx 1997.

30. See comparison with modern West Nile Virus: Lyons, Smith, Wagner, et al. 2004.

31. See, for example, McGuire 2014: notably, the huge asteroid impact ca 65.5 mya is generally blamed for wiping out the (non-avian) dinosaurs.

32. Firestone et al. 2007.

33. Holliday et al. 2014; Kerr 2008; Pinter et al. 2011; A. C. Scott et al. 2010; Surovell et al. 2009.

34. La Violette 2011.

35. Van der Plicht and Jull 2011.

CHAPTER FIVE

1. Ehlers and Gibbard 2004a.
2. Lambeck and Chappell 2001.
3. Allen, Hickler, et al. 2010.
4. Lister and Bahn 2007.
5. Bahn 2016; Cook 2013.
6. Chauvet, Brunel Deschamps, and Hillaire 2001.
7. Bahn 2016.
8. Stringer and Andrews 2012.
9. Higham, Douka, et al. 2014.
10. Reich et al. 2010.
11. Hublin et al. 2017.
12. Higham, Douka, et al. 2014.
13. Sankaraman et al. 2014.
14. Molyneaux 1695, 489.
15. Molyneaux 1695, 489–90.
16. G. F. Mitchell and Parkes 1949; Lister 1994.
17. Saarinen et al. 2016.
18. Gould 1974, pg191.
19. Lister, Edwards, et al. 2005; Hughes et al. 2006.
20. Millais 1897.
21. Barnosky 1985, 1986.
22. Kitchener 1987.
23. Chauvet, Brunel Deschamps, and Hillaire 2001; Guthrie 2005.
24. Van der Plicht, van der Molodin, et al. 2015.
25. Kahlke 1999; Stuart, Kosintsev, et al. 2004; Lister and Stuart 2019.
26. Van der Made and Tong 2008.
27. Stuart, Kosintsev, et al. 2004; Lister and Stuart 2019.
28. Geist 1998.
29. Dowd and Carden 2016.
30. Stuart, Kosintsev, et al. 2004.
31. Van der Plicht, van der Molodin, et al. 2015; Lister and Stuart 2019.
32. Saarinen et al. 2016.
33. Stuart and Lister 2012.
34. Orlando et al. 2003.
35. Deng et al. 2011.
36. Boeskorov et al. 2011.
37. Kubiak 1969.
38. Stuart and Lister 2012.
39. Bahn 2016; Chauvet, Brunel Deschamps, and Hillaire 2001.
40. Chauvet, Brunel Deschamps, and Hillaire 2001.

41. Stuart and Lister 2012; Lister and Stuart 2013.
42. Guthrie 2001.
43. Fischer 1809; Zhegallo et al. 2005.
44. Kosintsev et al. 2018.
45. Schvyreva 2015.
46. Kosintsev et al. 2018.
47. Lister and Bahn 2007; Lister 2014.
48. Pfizenmayer 1939.
49. Shapiro 2015.
50. Lister 2014.
51. Lister and Bahn 2007.
52. Bahn 2016.
53. Pitulko, Pavlova, and Nikolskiy 2017.
54. Wojtal et al., 2019, 163.
55. Lister and Bahn 2007.
56. Lister and Stuart 2019.
57. Lorenzen et al. 2011.
58. Lister and Stuart 2019.
59. Vartanyan, Garutt, and Sher 1993; Vartanyan, Arslanov, et al. 2008.
60. Dikov 1988; Gerasimov et al. 2006.
61. Rosenmüller 1794.
62. Stiller et al. 2014.
63. Pacher and Stuart 2009.
64. Fortes et al. 2016.
65. Chauvet, Brunel Deschamps, and Hillaire 2001.
66. Pacher and Stuart 2009.
67. Baca et al. 2016.
68. Münzel et al. 2011.
69. Stuart and Lister 2010.
70. Protopopov et al. 2016; Chernova et al. 2016.
71. Bahn 2016.
72. Stuart and Lister 2010.
73. Stuart and Lister 2014.
74. Stuart and Lister 2014.
75. Higham, Douka, et al. 2014.
76. Reich et al. 2010.
77. Sankaraman et al. 2014.
78. Stuart 1982, 1987.
79. Mazza and Bertini 2013.
80. Larramendi 2016.
81. Saarinen et al. 2016
82. Stuart 2005; Stuart and Lister 2007.

83. Bosscha Erdbrink, Brewer, and Mol 2001; Mol, de Vos, and van der Plicht 2007.
84. Iwase et al. 2012.
85. Saarinen et al. 2016.
86. Shapiro et al. 2004.
87. Soubrier et al., 2016.
88. Grange et al. 2018.
89. Reumer et al. 2003; Paijmans et al. 2017.
90. Campos, Willerslev, et al. 2010.
91. Crees and Turvey 2014.
92. Saarinen et al. 2016.
93. Crees, Carbone, et al. 2016.
94. Kahlke 1999.
95. Titov 2008; Kahlke 1999.
96. Jacobi and Higham 2011.
97. J. S. Hallam et al. 1973.
98. Jürgensen et al. 2017.

CHAPTER SIX

1. Jefferson 1799.
2. McDonald, Stafford, and Gnidovec 2015.
3. Quoted by G. Jefferson in Harris 2001, 3–6.
4. Froese et al. 2017.
5. Harris 2015.
6. Stock 1956.
7. Marcus 1960.
8. Stock 1956.
9. Harris 2001; Harris and G. Jefferson 1985; Stock 1956; Stock and Harris 1992.
10. Campbell in Harris 2001.
11. Coltrain et al. 2004.
12. Shaw in Harris 2001.
13. Harris 2015. See also La Brea Tar Pits & Museum, n.d.
14. Woodburne 2010; Bacon et al. 2015; Cione et al. 2015; Montes et al. 2015; Hoorn and Flantua 2015.
15. Ehlers and Gibbard 2004b.
16. J. W. Williams et al. 2004.
17. For distribution maps of extinct North American megafauna, see Grayson 2016.
18. Antón 2013.
19. Coltrain et al. 2004.
20. Antón 2013.
21. Antón 2013.
22. Fox-Dobbs, Leonard, and Koch 2008; D. R. Williams 2009.
23. Barnett, Barnes, et al. 2005.

24. Barnett, Yamaguchi, et al. 2006.
25. Guthrie 1990.
26. Stuart and Lister 2011.
27. Figueirido et al. 2010.
28. This is disputed in Figueirido et al. 2010.
29. Fox-Dobbs, Leonard, and Koch 2008.
30. Fox-Dobbs, Leonard, and Koch 2008.
31. Elias and Crocker 2008.
32. Schubert 2010.
33. Fuller et al. 2014.
34. Mead and Agenbroad 1992; Hofreiter et al. 2000.
35. Lull 1929.
36. Hansen 1978.
37. listed by Mead and Agenbroad 1992.
38. Hofreiter et al. 2000.
39. McDonald, Harington, and De Iuliis 2000.
40. McDonald 2005.
41. McDonald and Pelikan 2006.
42. Grayson 2016.
43. Stock 1920.
44. Gillette and Ray 1981.
45. Gillette and Ray 1981.
46. Gillette and Ray 1981, 205.
47. Conniff 2010.
48. Hedeen 2008, 158
49. Kentucky State Parks 2016.
50. Croghan in Hedeen 2008, 42.
51. Hedeen 2008, 42.
52. Franklin in Smyth 1970, 39–40.
53. Franklin in Smyth 1970, 40.
54. Saunders 1996.
55. Warren 1852.
56. Lepper et al. 1991.
57. Fisher in Lepper et al. 1991.
58. Smithsonian 1915, 28.
59. Woodman and Athfield 2009.
60. Graham, Haynes, et al. 1981.
61. Faith and Surovell 2009.
62. Woodman and Athfield 2009.
63. Sanchez et al. 2014.
64. Lister and Sher 2015.
65. Lister and Bahn 2007; Lister 2014.
66. Lister and Stuart 2010.

67. Agenbroad and Mead 1994.
68. Jim Mead, pers. comm., 2018.
69. Lister and Bahn, 2007; Haynes 2018.
70. Haury 1953.
71. Mead and Agenbroad 1992.
72. Mead and Agenbroad 1992.
73. Waters and Stafford 2007.
74. Sanchez et al. 2014
75. Graham, Belmecheria, et al. 2016.
76. S. D. Webb 1965.
77. Heintzman, Zazula, Cahill, et al. 2015.
78. Semprebon and Rivals 2010.
79. Zazula, Turner, et al. 2011.
80. Waters et al. 2015.
81. Grayson 2016.
82. McDonald and Bryson 2010.
83. Mayhew 1978.
84. Kropf, Mead, and Anderson 2007.
85. Grayson 2016.
86. Guthrie 1992.
87. Mead and Agenbroad 1992; Mead, Martin, et al. 1986.
88. Mead, Martin, et al. 1986.
89. Weinstock et al. 2005.
90. Heintzman, Zazula, and MacPhee 2017.
91. Guthrie 2003.
92. Guthrie 2006.
93. Guthrie 1990.
94. G. Jefferson in Harris 2001, 3–9.
95. Shapiro et al. 2004.
96. Grayson 2016.
97. Grayson 2016.
98. Lundelius 1961.
99. Tankersley 2011.
100. Grayson 2016.
101. Kurtén and Anderson 1980; Grayson 2016.
102. Kurtén and Anderson 1980.
103. Grayson 2016.
104. Grayson 2016.
105. Jürgensen et al. 2017.
106. King and Saunders 1986.
107. Meltzer 2010.
108. Goebel, Waters, and O'Rourke 2008; Potter 2008.
109. Bourgeon, Burke, and Higham 2017.

110. Ehlers and Gibbard 2004b.
111. Waters and Stafford 2007.
112. Sanchez et al. 2014.
113. Dillehay 1997. See Fiedel 1999 for an alternative view.
114. Rasmussen et al. 2014.
115. Pedersen, Ruter, and Schweger 2016.
116. Gilbert et al. 2008; Jenkins et al. 2012.
117. Llamas et al. 2016.
118. Holen et al. 2017.
119. Holen et al. 2017, 479.
120. Fox-Dobbs, Leonard, and Koch 2008; Schubert 2010.
121. Fox-Dobbs, Leonard, and Koch 2008.
122. Zazula, MacPhee, et al. 2014.
123. Faith and Surovell 2009.
124. Signor and Lipps 1982.
125. Grayson 2007, 185.
126. Davis and Shafer 2005; Gill et al. 2009; Gill 2013.

CHAPTER SEVEN

1. Falkner in Fernicola et al. 2009, 148.
2. Darwin 1845.
3. Woodburne 2010; Bacon et al. 2015; Cione et al. 2015; Montes et al. 2015.
4. Bargo, Toledo, and Vizcaíno 2006.
5. Piñero 1988, 158.
6. Piñero 1988, 158.
7. Cuvier 1812.
8. Fariña, Vizcaíno, and De Iuliis 2013.
9. Blanco and Czerwonogora 2003.
10. Fariña, Vizcaíno, and De Iuliis 2013.
11. Fariña and Blanco 1996.
12. Fariña and Blanco 1996, 1725.
13. Green and Kalthoff 2015, 645.
14. Bargo 2001.
15. Bargo, Toledo, and Vizcaíno 2006.
16. Carretero, García, and M. A. Dacar 2004.
17. Fariña, Vizcaíno, and De Iuliis 2013.
18. Bargo 2001, 189.
19. Bargo, De Iuliis, and Vizcaíno 2006, 57.
20. Fariña, Vizcaíno, and De Iuliis 2013.
21. Gazin 1956, 343.
22. Gazin 1953, 9.
23. Barnett and Sylvester 2010; Borrero and Martin 2012a, 2012b.

24. Moore 1978.
25. Villavicencio et al. 2015; Metcalf et al. 2016.
26. Markgraf 1985; Tonni et al. 2003.
27. Martin 2005.
28. Borrero and Martin 2012b, 103–4.
29. Borrero and Martin 2012b, 103–4.
30. Owen 1840.
31. Christiansen and Fariña in Fariña, Vizcaíno, and De Iuliis 2013, 207.
32. Fariña, Vizcaíno, and De Iuliis 2013.
33. Blanco and Rinderknecht 2008.
34. Owen 1857.
35. Bargo, Toledo, and Vizcaíno 2006.
36. Fariña, Vizcaíno, and De Iuliis 2013.
37. Cartelle, De Iuliis, and R. L. Ferreira 2009.
38. Vizcaíno et al. 2001; Fariña, Vizcaíno, and De Iuliis 2013; Lopes, Frank, et al. 2017.
39. Falkner in Fernicola et al. 2009, 148.
40. Welker et al. 2015.
41. Huxley 1865.
42. Gillette and Ray, 1981.
43. Fariña, 1995.
44. Fariña, Vizcaíno, and De Iuliis 2013.
45. Fariña, Vizcaíno, and De Iuliis 2013.
46. Fariña 1995.
47. Fariña, Vizcaíno, and De Iuliis. 2013; Zamorano, Scillato-Yané, and Zurita 2014.
48. Alexander et al. 1999; Fariña, Vizcaíno, and De Iuliis 2013, 245.
49. Alexander et al. 1999.
50. Blanco, Jones, and Rinderknecht 2009.
51. Arbour 2009.
52. Lister 2018.
53. Welker et al., 2015.
54. Bond et al. 2006.
55. Fernicola et al. 2009; Lister 2018.
56. Owen 1870b.
57. Fariña, Blanco, and Christiansen 2005; Fariña, Vizcaíno, and De Iuliis 2013.
58. Fariña, Blanco, and Christiansen 2005.
59. Fariña, Vizcaíno, and De Iuliis 2013, 202–3.
60. Lopes, Ribeiro, et al. 2013. C4 plants use a different method of carbon fixation in photosynthesis than the globally more abundant C3 plants, which is advantageous under such adverse conditions as drought, high temperatures, and reduced nitrogen availability.
61. Darwin 1845, 156.
62. Lister 2018.
63. Lopes, Ribeiro, et al. 2013.
64. Weinstock et al. 2005.

65. Fariña, Vizcaíno, and De Iuliis 2013.
66. Prado, Alberdi, et al. 2005; Ferretti 2010.
67. Mothé et al. 2017
68. Fariña, Vizcaíno, and De Iuliis 2013; Larramendi 2016.
69. Mothé et al. 2017; Sánchez, Prado, and Alberdi 2004.
70. Prevosti and Vizcaíno 2006; Prevosti and Martin 2013.
71. Fariña, Vizcaíno, and De Iuliis 2013.
72. K. J. Mitchell et al. 2016; Fariña, Vizcaíno, and De Iuliis 2013.
73. Prevosti and Martin 2013.
74. Fariña, Vizcaíno, and De Iuliis 2013.
75. Prado, Alberdi, et al. 2015
76. Dillehay 1997; Dillehay et al. 2015.
77. Politis et al. 2016.
78. Politis et al. 2016, 1.
79. Hulton et al. 2002; Ehlers and Gibbard 2004c.
80. Prado and Alberdi 2010.
81. Mayle et al. 2009.
82. Prado and Alberdi 2010.
83. Borrero 2009, 161.
84. Prado, Alberdi, et al. 2015.
85. Cozzuol et al. 2013.

CHAPTER EIGHT

1. Wells, Moriarty, and Williams 1984; Reed 2006.
2. Grün, Moriarty, and Wells 2001; Saltré, Rodríguez-Rey, et al. 2016.
3. Fraser and Wells 2006.
4. Grün, Moriarty, and Wells 2001, 58.
5. Johnson 2006, table 2.1; S. Webb 2013.
6. Roberts et al. 2001.
7. Wroe, Field, Archer, et al. 2013.
8. Cohen et al. 2015, 195.
9. Van der Kaars and De Deckker 2002.
10. Rule et al. 2012.
11. O'Connell and Allen 2015.
12. Clarkson et al. 2017; Marean 2017.
13. Clarkson et al. 2017, 306.
14. Clarkson et al. 2017, 306.
15. Clarkson et al. 2017, 309.
16. Clarkson et al. 2017, 306, 309; O'Connell et al. 2018, 8482; K. E. Westaway et al. 2017.
17. Turney et al. 2008.
18. Price 2008; Price and Piper 2009.
19. Vickers-Rich and Archbold 1991.

20. Gröcke 1997.
21. Owen 1870a.
22. Owen 1859b, 47, 48.
23. Tedford 1973.
24. Tedford 1973.
25. Roberts et al. 2001; Saltré, Rodríguez-Rey, et al. 2016.
26. Price 2012a, 2012b.
27. Price 2012b.
28. Price and Piper 2009.
29. Sharp 2014.
30. Tedford 1973.
31. Carey et al. 2011.
32. Price, Ferguson, et al. 2017.
33. Runnegar 1983.
34. M. Smith 2013.
35. Murray 1991.
36. Tasmanian Museum and Art Gallery 2013.
37. Gillespie, Wood, et al. 2014.
38. M. Westaway, Olley, and Grün 2017.
39. Owen 1874.
40. Banks, Colhoun, and van den Geer 1976; Murray 1991.
41. Owen 1873.
42. Janis, Buttrill, and Figueirido 2014.
43. Murray 1991.
44. Prideaux, Ayliffe, et al. 2009.
45. Murray 1991.
46. Turney et al. 2008.
47. Owen 1859a, 319.
48. Wells and Camens 2018.
49. Murray 1991.
50. Prideaux, Long, et al. 2007.
51. Wroe, McHenry, and Thomason 2005.
52. Wroe, Myers, et al. 2003.
53. Stephen Wroe, pers. comm., February 2020.
54. Wells and Camens 2018.
55. Arman and Prideaux 2016.
56. Murray 1991.
57. Pledge 1981; Ride et al. 1997.
58. Oskarsson et al. 2012; Letnic et al. 2012.
59. Turvey 2009b.
60. Molnar 2004; Fry et al. 2009; Wroe 2002.
61. Molnar 1991.
62. Owen 1859b. For subsequent papers, see, for example, Owen 1880.

63. Owen 1859b, 147, 148.
64. Hocknull et al. 2009, 13.
65. Fry et al. 2009.
66. Wroe 2002.
67. Price, Louys, et al. 2015, 101.
68. Price 2016, 23.
69. Molnar 1991; Willis and Molnar 1997.
70. Archer, Hand, and Godthelp 2000.
71. S. Webb 2013, 167.
72. M. J. Smith 1985.
73. Scanlon and Lee 2000.
74. Vickers-Rich 1987, 1991; Murray and Vickers-Rich 2004.
75. Stirling and Zeitz 1900.
76. Miller, Magee, Johnson, et al. 1999.
77. Miller, Magee, Smith, et al. 2016.
78. Grellet-Tinner, Spooner, and Worthy 2016.
79. Miller, Fogel, et al. 2017; Grellet-Tinner et al. 2017.
80. Grellet-Tinner 2017, 1–6.
81. See, for example, Field and Fullagar 2001; Field, Dodson, and Prosser 2002; Field, Wroe, et al. 2013; Field 2006; Gillespie and Brook 2006; Fillios, Field, and Charles 2010; Grün, Eggins, et al. 2010.
82. Wilkinson 1885.
83. Dodson and Field 2018.
84. Gillespie and Brook 2006.
85. Dodson and Field 2018.
86. Field 2006; Field, Fillios, and Wroe 2008.
87. Fillios, Field, and Charles 2010.
88. See Field, Fillios, and Wroe 2008.
89. Grün, Eggins, et al. 2010.
90. Dodson and Field 2018.
91. Roberts et al. 2001.
92. Clarkson et al. 2017; M. Westaway, Olley, and Grün 2017.
93. See, for example, Field, Fillios, and Wroe 2008, fig. 1.
94. Flannery and Roberts 1999, appendices 1 and 2; Sutton et al. 2009.
95. Flannery and Roberts 1999.
96. See, for example, Brook et al. 2013; Wroe, Field, Archer, et al. 2013a, 2013b.
97. Saltré, Brook, et al. 2015; Saltré, Rodríguez-Rey, et al. 2016.
98. Turney et al. 2008.
99. Clarkson et al. 2017.
100. Roberts et al. 2001; Hamm et al. 2016.
101. M. Westaway et al. 2017.
102. M. Westaway et al. 2017, 206.
103. Hamm et al. 2016.

CHAPTER NINE

1. Grandidier 1971; translation from Godfrey and Jungers 2003.
2. Crowley et al. 2017.
3. In Goodman and Jungers 2014.
4. Lamberton 1931.
5. Hansford and Turvey 2018.
6. Lamberton 1934a, 1934b.
7. Hume and Walters 2012, 24.
8. Flacourt 1658, in Hume and Walters 2012, 25.
9. Hume and Walters 2012.
10. Clarke et al. 2006.
11. Midgley and Illing 2009.
12. Arnold 1979.
13. Burleigh and Arnold 1986.
14. Crowley 2010.
15. Pedrono et al. 2013.
16. Brochu 2007.
17. Samonds et al. 2019.
18. Weston and Lister 2009.
19. Stuenes 1989; Rakotovao 2014.
20. Godfrey 1986; Burney and Ramilisonina 1998.
21. Simpson 1940.
22. Godfrey, Jungers, and Burney 2010.
23. Goodman and Jungers 2014, 136, 137.
24. Godfrey, Jungers, and Burney 2010.
25. Godfrey, Jungers, and Burney 2010, 356.
26. Godfrey and Jungers 2003, 255.
27. Rosenberger et al. 2015.
28. Wroe, Field, Fullagar, and Jermin 2004.
29. Meador et al. 2017.
30. Buckley 2013.
31. Burney, Burney, et al. 2004.
32. Hansford, Wright, et al. 2018.
33. Hansford, Wright, et al. 2018, 1.
34. Hansford, Wright, et al. 2018, 1.
35. Godfrey, Jungers, and Burney 2010, 363.
36. Burney, Burney, et al. 2004; Burney and Flannery 2005.
37. MacPhee and Marx 1997.
38. Martin 1984.
39. Virah-Sawmy, Willis, and Gillson 2010.
40. Crowley et al. 2017.

41. Crowley et al. 2017, 1.
42. Burns et al. 2016.
43. Burns et al. 2016, 98.
44. MacPhee and Burney 1991.
45. Perez et al. 2005.
46. Burney, Burney, et al. 2004.
47. Goodman and Jungers 2014.
48. Hansford, Wright, et al. 2018.
49. Burney and Ramilisonina 1998.
50. Burney, Robinson, and Burney 2003, 10800–805.
51. Godfrey, Jungers, and Burney 2010, 363.

CHAPTER TEN

1. For example, Allentoft et al. 2014.
2. Daugherty et al. 1990.
3. Worthy, Tennyson, et al. 2006.
4. Hand et al. 2015.
5. Wilmshurst et al. 2011.
6. Higham, Anderson, and Jacomb 1999.
7. Jacomb et al. 2014.
8. Holdaway, Allentoft, et al. 2014.
9. Quoted in Worthy and Holdaway 2002, 155.
10. Bunce, Worthy, Phillips, et al. 2009.
11. Dawson 2012.
12. Dawson 2012.
13. Dawson 2012.
14. Miskelly 2013; Bunce, Worthy, Phillips, et al. 2009; Holdaway, Allentoft, et al. 2014; Perry et al. 2014; Rawlence and Cooper 2013.
15. Worthy 2002.
16. Wood and Wilmshurst 2013.
17. Bunce, Worthy, Ford, et al. 2003; Worthy 2015.
18. Rawlence and Cooper 2013.
19. Perry et al. 2014; Holdaway, Allentoft, et al. 2014.
20. Anderson 1989.
21. Perry et al. 2014, 126.
22. Holdaway, pers. comm., 2017.
23. Holdaway, Allentoft, et al. 2014, 1.
24. Holdaway, Allentoft, et al. 2014, 1.
25. McWethy et al. 2009.
26. Furey 2006.
27. I. Smith 2013.

28. Holdaway and Jacomb 2000; Oskam et al. 2012.
29. McWethy et al. 2009.
30. Bunce, Szulkin, et al. 2005.
31. Holdaway 2015.

CHAPTER ELEVEN

1. Foster 1964.
2. Turvey 2009a.
3. Jaffe, Slater, and Alfaro 2011.
4. Steadman et al. 1991.
5. International Union for Conservation of Nature 2017.
6. A. W. White et al. 2010.
7. Hawkins et al. 2016.
8. A. W. White et al. 2010.
9. International Union for Conservation of Nature 2017.
10. Hocknull et al. 2009.
11. Brown et al. 2004.
12. Morwood et al. 2004.
13. Sutikna et al. 2016.
14. Van Den Bergh et al. 2008.
15. Benzi et al. 2007.
16. Benzi et al. 2007, 793.
17. Geer et al. 2010.
18. Köhler and Moyà-Solà 2009.
19. Ramis et al. 2002.
20. Bover, Antoni, and Alcover 2003; Bover, Quintanab, and Alcover 2008.
21. Steadman et al. 2005.
22. Burness, Diamond, and Flannery 2001.
23. MacPhee, Iturralde-Vinent, and Vázquez 2007.
24. Steadman et al. 2005.
25. MacPhee, Iturralde-Vinent, and Vázquez, 2007.
26. Biknevicius, McFarlane, and MacPhee 1993.
27. Agenbroad 1998, 2005; Agenbroad and Morris 1999.
28. Agenbroad 2005; Muhs et al. 2015.
29. Graham, Belmecheria, et al. 2016.
30. Guthrie 2006.

CHAPTER TWELVE

1. Klein 1984, 1994; Steele 2007; Faith 2014.
2. Faith 2014; Klein 1984.
3. Klein 1984.

4. Klein 1974; 1984; Steele 2007; Faith 2014.
5. Tryon et al., 2012.
6. O'Brien et al. 2016.
7. Tryon et al. 2012.
8. see Louys, Curnoe, and Tong, 2007.
9. Turvey et al. 2013.
10. Louys, Curnoe, and Tong, 2007.
11. Louys, Price, and O'Connor 2016.
12. see Stuart and Lister 2012; Turvey et al. 2013.
13. Turvey et al. 2013.
14. Bartosiewicz 2009.

CHAPTER THIRTEEN

1. Clarkson et al. 2017.
2. Lorenzen et al. 2011.
3. Campos, Kristensen, et al. 2010; Campos, Willerslev, et al. 2011.

APPENDIX

1. Walker 2005 gives detailed information on dating methods.
2. See the website for the Oxford Radiocarbon Accelerator Unit, https://c14.arch.ox.ac.uk.
3. Devièse and Higham in Kosintsev et al. 2018.
4. Lister and Stuart 2013.
5. Penkman et al. 2008.

REFERENCES

Agenbroad, L. D., 1998. "Pygmy (Dwarf) Mammoths of the Channel Islands of California." Hot Springs, SD, Mammoth Site.

———. 2005. "North American Proboscideans: Mammoths: The State of Knowledge 2003." *Quaternary International*: 126–28:73–92.

Agenbroad, L., J. Johnson, D. Morris, and T. Stafford Jr. 2005. "Mammoths and Humans as Late Pleistocene Contemporaries on Santa Rosa Island." In *Proceedings of the Sixth California Islands Symposium*, edited by D. Garcelon and C. Schwemm (Arcata, CA: National Park Service, Institute for Wildlife Studies), 3–7.

Agenbroad, L. D., and J. I. Mead, eds. 1994. The Hot Springs Mammoth Site. http://www.mammothsite.org.

Agenbroad, L. D., and D. P. Morris. 1999. "Giant Island/Pigmy Mammoths: The Late Pleistocene Prehistory of Channel Islands National Park." *National Parks Service Paleontological Research*. 4:27–31.

Agassiz, L. 1840. *Étude sur les glaciers: Neuchâtel*. En commission chez Jent et Gassmann, Libraires, à Soleure.

Alcover, J. A., A. Sans, and M. Palmer. 1998. "The Extent of Extinctions of Mammals on Islands." *Journal of Biogeography* 25:913–18.

Alexander, R. McNeill, R. A. Fariña, and S. F. Vizcaíno. 1999. "Tail Blow Energy and Carapace Fractures in a Large Glyptodont (Mammalia, Xenarthra)." *Zoological Journal of the Linnean Society*. 126:41–49.

Allen, J. R. M., T. Hickler, J. S. Singarayer, M. T. Sykes, P. J. Valdes, and B. Huntley. 2010. "Last Glacial Vegetation of Northern Eurasia." *Quaternary Science Reviews* 29:2604–18.

Allen, J. R. M., and B. Huntley. 2009. "Last Interglacial Palaeovegetation, Palaeoenvironments and Chronology: A New Record from Lago Grande di Monticchio, Southern Italy." *Quaternary Science Reviews* 28:1521–38.

Allen, J. R. M., W. A. Watts, and B. Huntley. 2000. "Weichselian Palynostratigraphy, Palaeovegetation and Palaeoenvironment: The Record from Lago Grande di Monticchio, Southern Italy." *Quaternary International* 73/74:91–110.

Allen, J. R. M., W. A. Watts, E. McGee, and B. Huntley. 2002. "Holocene Environmental Variability—the Record from Lago Grande di Monticchio, Italy." *Quaternary International* 88:69–80.

Allentoft, M. E., R. Heller, C. L. Oskam, E. D. Lorenzen, M. L. Halec, T. M. Gilbert, C. Jacomb, R. N. Holdaway, and M. Bunce. 2014. "Extinct New Zealand Megafauna Were Not in Decline before Human Colonization." *Proceedings of the National Academy of Sciences USA* 111:4922–27.

Alroy, J. 2001. "A Multispecies Overkill Simulation of the End-Pleistocene Megafaunal Mass Extinction." *Science* 292:1893–96.

Alvarez, L. W., W. Alvarez, F. Asaro, and H. V. Michel. 1980. "Extraterrestrial Cause for the Cretaceous–Tertiary Extinction." *Science* 208:1095–108.

Anderson, A. J. 1989. *Prodigious Birds: Moas and Moa-Hin Prehistoric New Zealand.* Cambridge: Cambridge University Press.

Antón, M. 2013. *Sabertooth*. Bloomington: Indiana University Press.

Arbour, V. M. 2009. "Estimating Impact Forces of Tail Club Strikes by Ankylosaurid Dinosaurs." *PLOS ONE* 4:e6738. https://doi.org/10.1371/journal.pone.0006738.

Archer, M., S. Hand, and H. Godthelp. 2000. *Australia's Lost World: Prehistoric Animals of Riversleigh*. Bloomington: Indiana University Press.

Archibald, J. D. 2012. "Dinosaur Extinction: Past and Present Perceptions." In *The Complete Dinosaur*, 2nd ed., edited by M. K. Brett-Surman, T. R. Holtz Jr., and J. O. Farlow (Bloomington: Indiana University Press), 1027–38.

———. 2014. "What the Dinosaur Record Says about Extinction Scenarios." In Keller and Kerr 2014, *Volcanism, Impacts, and Mass Extinctions: Causes and Effects: Geological Society of America Special Paper* 505, 213–24.

Archibald, J. D., and N. MacLeod. 2013. "The End-Cretaceous Extinction." *Grzimek's Animal Life Encyclopedia: Extinction*, 497–512.

Arman, S. D., and G. J. Prideaux. 2016. "Behaviour of the Pleistocene Marsupial Lion Deduced from Claw Marks in a Southwestern Australian Cave." *Scientific Reports* 6:21372. https://doi.org/10.1038/srep21372.

Arnold, E. N. 1979. "Indian Ocean Giant Tortoises: Their Systematic and Island Adaptations." *Philosophical Transactions of the Royal Society of London B* 286:127–45.

Baca, M., D. Popović, K. Stefaniak, A. Marciszak, M. Urbanowski, A. Nadachowski, and P. Mackiewicz. 2016. "Retreat and Extinction of the Late Pleistocene Cave Bear (*Ursus spelaeus sensu lato*)." *Naturwissenschaften* 103:92.

Bacon, C. D., D. Silvestro, C. Jaramillo, B. T. Smith, B. T., P. Chakrabarty, and A. Antonelli. 2015. "Biological Evidence Supports an Early and Complex Emergence of the Isthmus of Panama." *Proceedings of the National Academy of Sciences USA* 112:6110–15.

Bahn, P. 2016. *Images of the Ice Age*. Oxford: Oxford University Press.

Banks, M. R., E. A. Colhoun, and G. van de Geer. 1976. "Late Quaternary *Palorchestes azael* (Mammalia, Diprotodontidae) from Northwestern Tasmania." *Alcheringa* 1:159–66.

Bargo, M. S. 2001. "The Ground Sloth *Megatherium americanum*: Skull Shape, Bite Forces, and Diet." *Acta Palaeontologica Polonica* 46:173–92.

Bargo, M. S., G. De Iuliis, and S. F. Vizcaíno. 2006. "Hypsodonty in Pleistocene Ground Sloths." *Acta Palaeontologica Polonica* 51:53–61.

Bargo, M. S., N. Toledo, and S. F. Vizcaíno. 2006. "Muzzle of South American Pleistocene Ground Sloths (Xenarthra, Tardigrada)." *Journal of Morphology* 267:248–63.

Barnett, R., I. Barnes, M. J. Phillips, L. D. Martin, C. R. Harington, J. A. Leonard, and A. Cooper. 2005. "Evolution of the Extinct Sabretooths and the American Cheetah-Like Cat." *Current Biology* 15:589–90.

Barnett, R., and S. Sylvester. 2010. "Does the Ground Sloth, *Mylodon darwinii*, Still Survive in South America?" *Deposits Magazine* 23:8–11.

Barnett, R., N. Yamaguchi, I. Barnes, and A. Cooper. 2006. "The Origin, Current Diversity and Future Conservation of the Modern Lion (*Panthera leo*)." *Proceedings of the Royal Society B* 273:2119–25.

Barnosky, A. D. 1985. "Taphonomy and Herd Structure of the Extinct Irish Elk *Megaloceros giganteus*." *Science* 228:340–44.

———. 1986. "Big Game Extinction Caused by Late Pleistocene Climatic Change: Irish Elk (*Megaloceros giganteus*) in Ireland." *Quaternary Research* 25:128–35.

Barnosky, A. D., P. L. Koch, R. S. Feranec, S. L. Wing, and A. B. Shabel. 2004. "Assessing the Causes of Late Pleistocene Extinctions on the Continents." *Science* 306:70–75.

Barnosky, A. D., and E. L. Lindsey. 2010. "Timing of Quaternary Megafaunal Extinction in South America in Relation to Human Arrival and Climate Change." *Quaternary International* 217:10–29.

Bartosiewicz, L. 2009. "A Lion's Share of Attention: Archaeology and the Historical Record." *Acta Archaeologica Academiae Scientarium Hungaricae* 60:275–89.

Benton, M. J. 2015. *When Life Nearly Died: The Greatest Mass Extinction of All Time*. London: Thames and Hudson.

Benzi, V., L. Abbazzi, P. Bartolomei, M. Esposito, C. Fasso, O. Fonzo, R. Giampieri, F. Murgia, and J.-L. Reyss. 2007. "Radiocarbon and U-Series Dating of the Endemic Deer *Praemegaceros cazioti* (Deperet) from 'Grotta Juntu,' Sardinia." *Journal of Archaeological Science* 34:790–94.

Berger, J., J. E. Swenson, and I. L. Persson. 2001. "Recolonizing Carnivores and Naïve Prey: Conservation Lessons from Pleistocene Extinctions." *Science* 291:1036–39.

Biknevicius, A. R., D. A. McFarlane, and R. D. E. MacPhee. 1993. "Body Size in *Amblyrhiza inundata* (Rodentia: Caviomorpha), an Extinct Megafaunal Rodent from the Anguilla Bank, West Indies: Estimates and Implications." *American Museum Novitates* 3079:25 pp.

Blanco, R. E., and A. Czerwonogora. 2003. "The Gait of *Megatherium* Cuvier 1796 (Mammalia, Xenarthra, Megatheriidae)." *Senckenbergiana Biologica* 83:61–68.

Blanco, R. E., W. W. Jones, and A. Rinderknecht. 2009. "The Sweet Spot of a Biological Hammer: The Centre of Percussion of Glyptodont (Mammalia: Xenarthra) Tail Clubs." *Proceedings: Biological Sciences* 276:3971–78.

Blanco, R. E., and A. Rinderknecht. 2008. "Estimation of Hearing Capabilities of Pleistocene Ground Sloths (Mammalia, Xenarthra) from Middle Ear Anatomy." *Journal of Vertebrate Paleontology* 28:274–76.

Bocherens, H. A. Bridault, D. G. Drucker, M. Hofreiter, S. C. Münzel, M. Stiller, and J. van der Plicht. 2013. "The Last of Its Kind? Radiocarbon, Ancient DNA and Stable Isotope Evidence from a Late Cave Bear (*Ursus spelaeus* ROSENMÜLLER, 1794) from Rochedane (France)." *Quaternary International* 339–40:179–88.

Boeskorov, G, P. A. Lazarev, A. V. Sher, S. P. Davydov, N. T. Bakulina, M. V. Shchelchkova, J. Binladen, E. Willerslev, B. Buigues, and A. N. Tikhonov. 2011. "Woolly Rhino Discovery in the Lower Kolyma River." *Quaternary Science Reviews* 30:2262–72.

Bond, M., M. A. Reguero, S. F. Vizcaíno, and S. A. Marenssi. 2006. "A New 'South American Ungulate' (Mammalia: Litopterna) from the Eocene of the Antarctic Peninsula." *Geological Society, London, Special Publications* 258:163–76.

Borrero, L. A. 2008. "Extinction of Pleistocene Megamammals in South America: The Lost Evidence." *Quaternary International* 185:69–74.

———. 2009. "The Elusive Evidence: The Archeological Record of the South American Extinct Megafauna." In *American Megafaunal Extinctions at the End of the Pleistocene*, edited by G. Haynes (Dordrecht, Netherlands: Springer), 145–68.

Borrero, L. A., and F. M. Martin. 2012a. "Taphonomic Observations on Ground Sloth Bone and Dung from Cueva del Milodón, Ultima Esperanza, Chile: 100 Years of Research History." *Quaternary International* 278:3–11.

———. 2012b. "Ground Sloths and Humans in Southern Fuego-Patagonia: Taphonomy and Archaeology." *World Archaeology* 44:102–17.

Bosscha Erdbrink, D. P., J. G. Brewer, and D. Mol. 2001. "Some Remarkable Weichselian Elephant Remains." *Deinsea* 8:21–26.

Bourgeon, L., A. Burke, and T. Higham. 2017. "Earliest Human Presence in North America Dated to the Last Glacial Maximum: New Radiocarbon Dates from Bluefish Caves, Canada." *PLOS ONE* 12 (1): e0169486. https://doi.org/10.1371/journal.pone.0169486.

Bover, P., J. Antoni, and A. Alcover. 2003. "Understanding Late Quaternary Extinctions: The Case of *Myotragus balearicus* (Bate, 1909)." *Journal of Biogeography* 30:771–81.

Bover, P., J. Quintanab, and J. A. Alcover. 2008. "Three Islands, Three Worlds: Paleogeography and Evolution of the Vertebrate Fauna from the Balearic Islands." *Quaternary International* 182:135–44.

Brochu, C. A. 2007. "Morphology, Relationships, and Biogeographical Significance of an Extinct Horned Crocodile (Crocodylia, Crocodylidae) from the Quaternary of Madagascar." *Zoological Journal of the Linnean Society* 150:835–63.

Bronk Ramsey, C., T. F. G. Higham, D. C. Owen, A. W. G. Pike, and R. E. M. Hedges. 2002. "Radiocarbon Dates from the Oxford AMS System: Datelist 31." *Archaeometry* 44 (3) Supplement 1:1–149.

Brook, B. W., C. J. A. Bradshaw, A. Cooper, C. N. Johnson, T. H. Worthy, M. Bird, R. Gillespie, and R. G. Roberts. 2013. "Lack of Chronological Support for Stepwise Prehuman Extinctions of Australian Megafauna." *Proceedings of the National Academy of Sciences USA* 110:E3368.

Brown, P., M. J. Morwood, T. Sutikna, R. P. Soejono, J. E. W. Wayhu Saptomo, and R. Awe Due. 2004. "A New Small-Bodied Hominin from the Late Pleistocene of Flores, Indonesia." *Nature* 431:1055–61.

Brusatte, S. L., R. J. Butler, A. Prieto-Márquez, and M. A. Norell. 2012. "Dinosaur Morphological Diversity and the End-Cretaceous Extinction." *Nature Communications* 3:804.

Buckley, M. 2013. "A Molecular Phylogeny of *Plesiorycteropus* Reassigns the Extinct Mammalian Order 'Bibymalagasia.'" *PLOS ONE*: 8 (3): e59614. https://doi.org/10.1371/journal.pone.0059614.

Bunce, M., M. Szulkin, H. R. L. Lerner, I. Barnes, B. Shapiro, A. Cooper, and R. N. Holdaway. 2005. "Ancient DNA Provides New Insights into the Evolutionary History of New Zealand's Extinct Giant Eagle." *PLOS Biol*: 3 (1): e9.

Bunce, M., T. H. Worthy, T. Ford, W. Hoppitt, E. Willerslev, A. Drummond, and A. Cooper. 2003. "Extreme Reversed Sexual Size Dimorphism in the Extinct New Zealand Moa *Dinornis*." *Nature*: 425:172–75.

Bunce, M., T. H. Worthy, M. J. Phillips, R. N. Holdaway, E. Willerslev, J. Haile, B. Shapiro, et al. 2009. "The Evolutionary History of the Extinct Ratite Moa and New Zealand Neogene Paleogeography." *Proceedings of the National Academy of Sciences USA* 106:20646–51.

Burleigh, R., and E. N. Arnold. 1986. "Age and Dietary Differences of Recently Extinct Indian Ocean Tortoises (*Geochelone* s. lat.) Revealed by Carbon Isotope Analysis." *Proceedings of the Royal Society of London B* 227:137–44.

Burney, D. A. 1999. "Rates, Patterns, and Processes of Landscape Transformation and Extinction in Madagascar." In *Extinction in Near Time*, edited by R. D. E. MacPhee (New York: Kluwer/Plenum), 145–64.

Burney, D. A., L. P. Burney, L. R. Godfrey, W. L. Jungers, S. M. Goodman, H. T. Wright, and A. J. Timothy Jull. 2004. "A Chronology for Late Prehistoric Madagascar." *Journal of Human Evolution* 47:25–63.

Burney, D. A., and T. F. Flannery. 2005. "Fifty Millenia of Catastrophic Extinctions after Human Contact." *Trends in Ecology and Evolution* 20:395–401.

Burney, D. A., and Ramilisonina. 1998. "The Kilopilopitsofy, Kidoky, and Bokyboky: Accounts of Strange Animals from Belo-sur-mer, Madagascar, and the Megafaunal 'Extinction Window.'" *American Anthropologist* 100:957–66.

Burney, D. A., G. S. Robinson, and L. P. Burney. 2003. "*Sporormiella* and the Late Holocene Extinctions in Madagascar." *Proceedings of the National Academy of Sciences USA* 100:10800–10805.

Burns, S. J., L. R. Godfrey, P. Faina, D. McGee, B. Hardt, L. Ranivoharimanana, and J. Randrianasy. 2016. "Rapid Human-Induced Landscape Transformation in Madagascar at the End of the First Millennium of the Common Era." *Quaternary Science Reviews* 134:92–99.

Burness, G. P., J. Diamond, and T. Flannery. 2001. "Dinosaurs, Dragons, and Dwarfs: The Evolution of Maximal Body Size." *Proceedings of the National Academy of Sciences USA* 98:14518–23.

Campos, P. F., T. Kristensen, L. Orlando, A. V. Sher, M. V. Kholodova, A. Götherström, M. Hofreiter, et al. 2010. "Ancient DNA Sequences Point to a Large Loss of Mitochondrial Genetic Diversity in the Saiga Antelope (*Saiga tatarica*) since the Pleistocene." *Molecular Ecology* 19:4863–75.

Campos, P. F., E. Willerslev, A. V. Sher, Ludovic Orlando, Erik Axelsson, Alexei Tikhonov, Kim Aaris-Sørensen, et al. 2011. "Ancient DNA Analyses Exclude Humans as the Driving

Force behind Late Pleistocene Musk Ox (*Ovibos moschatus*) Population Dynamics." *Proceedings of the National Academy of Sciences USA* 107:5675–80.

Carey, S. P., A. B. Camens, M. L. Cupper, R. Grün, J. C. Hellstrom, S. W. McKnight, I. Mclennan, et al. 2011. "A Diverse Pleistocene Marsupial Trackway Assemblage from the Victorian Volcanic Plains, Australia." *Quaternary Science Reviews* 30:591–610.

Cartelle, C., G. De Iuliis, and R. L. Ferreira. 2009. "Systematic Revision of Tropical Brazilian Scelidotheriine Sloths (Xenarthra, Mylodontoidea)." *Journal of Vertebrate Paleontology* 29:555–66.

Carretero, D. D., A. García, and M. A. Dacar. 2004. "First Data on Differential Use of the Environment by Pleistocene Megafauna Species (San Juan, Argentina)." *Current Research in the Pleistocene* 21:91–92.

Chauvet, J.-M., E. Brunel Deschamps, and C. Hillaire. 2001. *Chauvet Cave: The Discovery of the World's Oldest Paintings*. London: Thames and Hudson.

Chernova, O. F., I. V. Kirillova, B. Shapiro, F. K. Shidlovskiy, A. E. R. Soares, V. A. Levchenko, and F. Bertuch. 2016. "Morphological and Genetic Identification and Isotopic Study of the Hair of a Cave Lion (*Panthera spelaea Goldfuss*, 1810) from the Malyi Anyui River (Chukotka, Russia)." *Quaternary Science Reviews* 142:61–73.

Cione, A. L., G. M. Gasparini, E. Soibelzon, L. H. Soibelzon, and E. P. Tonni. 2015. "The Great American Biotic Interchange: A South American Perspective." In *Springer Briefs in Earth System Sciences*. 97 pp.

Clarke, S. J., G. H. Miller, M. L. Fogel, A. R. Chivas, and C. V. Murray-Wallace. 2006. "The Amino Acid and Stable Isotope Biogeochemistry of Elephant Bird (*Aepyornis*) Eggshells from Southern Madagascar." *Quaternary Science Reviews* 25:2343–56.

Clarkson, C., Z. Jacobs, B. Marwick, R. Fullagar, L. Wallis, M. Smith, R. G. Roberts, et al. 2017. "Human Occupation of Northern Australia by 65,000 Years Ago." *Nature*: 547:306–10.

Cohen, T., J. D. Jansen, L. A. Gliganic, J. R. Larsen, G. C. Nanson, J.-H. May, B. G. Jones, and D. M. Price. 2015. "Hydrological Transformation Coincided with Megafaunal Extinction in Central Australia." *Geology* 43:195–98.

Coltrain, J. B., J. M. Harris, T. E. Cerling, J. R. Ehleringer, M. D. Dearing, J. Ward, and J. Allen. 2004. "Rancho La Brea Stable Isotope Biogeochemistry and Its Implications for the Palaeoecology of Late Pleistocene, Coastal Southern California." *Palaeogeography, Palaeoclimatology, Palaeoecology* 205:199–219.

Conniff, R. 2010. "Mammoths and Mastodons: All American Monsters." *Smithsonian Magazine*, April 2010, http://www.smithsonianmag.com/science-nature/mammoths-and-mastodons-all-american-monsters-8898672/#FKxoQO1WKmP9JBmU.99.

Cook, J. 2013. *Ice Age Art: Arrival of the Modern Mind*. London: British Museum Press.

Cooper, A., C. Turney, K. A. Hughen, B. W. Brook, H. G. McDonald, and C. J. A. Bradshaw. 2015. "Abrupt Warming Events Drove Late Pleistocene Holarctic Megafaunal Turnover." *Science* 349:602–6.

Cozzuol, M. A., C. L. Clozato, E. C. Holanda, F. H. G. Rodrigues, S. Nienow, B. De Thoisy, A. F. Rodrigo, and F. R. Redondo. 2013. "A New Species of Tapir from the Amazon." *Journal of Mammalogy* 94:1331–45.

Crees, J. J., C. Carbone, C., R. S. Sommer, N. Benecke, and S. T. Turvey. 2016. "Millennial-Scale Faunal Record Reveals Differential Resilience of European Large Mammals to Human Impacts across the Holocene." *Proceedings of the Royal Society B: Biological Sciences* 283:2015–152.

Crees, J. J., and S. T. Turvey. 2014. "Holocene Extinction Dynamics of *Equus hydruntinus*, a Late-Surviving European Megafaunal Mammal." *Quaternary Science Reviews* 91:16–29.

Crowley, B. E. 2010. "A Refined Chronology of Prehistoric Madagascar and the Demise of the Megafauna." *Quaternary Science Reviews* 29:2591–603.

Crowley, B. E., L. R. Godfrey, R. J. Bankoff, G. H. Perry, B. J. Culleton, D. J. Kennett, M. R. Sutherland, K. A. Samonds, and D. A. Burney. 2017. "Island-Wide Aridity Did Not Trigger Recent Megafaunal Extinctions in Madagascar." *Ecography* 39:001–012. https://doi.org/10.1111/ecog.02376.

Cuvier, G. 1812. *Recherches sur les Ossemens Fossiles de Quadrupèdes, où l'on Rétablit les Caractères de Plusieurs Espèces d'Animaux que les Révolutions du Globe Paroissent Avoir Détruites*. 4 vols.

Czaplewski, N. J., and C. Cartelle. 1998. "Pleistocene Bats from Cave Deposits in Bahia, Brazil." *Journal of Mammalogy* 79:784–803.

Darwin, C. R. 1845. *Journal of Researches into the Natural History and Geology of the Countries Visited during the Voyage of* H.M.S. Beagle *Round the World, under the Command of Capt. Fitz Roy, R.N.* 2nd ed. London: John Murray.

———. 1859. *On the Origin of Species by Means of Natural Selection, or the Preservation of Favoured Races in the Struggle for Life*. London: John Murray.

Daugherty, C. H., A. Cree, J. M. Hay, and M. B. Thompson. 1990. "Neglected Taxonomy and Continuing Extinctions of Tuatara (*Sphenodon*)." *Nature* 347:177–79.

Davis, O. K., and D. S. Shafer. 2005. "*Sporormiella* Fungal Spores, a Palynological Means of Detecting Herbivore Density." *Palaeogeography, Palaeoclimatology, Palaeoecology* 237:40–50. https://doi.org/10.1016/j.palaeo.2005.11.028.

Dawson, G. 2012. "On Richard Owen's Discovery, in 1839, of the Extinct New Zealand Moa from Just a Single Bone." *BRANCH: Britain, Representation and Nineteenth-Century History*. Edited by Dino Franco Felluga. http://www.branchcollective.org/?ps_articles=gowan-dawson-on-richard-owens-discovery-in-1839-of-the-extinct-new-zealand-moa-from-just-a-single-bone.

De Beaulieu, J.-L., and M. Reille. 1992. "The Last Climatic Cycle at La Grande Pile (Vosges, France): A New Pollen Profile." *Quaternary Science Reviews* 11:431–38.

Deng, T., X. Wang, M. Fortelius, Q. Li, Y. Wang, Z. J. Tseng, G. T. Takeuchi, J. E. Saylor, L. K. Säilä, and G. Xie. 2011. "Out of Tibet: Pliocene Woolly Rhino Suggests High Plateau Origin of Ice Age Megaherbivores." *Science* 333:1285–88.

Devièse, T., T. W. Stafford Jr., M. R. Waters, C. Wathen, D. Comeskey, L. Becerra-Valdivia, and T. Higham. 2018. "Increasing Accuracy for the Radiocarbon Dating of Sites Occupied by the First Americans." *Quaternary Science Reviews* 198:171–80.

Dikov, N. N. 1988. "The Earliest Sea Mammal Hunters of Wrangell Island." *Arctic Anthropology* 25:80–93.

Dillehay, T. D. 1997. *Monte Verde: A Late Pleistocene Settlement in Chile: Volume 2. The Archaeological Context and Interpretation*. Washington, DC: Smithsonian Press.

Dillehay, T. D., C. Ocampo, J. Saavedra, A. O. Sawakuchi, R. M. Vega, M. Pino, M. B. Collins, et al. 2015. "New Archaeological Evidence for an Early Human Presence at Monte Verde, Chile." *PLOS ONE* 10:e0141923. https://doi.org/10.1371/journal.pone.0141923.

Dodson, J., and J. Field. 2018. "What Does the Occurrence of *Sporormiella* (*Preussia*) Spores Mean in Australian Fossil Sequences?" *Journal of Quaternary Science* 33:380–92. https://doi.org/10.1002/jqs.3020.

Donlan, J., H. Greene, J. Berger, C. E. Bock, J. H. Bock, D. A. Burney, J. A. Estes, et al. 2005. "Re-wilding North America." *Nature* 436:913–14.

Dowd, M., and R. Carden. 2016. "First Evidence of a Late Upper Palaeolithic Human Presence in Ireland." *Quaternary Science Reviews* 139:158–63.

Dulvy, N. K., J. K. Pinnegar, and J. D. Reynolds. 2009. "Holocene Extinctions in the Sea." *Holocene Extinctions*., edited by S. T. Turvey (Oxford: Oxford University Press), 129–50.

Ehlers, J., and P. L. Gibbard, eds. 2004a. *Quaternary Glaciations: Extent and Chronology: Part 1: Europe*. Developments in Quaternary Science. Amsterdam: Elsevier.

———, eds. 2004b. *Quaternary Glaciations: Extent and Chronology: Part 2: North America*. Developments in Quaternary Science. Amsterdam: Elsevier.

———, eds. 2004c. *Quaternary Glaciations: Extent and Chronology: Part 3: South America, Asia, Australasia, Antarctica*. Developments in Quaternary Science. Elsevier: Amsterdam.

Elias, S. A., and B. Crocker. 2008. "The Bering Land Bridge: A Moisture Barrier to the Dispersal of Steppe–Tundra Biota?" *Quaternary Science Reviews* 27:2473–83.

EPICA Community Members. 2004. "Eight Glacial Cycles from an Antarctic Ice Core." *Nature* 429:623–28.

Ersmark, E., L. Orlando, E. Sandoval-Castallanos, I. Barnes, R. Barnett, A. Stuart, A. Lister, and L. Dalén. 2015. "Population Demography and Genetic Diversity in the Pleistocene Cave Lion." *Open Quaternary* 1 (1), part 4. https://doi.org/10.5334/oq.aa.

Erwin, D. H. 2006. *Extinction: How Life on Earth Nearly Ended 250 Million Years Ago*. Princeton, NJ: Princeton University Press.

Faith, J. T. 2014. "Late Pleistocene and Holocene Mammal Extinctions on Continental Africa." *Earth-Science Reviews* 128:105–21.

Faith, J. T., and T. Surovell. 2009. "Synchronous Extinction of North America's Pleistocene Mammals." *Proceedings of the National Academy of Sciences, USA* 106:20641–45.

Fariña, R. A. 1995. "Limb Bone Strength and Habits in Large Glyptodonts." *Lethaia* 28:189–96.

Fariña, R. A., and R. E. Blanco. 1996. "*Megatherium* the Stabber." *Proceedings of the Royal Society B*. 263:1725–29.

Fariña, R. A., R. E. Blanco, and P. Christiansen. 2005. "Swerving as the Escape Strategy of *Macrauchenia patachonica* (Mammalia; Litopterna)." *Ameghiniana* 42:751–60.

Fariña, R. A., S. F. Vizcaíno, and G. De Iuliis. 2013. *Megafauna: Giant Beasts of Pleistocene South America*. Bloomington: Indiana University Press.

Feranec, R. S., and A. L. Kozlowski. 2010. "AMS Radiocarbon Dates from Pleistocene and Holocene Mammals Housed in the New York State Museum, Albany, New York, USA." *Radiocarbon* 52:205–8.

Fernández, J., V. Markgraf, H. O. Panarello, M. Albero, F. E. Angiolini, S. Valencio, and M. Arriaga. 1991. "Late Pleistocene/Early Holocene Environments and Climates, Fauna, and Human Occupation in the Argentine Altiplano." *Geoarchaeology* 6:251–72.

Fernicola, J. C., S. F. Vizcaíno, and G. De Iuliis. 2009. "The Fossil Mammals Collected by Charles Darwin in South America during His Travels on Board the *HMS Beagle*." *Revista de la Asociación Geológica Argentina* 64:147–59.

Ferretti, M. P. 2010. "Anatomy of *Haplomastodon chimborazi* (Mammalia, Proboscidea) from the Late Pleistocene of Ecuador and Its Bearing on the Phylogeny and Systematics of South American Gomphotheres." *Geodiversitas* 32:663–721.

Fiedel, S. J. 1999. "Artifact Provenience at Monte Verde: Confusion and Contradictions." *Discovering Archaeology*. Special report. 1, 12.

Fiedel, S. J., and G. Haynes. 2004. "A Premature Burial: Comments on Grayson and Meltzer's 'Requiem for Overkill.'" *Journal of Archaeological Science* 31:121–31.

Field, J. 2006. "Trampling through the Pleistocene: Does Taphonomy Matter at Cuddie Springs?" *Australian Archaeology* 63:9–20.

Field, J., J. Dodson, and I. Prosser. 2002. "A Late Pleistocene Vegetation History from the Australian Semi-arid Zone." *Quaternary Science Reviews* 21:1005–19.

Field, J., M. Fillios, and S. Wroe. 2008. "Chronological Overlap between Humans and Megafauna in Sahul (Pleistocene Australia-New Guinea): A Review of the Evidence." *Earth-Science Reviews* 89:97–115.

Field, J., and R. Fullagar. 2001. "Archaeology and Australian Megafauna." *Science* 294:7.

Field, J., S. Wroe, C. N. Trueman, J. Garvey, and S. Wyatt-Spratt. 2013. "Looking for the Archaeological Signature in Australian Megafaunal Extinctions." *Quaternary International* 285:76–88. https://doi.org/10.1016/j.quaint.2011.04.013.

Figueirido, B., J. A. Pérez-Claros, V. Torregrosa, A. Martín-Serra, and P. Palmqvist. 2010. "Demythologizing *Arctodus simus*, the 'Short-Faced' Long-Legged and Predaceous Bear That Never Was." *Journal of Vertebrate Paleontology* 30:262–75.

Fillios, M., J. Field, and B. Charles. 2010. "Investigating Human and Megafauna Co-occurrence in Australian Prehistory: Mode and Causality in Fossil Accumulations at Cuddie Springs." *Quaternary International* 211:123–43.

Firestone, R. B., A. West, J. P. Kennett, L. Becker, T. E. Bunch, Z. S. Revay, P. H. Schultz, et al. 2007. "Evidence for an Extraterrestrial Impact 12,900 Years Ago That Contributed to the Megafaunal Extinctions and the Younger Dryas Cooling." *Proceedings of the National Academy of Sciences USA* 104:16016–21.

Fischer, G. 1809. "Sur l'Elasmotherium et le Trogonthérium." In *Memoires de la Société Impériale des Naturalistes de Moscou*, book 2 (Moscow: Imprimerie de l'Université Impériale), 255.

Flacourt, E. de. 1658. *Histoire de la Grande Isle Madagascar*. Paris: Chez G. de Lynes.

Flannery, T. F., and R. G. Roberts. 1999. "Late Quaternary Extinctions in Australasia: An Overview." In *Extinctions in Near Time: Causes, Contexts, and Consequences*, edited by R. MacPhee (New York: Kluwer Academic/Plenum), 239–55.

Fortes, G. G., A. Grandal-d'Anglade, B. Kolbe, D. Fernandes, I. N. Meleg, A. García-Vázquez, A. C. Pinto-Llona, S. Constantin, T. J. de Torres, J. E. Ortiz, C. Frischauf, G. Rabeder,

M. Hofreiter, and A. Barlow. 2016. "Ancient DNA Reveals Differences in Behaviour and Sociality between Brown Bears and Extinct Cave Bears." *Molecular Ecology* 25:4907–18.

Foster, J. B. 1964. "The Evolution of Mammals on Islands." *Nature* 202:234–35.

Fox-Dobbs, K., J. A. Leonard, and P. L. Koch. 2008. "Pleistocene Megafauna from Eastern Beringia: Paleoecological and Paleoenvironmental Interpretations of Stable Carbon and Nitrogen Isotope and Radiocarbon Records." *Palaeogeography, Palaeoclimatology, Palaeoecology* 261:30–46.

Fraser, R., and R. T. Wells. 2006. "Palaeontological Excavation and Taphonomic Investigation of the Late Pleistocene Fossil Deposit in Grant Hall, Victoria Fossil Cave, Naracoorte, South Australia." *Alcheringa* 30:147–61.

Frison, G. C. 2000. "A ^{14}C date on a Late-Pleistocene Camelops at the Casper-Hell Gap Site, Wyoming." *Current Research in the Pleistocene* 17:28–29.

Froese, D., M. Stiller, P. D. Heintzman, A. V. Reyes, G. D. Zazula, A. E. Soares, M. Meyer, et al. 2017. "Fossil and Genomic Evidence Constrains the Timing of Bison Arrival in North America." *Proceedings of the National Academy of Sciences USA* 114:3457–62.

Fry, B. G., S. Wroe, W. Teeuwisse, M. J. P. van Osch, K. Moreno, J. Ingle, C. McHenry, et al. 2009. "A Central Role for Venom in Predation by *Varanus komodoensis* (Komodo Dragon) and the Extinct Giant *Varanus* (*Megalania*) *priscus*." *Proceedings of the National Academy of Sciences USA* 106:8969–74.

Fuller, B. T., S. H. Fahrni, J. M. Harris, A. B. Farell, J. B. Coltrain, L. M. Gerhart, J. K. Ward, R. E. Taylor, and J. R. Southon. 2014. "Ultrafiltration for Asphalt Removal from Bone Collagen for Radiocarbon Dating and Isotopic Analysis of Pleistocene Fauna at the Tar Pits of Rancho La Brea, Los Angeles, California." *Quaternary Geochronology* 22:85–98.

Furey, L. 2006. *Maori Gardening: An Archaeological Perspective*. Wellington, NZ: Department of Conservation.

Galetti, M., M. Moleón, P. Jordano, M. M. Pires, P. R. Guimarães Jr, T. Pape, E. Nichols, et al. 2018. "Ecological and Evolutionary Legacy of Megafauna Extinctions." *Biological Reviews* 93:845–62.

Gazin, C. L. 1956. "Exploration for the Remains of Giant Ground Sloths in Panama." *Smithsonian Institution Annual Report*, 341–54.

Geer, A. van der, G. Lyras, J. De Vos, and M. Dermitzakis. 2010. *Evolution of Island Mammals: Adaptation and Extinction of Placental Mammals on Islands*. Wiley-Blackwell.

Geist, V. 1998. *Deer of the World: Their Evolution, Behaviour and Ecology*. Mechanicsburg, PA: Stackpole Books.

Gerasimov, D. V., E. Yu Giria, V. V. Pitul'ko, and A. N. Tikhonov. 2006. "New Materials for the Interpretation of the Chertov Ovrag Site on Wrangel Island." Chapter 10 of *Archaeology in Northeast Asia: On the Pathway to Bering Strait*, edited by Don E. Dumond and Richard L. Bland (N.p.: Beringia Program).

Gibbard, P. L., and J. Lewin. 2016. "Partitioning the Quaternary." *Quaternary Science Reviews* 151:127–39.

Gibbard, P. L., and T. van Kolfschoten. 2005. "The Quaternary System (The Pleistocene and Holocene Series)." In *A Geologic Time Scale*, edited by F. Gradstein, J. Ogg, and A. Smith (Cambridge: Cambridge University Press), 441–52.

Gilbert, M. T. P., D. L. Jenkins, A. Götherström, N. Naveran, J. J. Sanchez, M. Hofreiter, P. F. Thomsen, et al. 2008. "DNA from Pre-Clovis Human Coprolites in Oregon, North America." *Science* 320:786–89.

Gill, J. L. 2013. "Ecological Impacts of the Late Quaternary Megaherbivore Extinctions." *New Phytologist* 201:1163–69.

Gill, J. L., J. W. Williams, S. T. Jackson, K. G. Lininger, and G. S. Robinson. 2009. "Pleistocene Megafaunal Collapse, Novel Plant Communities, and Enhanced Fire Regimes in North America." *Science* 326:1100–103.

Gillespie, R., and B. W. Brook. 2006. "Is There a Pleistocene Archaeological Site at Cuddie Springs?" *Archaeology in Oceania* 41:1–11.

Gillespie, R., A. B. Camens, T. H. Worthy, N. J. Rawlence, C. Reid, F. Bertuch, V. Levchenko, and A. Cooper. 2012. "Man and Megafauna in Tasmania: Closing the Gap." *Quaternary Science Reviews* 37:38–47.

Gillespie, R., R. Wood, S. Fallon, T. W. Stafford Jr., and J. Southon. 2014. "New ^{14}C Dates for Spring Creek and Mowbray Swamp Megafauna: XAD-2 Processing." *Archaeology in Oceania* 50:43–48. https://doi.org/10.1002/arco.5045.

Gillette, D., and C. Ray. 1981. "Glyptodonts of North America." *Smithsonian Contributions to Paleobiology* 40:1–255.

Gilmour, D. M., V. L. Butler, J. E. O'Connor, E. B. Davis, B. J. Culleton, D. J. Kennett, and G. Hodgins. 2015. "Chronology and Ecology of Late Pleistocene Megafauna in the Northern Willamette Valley, Oregon." *Quaternary Research* 83:127–36.

Godfrey, L. 1986. "The Tale of the Tsy-aomby-aomby." *Sciences* 1986:49–51.

Godfrey, L. R., and W. L. Jungers. 2003. "The Extinct Sloth Lemurs of Madagascar." *Evolutionary Anthropology* 12:252–63.

Godfrey, L., W. L. Jungers, and D. A. Burney. 2010. "Subfossil Lemurs of Madagascar." In *Cenozoic Mammals of Africa*, edited by L. Werdelin and W. Sanders (Berkeley: University of California Press), 351–67.

Goebel, T., M. R. Waters, and D. H. O'Rourke. 2008. "The Late Pleistocene Dispersal of Modern Humans in the Americas." *Science* 319:1497–502.

Goodman, S. M., and W. L. Jungers. 2014. *Extinct Madagascar: Picturing the Island's Past.* Chicago: University of Chicago Press.

Gould, S. J. 1974. "The Origin and Function of 'Bizarre' Structures: Antler Size and Skull Size in the 'Irish Elk' *Megaloceros giganteus*." *Evolution* 28:191–220.

Graham, R. W., S. Belmecheria, S. K. Choy, B. J. Culleton, Lauren J. Davies, Duane Froese, Peter D. Heintzman, et al. 2016. "Timing and Causes of Mid-Holocene Mammoth Extinction on St. Paul Island, Alaska." *Proceedings of the National Academy of Sciences USA* 113:9310–14.

Graham, R.W., C. V. Haynes, D. L. Johnson, and M. Kay. 1981. "A Clovis-Mastodon Association in Eastern Missouri." *Science* 213:1115–17.

Graham, R. W., and E. L. Lundelius. 1984. "Coevolutionary Disequilibrium and Pleistocene Extinctions." In *Quaternary Extinctions: A Prehistoric Revolution*, edited by P. S. Martin and R. G. Klein (Tucson: University of Arizona Press), 223–49.

Grandidier, A. 1971. *Souvenirs de Voyages d'Alfred Grandidier, 1865–1870.* Tanarive, Madagascar: Association Malgache d'Archéologie.

Grange, T., J.-P. Brugal, L. Flori, M. Gautier, A. Uzunidis, and E.-M. Geig. 2018. "The Evolution and Population Diversity of Bison in Pleistocene and Holocene Eurasia: Sex Matters." *Diversity* 10:65. https://doi.org/10.3390/d10030065.

Grayson, D. K. 1984a. "Explaining Pleistocene Extinctions: Thoughts on the Structure of Debate." In *Quaternary Extinctions: A Prehistoric Revolution*, edited by P. S. Martin and R. G. Klein (Tucson: University of Arizona Press), 807–23.

———. 1984b. "Nineteenth-Century Explanations of Pleistocene Extinctions: A Review and Analysis." In *Quaternary Extinctions: A Prehistoric Revolution*, edited by P. S. Martin and R. G. Klein (Tucson: University of Arizona Press), 5–39.

———. 2007. "Deciphering North American Pleistocene Extinctions." *Journal of Anthropological Research* 63:185–213.

———. 2016. *Giant Sloths and Sabertooth Cats: Extinct Mammals and the Archaeology of the Ice Age Great Basin*. Salt Lake City: University of Utah Press.

Grayson, D. K., and D. J. Meltzer. 2002. "Clovis Hunting and Large Mammal Extinction: A Critical Review of the Evidence." *Journal of World Prehistory* 16 (4): 313–59.

———. 2003. "A Requiem for North American Overkill." *Journal of Archaeological Science* 30:585–93.

———. 2004. "North American Overkill Continued?" *Journal of Archaeological Science* 31:133–36.

Green, J. L., and D. C. Kalthoff. 2015. "Xenarthran Dental Microstructure and Dental Microwear Analyses, with New Data for *Megatherium americanum* (Megatheriidae)." *Journal of Mammalogy* 96:645–57. https://doi.org/10.1093/jmammal/gyv045.

Grellet-Tinner, G., N. A. Spooner, W. D. Handley, and T. H. Worthy. 2017. "The Genyornis Egg: Response to Miller et al.'s Commentary on Grellet-Tinner et al., 2016." *Quaternary Science Reviews* 161:128–33.

Grellet-Tinner, G., N. A. Spooner, and T. H. Worthy. 2016. "Is the 'Genyornis' Egg of Amihirung or Another Extinct Bird from the Australian Dreamtime?" *Quaternary Science Reviews* 133:147–64.

Gröcke, D. R., 1997. "Distribution of C3 and C4 Plants in the Late Pleistocene of South Australia Recorded by Isotope Biogeochemistry of Collagen in Megafauna." *Australian Journal of Botany* 45:607–17.

Grün, R., S. Eggins, M. Aubert, N. Spooner, A. W. G. Pike, and W. Müller. 2010. "ESR and U-Series Analyses of Faunal Material from Cuddie Springs, NSW, Australia: Implications for the Timing of the Extinction of the Australian Megafauna." *Quaternary Science Reviews* 29:596–610.

Grün, R., K. C. Moriarty, and R. T. Wells. 2001. "Electron Spin Resonance Dating of the Fossil Deposits in the Naracoorte Caves, South Australia." *Journal of Quaternary Science* 16:49–59.

Guthrie, R. D. 1984. "Mosaics, Allelochemics, and Nutrients: An Ecological Theory of Late Pleistocene Megafaunal Extinctions." In *Quaternary Extinctions*, edited by P. S. Martin and R. G. Klein (Tucson: University of Arizona Press), 259–98.

———. 1990. *Frozen Fauna of the Mammoth Steppe: The Story of Blue Babe*. Chicago: University of Chicago Press.

———. 1992. "New Paleoecological and Paleoethological Information on the Extinct Helmeted Muskox from Alaska." *Annales Zoologici Fennici* 28:175–86.

———. 2001. "Origin and Causes of the Mammoth Steppe: A Story of Cloud Cover, Woolly Mammoth Tooth Pits, Buckles, and Inside-Out Beringia." *Quaternary Science Reviews* 20:549–74.

———. 2003. "Rapid Body Size Decline in Alaskan Pleistocene Horses before Extinction." *Nature* 426:169–71.

———. 2005. *The Nature of Paleolithic Art*. Chicago: University of Chicago Press.

———. 2006. "New Carbon Dates Link Climatic Change with Human Colonization and Pleistocene Extinctions." *Nature* 441:207–9.

Hallam, A. 2005. *Catastrophes and Lesser Calamities: The Causes of Mass Extinctions*. Oxford: Oxford University Press.

Hallam, A., and P. B. Wignall. 1997. *Mass Extinctions and Their Aftermath*. Oxford: Oxford University Press.

Hallam, J. S., B. J. N. Edwards, B. Barnes, and A. J. Stuart. 1973. "The Remains of a Late Glacial Elk with Associated Barbed Points from High Furlong, Near Blackpool, Lancashire." *Proceedings of the Prehistoric Society* 39:100–128.

Hamm, G., P. Mitchell, L. J. Arnold, G. J. Prideaux, D. Questiaux, N. A. Spooner, V. A. Levchenko, et al. 2016. "Cultural Innovation and Megafauna Interaction in the Early Settlement of Arid Australia." *Nature* 539:280–83.

Hand, S. J., D. E. Lee, T. H. Worthy, M. Archer, J. P. Worthy, A. J. D. Tennyson, S. W. Salisbury, et al. 2015. "Miocene Fossils Reveal Ancient Roots for New Zealand's Endemic *Mystacina* (Chiroptera) and Its Rainforest Habitat." *PLOS ONE* 10 (6): e0128871. https://doi.org/10.1371/journal.pone.0128871.

Hansen, R. M. 1978. "Shasta Ground Sloth Food Habits, Rampart Cave." *Arizona Paleobiology* 4:302–19.

Hansford, J. P., and S. T. Turvey. 2018. "Unexpected Diversity within the Extinct Elephant Birds (Aves: Aepyornithidae) and a New Identity for the World's Largest Bird." *Royal Society Open Science* 5:181295. http://doi.org/10.1098/rsos.181295.

Hansford, J., P. C. Wright, A. Rasoamiaramanana, V. R. Pérez, L. R. Godfrey, D. Errickson, T. Thompson, and S. T. Turvey. 2018. "Early Holocene Human Presence in Madagascar Evidenced by Exploitation of Avian Megafauna." *Science Advances* 4:eaat6925. http://advances.sciencemag.org/content/4/9/eaat6925.

Harris, J. M., ed. 2001. "Rancho La Brea: Death Trap and Treasure Trove." *TERRA* 38, no. 2 (April/May/June 2001): 62 pp.

———, ed. 2015. "La Brea and Beyond, the Paleontology of Asphalt-Preserved Biotas." Science Series No. 42. Natural History Museum of Los Angeles County, Los Angeles, CA. 174 pp.

Harris, J. M., and G. T. Jefferson. 1985. "Rancho La Brea: Treasures of the Tar Pits." Science Series No. 31. Natural History Museum of Los Angeles County, Los Angeles, CA.

Haury, E. W. 1953. "Artifacts with Mammoth Remains, Naco, Arizona." *American Antiquity* 19:1–14.

Hawkins, S., T. H. Worthy, S. Bedford, M. Spriggs, G. Clark, G. Irwin, S. Best, and P. Kirch. 2016. "Ancient Tortoise Hunting in the Southwest Pacific." *Scientific Reports* 6:38317.

Haynes, G. 2018. "North American Megafauna Extinction: Climate or Overhunting?" In *Encyclopedia of Global Archaeology*, living edition, edited by C. Smith (N.p.: Springer International Publishing). https://doi.org/10.1007/978-3-319-51726-1_1853-2.

Hays, J. D., J. Imbrie, and N. J. Shackleton. 1976. "Variations in the Earth's Orbit: Pacemaker of the Ice Ages." *Science* 194 (4270): 1121–32.

Hedeen, S. 2008. *Big Bone Lick: The Cradle of American Paleontology*. Lexington: University Press of Kentucky.

Heintzman, P. D., G. D. Zazula, J. A. Cahill, A. V. Reyes, R. D. MacPhee, and B. Shapiro. 2015. "Genomic Data from Extinct North American Camelops Revise Camel Evolutionary History." *Molecular Biology and Evolution* 32:2433–40.

Heintzman, P. D., G. D. Zazula, and R. D. E. MacPhee. 2017. "A New Genus of Horse from Pleistocene North America." eLife 6:e29944.

Hershkovitz, I., G. W. Weber, R. Quam, M. Duval, R. Grün, L. Kinsley, A. Ayalon, et al. 2018. "The Earliest Modern Humans outside Africa." *Science* 359:456–59.

Higham, T., A. Anderson, and C. Jacomb. 1999. "Dating the First New Zealanders: The Chronology of Wairau Bar." *Antiquity* 73:420–27.

Higham, T., K. Douka, R. Wood, C. B. Ramsey, F. Brock, L. Basell, M. Camps, A. Arrizabalaga, et al. 2014. "The Timing and Spatiotemporal Patterning of Neanderthal Disappearance." *Nature* 512:306–9.

Hockett, B., and E. Dillingham. 2004. *Paleontological Investigations at Mineral Hill Cave. Contributions to the Study of Cultural Resources Technical Report 18*. Reno, NV: US Department of the Interior, Bureau of Land Management.

Hocknull, S. A., P. J. Piper, G. D. van den Bergh, R. A. Due, M. J. Morwood, and I. Kurniawan. 2009. "Dragon's Paradise Lost: Palaeobiogeography, Evolution and Extinction of the Largest-Ever Terrestrial Lizards (Varanidae)." *PLOS ONE* 4 (9): e7241. https://doi.org/10.1371/journal.pone.0007241.

Hofreiter, M., H. N. Poinar, W. G. Spaulding, K. Bauer, P. S. Martin, G. Possnert, and S. Pääbo. 2000. "A Molecular Analysis of Ground Sloth Diet through the Last Glaciation." *Molecular Ecology* 9:1975–84.

Holdaway, R. N. 2015. *Pyramid Valley and Beyond; Discovering the Prehistoric Birdlife of North Canterbury, New Zealand*. Christchurch: Turnagra Press.

Holdaway, R. N., M. E. Allentoft, C. Jacomb, C. L. Oskam, N. R. Beavan, and M. Bunce. 2014. "An Extremely Low-Density Human Population Exterminated New Zealand Moa." *Nature Communications* 5, article 5436. https://doi.org/10.1038/ncomms6436.

Holdaway, R. N., and C. Jacomb. 2000. "Rapid Extinction of the Moas (Aves: Dinornithiformes): Model, Test, and Implications." *Science* 287:2250–54.

Holen, S. R., T. A. Deméré, D. C. Fisher, R. Fullagar, J. B. Paces, G. T. Jefferson, J. M. Beeton, R. A. Cerutti, A. N. Rountrey, L. Vescera, and K. A. Holen. 2017. "A 130,000-Year-Old Archaeological Site in Southern California, USA." *Nature* 544:479–83.

Holliday, V. T., T. Surovell, D. J. Meltzer, D. K. Grayson, and M. Boslough. 2014. "The Younger Dryas Impact Hypothesis: A Cosmic Catastrophe." *Journal of Quaternary Science* 29:515–30.

Hoorn, C., and S. Flantua. 2015. "An Early Start for the Panama Land Bridge." *Science* 348:186–87.
Hubbe, A., M. Hubbe, and W. A. Neves. 2009. "New Late-Pleistocene Dates for the Extinct Megafauna of Lagoa Santa, Brazil." *Current Research in the Pleistocene* 26:154–56.
———. 2013. "The Brazilian Megamastofauna of the Pleistocene/Holocene Transition and Its Relationship with the Early Human Settlement of the Continent." *Earth Science Reviews* 118:1–10.
Hublin, J.-J., A. Ben-Ncer, S. E. Bailey, S. E. Freidline, S. Neubauer, M. M. Skinner, I. Bergmann, et al. 2017. "New Fossils from Jebel Irhoud, Morocco and the Pan-African Origin of *Homo sapiens*." *Nature* 546:289–92.
Hughes, S., T. J. Hayden, C. J. Douady, C. Tougard, M. Germonpré, A. Stuart, L. Lbova, et al. 2006. "Molecular Phylogeny of the Extinct Giant Deer, *Megaloceros giganteus*." *Molecular Phylogenetics and Evolution* 40:285–91.
Hulton, N. R. J., R. S. Purves, R. D. McCulloch, D. E. Sugden, and M. J. Bentley. 2002. "The Last Glacial Maximum and Glaciation in Southern South America." *Quaternary Science Reviews* 21:233–41.
Hume, J. P., and M. Walters. 2012. *Extinct Birds*. London: T. & A. D. Poyser.
Huntley, B., J. R. M. Allen, Y. C. Collingham, T. Hickler, A. M. Lister, J. Singarayer, A. J. Stuart, M. T. Sykes, and P. J. Valdes. 2013. "Millennial Climatic Fluctuations Are Key to the Structure of Last Glacial Ecosystems." *PLOS ONE* 8 (4): e61963. http://doi.org/10.1371/journal.pone.0061963.
Huxley, T. H. 1865. "On the Osteology of the Genus *Glyptodon*." *Philosophical Transactions of the Royal Society of London* 155:31–70.
International Union for Conservation of Nature. 2017. *IUCN Red List of Threatened Species*. https://www.iucnredlist.org.
Iwase, A., J. Hashizume, M. Izuho, K. Takahashi, and H. Sato. 2012. "Timing of Megafaunal Extinction in the Late Late Pleistocene on the Japanese Archipelago." *Quaternary International* 255:114–24.
Jacobi, R., and T. Higham. 2011. "The Later Upper Palaeolithic Recolonisation of Britain: New Results from AMS Radiocarbon Dating." Chapter 12 of *The Ancient Human Occupation of Britain*, edited by N. M. Ashton, S. G. Lewis, and C. B. Stringer (Amsterdam: Elsevier), 226.
Jacomb, C., R. N. Holdaway, M. E. Allentoft, M. Bunce, C. L. Oskam, R. Walter, and E. Brooks. 2014. "High-Precision Dating and Ancient DNA Profiling of Moa (Aves: Dinornithiformes) Eggshell Documents a Complex Feature at Wairau Bar and Refines the Chronology of New Zealand Settlement by Polynesians." *Journal of Archaeological Science* 50:24–30.
Jaffe, A. L., G. J. Slater, and M. E. Alfaro. 2011. "The Evolution of Island Gigantism and Body Size Variation in Tortoises and Turtles." *Biology Letters* 7:558–61.
Janis, C. M., K. Buttrill, and B. Figueirido. 2014. "Locomotion in Extinct Giant Kangaroos: Were Sthenurines Hop-less Monsters?" *PLOS ONE* 9(10): e109888. https://doi.org/10.1371/journal.pone.0109888.

Janzen, D. H., and P. S. Martin. 1982. "Neotropical Anachronisms: The Fruits the Gomphotheres Ate." *Science* 215:19–27.

Jefferson, T. 1799. "A Memoir on the Discovery of Certain Bones of a Quadruped of the Clawed Kind in the Western Parts of Virginia." *Transactions of the American Philosophical Society* 4:246–60.

Jenkins, D. L., L. G. Davis, T. W. Stafford Jr., P. F. Campos, B. Hockett, G. T. Jones, L. S. Cummings, et al. 2012. "Clovis Age Western Stemmed Projectile Points and Human Coprolites at the Paisley Caves." *Science* 337:223–28.

Johnson, C. N. 2002. "Determinants of Loss of Mammal Species during the Late Quaternary 'Megafauna' Extinctions: Life History and Ecology, but Not Body Size." *Proceedings of the Royal Society, London B* 269:2221–27.

———. 2006. *Australia's Mammal Extinctions: A 50,000 Year History*. Melbourne: Cambridge University Press.

Jürgensen, J, D. G. Drucker, A. J. Stuart, M. Schneider, B. Buuveibaatar, and H. Bocherens. 2017. "Diet and Habitat of the Saiga Antelope during the Late Quaternary Using Stable Carbon and Nitrogen Isotope Ratios." *Quaternary Science Reviews* 160:150–61.

Kahlke, R.-D. 1999. *The History of the Origin, Evolution and Dispersal of the Late Pleistocene Mammuthus-Coelodonta Faunal Complex in Eurasia (Large Mammals)*. Rapid City, SD: Fenske Companies.

Keller, G., T. Adatte, S. Gardin, A. Bartolini, and S. Bajpai. 2008. "Main Deccan Volcanism Phase Ends Near the K–T Boundary: Evidence from the Krishna-Godavari Basin, SE India." *Earth and Planetary Science Letters* 268:293–311.

Keller, G., J. Punekar, and P. Mateo. 2015. "Upheavals during the Late Maastrichtian: Volcanism, Climate and Faunal Events Preceding the End-Cretaceous Mass Extinction." *Palaeogeography, Palaeoclimatology, Palaeoecology* 441:137–51. https://doi.org/10.1016/j.palaeo.2015.06.034.

Keller G., A. Sahni, and S. Bajpai. 2009. "Deccan Volcanism, the KT Mass Extinction and Dinosaurs." *Journal of Bioscience* 34:709–28. https://doi.org/10.1007/s12038-009-0059-6.

Kentucky State Parks. *Big Bone Lick State Historic Site*. 2016. http://parks.ky.gov/!userfiles/big-bone-lick/big-bone-lick-history.pdf.

Kerr, R. A. 2008. "Experts Find No Evidence for a Mammoth-Killer Impact." *Science* 319:1331–32.

King, J. E., and J. J. Saunders. 1986. "*Geochelone* in Illinois and the Illinoian-Sangamonian Vegetation of the Type Region." *Quaternary Research* 25:89–99.

Kirillova, I. V., O. G. Zanina, O. F. Chernova, E. G. Lapteva, S. S. Trofimova, V. S. Lebedev, A. V. Tuinov, et al. 2015. "An Ancient Bison from the Mouth of the Rauchua River (Chukotka, Russia)." *Quaternary Research* 84:232–45.

Kitchener, A. 1987. "Fighting Behaviour of the Extinct Irish Elk." *Modern Geology* 11:1–28.

Klein, R. G. 1974. "On the Taxonomic Status, Distribution and Ecology of the Blue Antelope, *Hippotragus leucophaeus*." *Annals of the South African Museum* 65:99–143.

———. 1984. "Mammalian Extinctions and Stone Age People in Africa." In *Quaternary Extinctions: A Prehistoric Revolution*, edited by P. S. Martin and R. G. Klein (Tucson: University of Arizona Press), 553–73.

———. 1994. "The Long-Horned African Buffalo (*Pelorovis antiquus*) Is an Extinct Species." *Journal of Archaeological Science* 21:725–33.

Koch, P. L., and A. D. Barnosky. 2006. "Late Quaternary Extinctions: State of the Debate." *Annual Review of Ecology, Evolution, and Systematics* 37:215–50.

Köhler, M., and S. Moyà-Solà. 2009. "Physiological and Life History Strategies of a Fossil Large Mammal in a Resource-Limited Environment." *Proceedings of the National Academy of Sciences USA* 106:20354–58.

Kosintsev, P., K. J. Mitchell, T. Devièse, J. van der Plicht, M. Kuitems, E. Petrova, A. Tikhonov, et al. 2018. "Evolution and Extinction of the Giant Rhinoceros *Elasmotherium sibiricum* Sheds Light on Late Quaternary Megafaunal Extinctions." *Nature Ecology and Evolution* 3:31–38.

Kropf, M., J. I. Mead, and R. S. Anderson. 2007. "Dung, Diet, and the Paleoenvironment of the Extinct Shrub-Ox (*Euceratherium collinum*) on the Colorado Plateau, USA." *Quaternary Research* 67:143–51.

Kubiak, H. 1969. "Uber die Bedeutung der Kadaver des Wollhaarnashorns von Starunia." *Berichte den Deutschen Gesellschaft für Geologische Wissenschaften. Geologie und Paläontologie* 14:345–47.

Kurtén, B., and E. Anderson. 1980. *Pleistocene Mammals of North America*. New York: Columbia University Press.

La Brea Tar Pits & Museum. n.d. "Experience the Tar Pits." https://tarpits.org/la-brea-tar-pits/current-excavations.

Lambeck, K., and J. Chappell. 2001. "Sea Level Change through the Last Glacial Cycle." *Science* 292:679–86.

Lamberton, C. 1931. "Contribution à l'étude anatomique des *Aepyornis*." *Bulletinde L'Académie Malgache* 13:151–74, plus plates.

———. 1934a. "Contribution à la connaissance de la faune subfossile de Madagascar: Lémuriens et ratites." *Mémoires de l'Académie Malgache* 17:1–168.

———. 1934b. "Ratites subfossiles de Madagascar: Les Mullerornithidae." *Mémoires de L'Académie Malgache* 17:123–68, plus plates.

Larramendi, A. 2016. "Shoulder Height, Body Mass, and Shape of Proboscideans." *Acta Palaeontologica Polonica* 61:537–74.

La Violette, P. A. 2011. "Evidence for a Solar Flare Cause of the Pleistocene Mass Extinction." *Radiocarbon* 53 (2): 303–23.

Lepper, B. T., T. A. Frolking, D. C. Fisher, G. Goldstein, J. E. Sanger, D. A. Wymer, J. G. Ogden III, and P. E.Hooge. 1991. "Intestinal Contents of a Late Pleistocene Mastodont from Midcontinental North America." *Quaternary Research* 36:120–25.

Letnic, M., M. Fillios, and M. S. Crowther. 2012. "Could Direct Killing by Larger Dingoes Have Caused the Extinction of the Thylacine from Mainland Australia?" *PLOS ONE* 7 (5): e34877. https://doi.org/10.1371/journal.pone.0034877.

Lister, A. M. 1994. "The Evolution of the Giant Deer, *Megaloceros giganteus* (Blumenbach)." *Zoological Journal of the Linnean Society* 112:65–100.

———. 2014. *Mammoths: Ice Age Giants*. London: Natural History Museum.

———. 2018. *Darwin's Fossils: Discoveries that Shaped the Theory of Evolution*. London: Natural History Museum.

Lister, A. M., and P. G. Bahn. 2007. *Mammoths: Giants of the Ice Age*. London: Frances Lincoln.

Lister, A. M., C. J. Edwards, D. A. W. Nock, M. Bunce, I. A. van Pijlen, D. G. Bradley, M. G. Thomas, and I. Barnes. 2005. "The Phylogenetic Position of the 'Giant Deer' *Megaloceros giganteus*." *Nature* 438:850–53.

Lister, A. M., and A. V. Sher. 2015. "Evolution and Dispersal of Mammoths across the Northern Hemisphere." *Science* 350:806–9.

Lister, A. M., and A. J. Stuart. 2010. "The West Runton Mammoth (*Mammuthus trogontherii*) and Its Evolutionary Significance." *Quaternary International* 228 (1–2): 180–209.

———. 2013. "Extinction Chronology of the Woolly Rhinoceros *Coelodonta antiquitatis*: Reply to Kuzmin." *Quaternary Science Reviews* 62:144–46.

———. 2019. "The Extinction of the Giant Deer *Megaloceros giganteus* (Blumenbach): New Radiocarbon Evidence." *Quaternary International* 500:185–203. https://doi.org/10.1016/j.quaint.2019.03.025.

Llamas, B., L. Fehren-Schmitz, G. Valverde, J. Soubrier, S. Mallick, N. Rohland, S. Nordenfelt, et al. 2016. "Ancient Mitochondrial DNA Provides High-Resolution Time Scale of the Peopling of the Americas." *Science Advances* 2:e1501385. https://doi.org/10.1126/sciadv.1501385.

Lopes, R. P., H. T. Frank, F. S. de Carvalho Buchmann, and F. Caron. 2017. "*Megaichnus* igen. nov.: Giant Paleoburrows Attributed to Extinct Cenozoic Mammals from South America" *Ichnos* 24 (2): 133–45. https://doi.org/10.1080/10420940.2016.1223654.

Lopes, R. P., A. M. Ribeiro, S. R. Dillenburg, and S. L. Schultz. 2013. "Late Middle to Late Pleistocene Paleoecology and Paleoenvironments in the Coastal Plain of Rio Grande do Sul State, Southern Brazil, from Stable Isotopes in Fossils of *Toxodon* and *Stegomastodon*." *Palaeogeography, Palaeoclimatology, Palaeoecology* 369:385–94.

Lorenzen, E., D. Nogués-Bravo, L. Orlando, J. Weinstock, J. Binladen, K. A. Marske, A. Ugan, et al. 2011. "Species-Specific Responses of Late Quaternary Megafauna to Climate and Humans." *Nature* 479:359–64.

Louys, J., D. Curnoe, and H. Tong. 2007. "Characteristics of Pleistocene Megafauna Extinctions in Southeast Asia." *Palaeogeography, Palaeoclimatology, Palaeoecology* 243:152–73.

Louys, J., G. J. Price, and S. O'Connor. 2016. "Direct Dating of Pleistocene *Stegodon* from Timor Island, East Nusa Tenggara." *PeerJ* 4:e1788. https://doi.org/10.7717/peerj.1788.

Lull, R. S. 1929. "A Remarkable Ground Sloth." *Memoirs of the Peabody Museum of Natural History* 3:1–39.

Lundelius, E. 1961. "*Mylohyus nasutus*: Long-Nosed Peccary of the Texas Pleistocene." *Bulletin of the Texas Memorial Museum*, vol. 1. 40 pp.

Lyell, C. 1830–1833. "*Principals of Geology* 3 vol. London: John Murray.

Lyons, S. K., F. A. Smith, and J. H. Brown. 2004. "Of Mice, Mastodons and Men: Human Mediated Extinctions on Four Continents." *Evolutionary Ecology Research* 6:339–58.

Lyons, S. K., F. A. Smith, P. J. Wagner, E. P. White, and J. H. Brown. 2004. "Was a 'Hyperdisease' Responsible for the Late Pleistocene Megafaunal Extinction?" *Ecology Letters* 7:859–68.

MacLeod, N. 2013. *The Great Extinctions: What Causes Them and How They Shape Life*. London: Natural History Museum Publications.

MacPhee, R. D. E., and D. A. Burney. 1991. "Dating of Modified Femora of Extinct Dwarf *Hippopotamus* from Southern Madagascar: Implications for Constraining Human Colonization and Vertebrate Extinction Events." *Journal of Archaeological Science* 18:695–706.

MacPhee, R. D. E., M. A. Iturralde-Vinent, and O. J. Vázquez. 2007. "Prehistoric Sloth Extinctions in Cuba: Implications of a New 'Last' Appearance Date." *Caribbean Journal of Science* 43:94–98.

MacPhee, R. D. E., and P. A. Marx. 1997. "Humans, Hyperdisease, and First-Contact Extinctions." In *Natural Change and Human Impact in Madagascar*, edited by S. M. Goodman and B. D. Patterson (Washington, DC: Smithsonian Institution Press), 169–217.

MacPhee, R. D. E., A. N. Tikhonov, D. Mol, C. de Marliave, J. van der Plicht, A. D. Greenwood, C. Flemming, and L. Agenbroad. 2002. "Radiocarbon Chronologies and Extinction Dynamics of the Late Quaternary Mammalian Megafauna of the Taimyr Peninsula, Russian Federation." *Journal of Archaeological Science* 29 (10): 1017–42.

Marcus, L. F. 1960. "A Census of the Abundant Large Pleistocene Mammals from Rancho La Brea." *Contributions in Science: Los Angeles County Museum of Natural History* 38:1–11.

Made, J. van der, and Tong, H. W. 2008. "Phylogeny of the Giant Deer with Palmate Brow Tines: *Megaloceros* from West and *Sinomegaceros* from East Eurasia." *Quaternary International* 179: 135–62.

Mann Butler, A.M. 1834. *A History of the Commonwealth of Kentucky*. Louisville, KY: Wilcox, Dickerman.

Marean, C.W. 2017. "Early Signs of Human Presence in Australia." *Nature* 547:285–86.

Markgraf, V. 1985. "Late Pleistocene Faunal Extinctions in Southern Patagonia." *Science* 288:1110–12.

Martin, P. S. 1967. "Prehistoric Overkill." In *Pleistocene Extinctions: The Search for a Cause*, edited by P. S. Martin and H. E. Wright (New Haven, CT: Yale University Press), 75–120.

———. 1973. "The Discovery of America." *Science* 179:969–74.

———. 1984. "Prehistoric Overkill: The Global Model." In *Quaternary Extinctions: A Prehistoric Revolution*, edited by P. S. Martin and R. G. Klein (Tucson: University of Arizona Press), 354–403.

———. 2005. *Twilight of the Mammoths: Ice Age Extinctions and the Rewilding of America*. Berkeley: University of California Press.

Martin, P. S., and R. G. Klein, eds. 1984. *Quaternary Extinctions: A Prehistoric Revolution*. Tucson: University of Arizona Press.

Martin, P. S., and D. W. Steadman. 1999. "Prehistoric Extinctions on Islands and Continents." In *Extinctions in Near Time: Causes, Contexts and Consequences*, edited by R. D. E. MacPhee (New York: Kluwer Academic/Plenum), 17–55.

Mayhew, D. F. 1978. "Reinterpretation of the Extinct Beaver *Trogontherium* (Mammalia, Rodentia)." *Philosophical Transactions of the Royal Society of London: Series B, Biological Sciences* 281:407–38.

Mayle, F. E., M. J. Burn, M. Power, and D. H. Urrego. 2009. "Vegetation and Fire at the Last Glacial Maximum in Tropical South America." Chapter 4 of *Past Climate Variability in*

South America and Surrounding Regions, Developments in Paleoenvironmental Research, edited by F. Vimeux, F. Sylvestre, and M Khodri (Dordrecht: Springer). https://doi.org/10.1007/978-90-481-2672-9_4.

Mazza, P., and A. Bertini. 2013. "Were Pleistocene Hippopotamuses Exposed to Climate-Driven Body Size Changes?" *Boreas* 42:194–209.

McDonald, H. G. 2005. "The Paleoecology of Extinct Xenarthrans and the Great American Biotic Interchange." *Bulletin of the Florida Museum of Natural History* 45:313–33.

McDonald, H. G., and R. A. Bryson. 2010. "Modeling Pleistocene Local Climatic Parameters Using Macrophysical Climate Modeling and the Paleoecology of Pleistocene Megafauna." *Quaternary International* 217:131–37.

McDonald, H. G., C. R. Harington, and G. De Iuliis. 2000. "The Ground Sloth *Megalonyx* from Pleistocene Deposits of the Old Crow Basin, Yukon, Canada." *Arctic* 53:213–20.

McDonald, H. G., and S. Pelikan. 2006. "Mammoths and Mylodonts: Exotic Species from Two Different Continents in North American Pleistocene Faunas." *Quaternary International* 142–43:229–41.

McDonald, H. G., T. W. Stafford Jr., and D. M. Gnidovec. 2015. "Youngest Radiocarbon Age for Jefferson's Ground Sloth, *Megalonyx jeffersonii* (Xenarthra, Megalonychidae)." *Quaternary Research* 83:355–59.

McGuire, W. 2014. *Global Catastrophes: A Very Short Introduction*. Oxford: Oxford University Press.

McWethy, D. B., C. Whitlock, J. M. Wilmshurst, M. S. McGlone, and X. Li. 2009. "Rapid Deforestation of South Island, New Zealand, by Early Polynesian Fires." *Holocene* 19: 883–97.

Mead, J. I., and L. D. Agenbroad. 1992. "Isotope Dating of Pleistocene Dung Deposits from the Colorado Plateau, Arizona and Utah." *Radiocarbon* 34:1–19.

Mead, J. I., P. S. Martin, R. C. Euler, A. Long, A. J. T. Jull, L. J. Toolin, D. J. Donahue, and T. W. Linick. 1986. "Extinction of Harrington's Mountain Goat." *Proceedings of the National Academy of Sciences USA* 83:836–39.

Meador, L. R., L. R. Godfrey, J. C. Rakotondramavo, L. Ranivoharimanana, A. Zamora, M. R. Sutherland, and M. T. Irwin. 2017. "*Cryptoprocta spelea* (Carnivora: Eupleridae): What Did It Eat and How Do We Know?" *Journal of Mammalian Evolution* 26:237–51. https://doi.org/10.1007/s10914-017-9391-z.

Meltzer, D. 2010. *First Peoples in a New World: Colonizing Ice Age America*. Berkeley: University of California Press.

Metcalf, J. L., C. Turney, R. Barnett, F. Martin, S. C. Bray, J. T. Vilstrup, L. Orlando, et al. 2016. "Synergistic Roles of Climate Warming and Human Occupation in Patagonian Megafaunal Extinctions during the Last Deglaciation." *Science Advances* 2:e1501682.

Midgley, J. J., and N. Illing. 2009. "Were Malagasy *Uncarina* Fruits Dispersed by the Extinct Elephant Bird?" *South African Journal of Science* 105:467–69.

Millais, J. G. 1897. *British Deer and Their Horns*. Henry Sotheran.

Miller, G. H., M. L. Fogel, J. W. Magee, and S. J. Clarke. 2017. "The *Genyornis* Egg: A Commentary on Grellet-Tinner et al., 2016." *Quaternary Science Reviews* 161:123–27. http://doi.org/10.1016/j.quascirev.2016.12.004.

Miller, G. H., J. W. Magee, B. J. Johnson, M. L. Fogel, N. A. Spooner, M. T. McCulloch, and L. K. Ayliffe. 1999. "Pleistocene Extinction of *Genyornis newtoni*: Human Impact on Australian Megafauna." *Science* 283:205–8.

Miller, G. H., J. Magee, M. Smith, N. Spooner, A. Baynes, S. Lehman, M. Fogel, et al. 2016. "Human Predation Contributed to the Extinction of the Australian Megafaunal Bird *Genyornis newtoni* ~ 47 ka." *Nature Communications* 7:10496. https://doi.org/10.1038/ncomms10496.

Miskelly, C. M., ed. 2013. *New Zealand Birds Online*. www.nzbirdsonline.org.nz.

Mitchell, G. F., and H. M. Parkes. 1949. "The Giant Deer in Ireland." *Proceedings of the Royal Irish Academy* 52B: 291–314.

Mitchell, K. J., S. C. Bray, P. Bover, L. Soibelzon, B. W. Schubert, F. Prevosti, A. Prieto, F. Martin, J. Austin, and A. Cooper. 2016. "Ancient Mitochondrial DNA Reveals Convergent Evolution of Giant Short-Faced Bears (Tremarctinae) in North and South America." *Biology Letters* 12:20160062. http://doi.org/10.1098/rsbl.2016.0062.

Mol, D., J. de Vos, and J. van der Plicht. 2007. "The Presence and Extinction of *Elephas antiquus* Falconer and Cautley 1847, in Europe." *Quaternary International* 169–70:149–53.

Molnar, R. E. 1991. "Fossil Reptiles in Australia." In *Vertebrate Palaeontology of Australia*, edited by P. Vickers-Rich, J. M. Monaghan, R. F. Baird, and T. H. Rich (Melbourne: Monash University Publications), 604–702.

———. 2004. *Dragons in the Dust: The Paleobiology of the Giant Monitor Lizard Megalania*. Bloomington: Indiana University Press.

Molyneaux, T. 1695–97. "A Discourse Concerning the Large Horns Frequently Found under Ground in Ireland, Concluding from Them That the Great American Deer, Call'd a Moose, Was Formerly Common in That Island: With Remarks on Some Other Things Natural to That Country." *Philosophical Transactions of the Royal Society of London* 19: 489–512.

Montes, C., A. Cardona, C. Jaramillo, A. Pardo, J. C. Silva, V. Valencia, C. Ayala, et al. 2015. "Middle Miocene Closure of the Central American Seaway." *Science* 348:226–29.

Moore, D. M. 1978. "Post-glacial Vegetation in the South Patagonian Territory of the Giant Ground Sloth, *Mylodon*." *Botanical Journal of the Linnean Society* 77:177–202.

Morwood, M. J., R. P. Soejono, R. G. Roberts, T. Sutikna, C. S. M. Turney, K. E. Westaway, W. J. Rink, et al. 2004. "Archaeology and Age of a New Hominin from Flores in Eastern Indonesia." *Nature* 431:1087–91.

Mothé, D., L. S. Avilla, L. Asevedo, L. Borges-Silva, M. Rosas, R. Labarca-Encina, R. Souberlich, et al. 2017. "Sixty Years after 'The Mastodonts of Brazil': The State of the Art of South American Proboscideans (Proboscidea, Gomphotheriidae)." *Quaternary International* 443:52–64.

Muhs, D. R., K. R. Simmons, L. T. Groves, J. P. McGeehin, R. R. Schumann, and L. D. Agenbroad. 2015. "Late Quaternary Sea-Level History and the Antiquity of Mammoths (*Mammuthus exilis* and *Mammuthus columbi*), Channel Islands National Park, California, USA." *Quaternary Research* 83:502–21. http://doi.org/10.1016/j.yqres.2015.03.001.

Münzel, S. C., M. Stiller, M. Hofreiter, A. Mittnik, N. J. Conard, and H. Bocherens. 2011. "Pleistocene Bears in the Swabian Jura (Germany): Genetic Replacement, Ecological Displacement, Extinctions and Survival." *Quaternary International* 245:1–13.

Murray, P. 1991. "The Pleistocene Megafauna of Australia." In *Vertebrate Palaeontology of Australia*, edited by P. Vickers-Rich, J. M. Monaghan, R. F. Baird, and T. H. Rich (Melbourne: Monash University Publications), 1071–164.

Murray, P., and P. Vickers-Rich. 2004. *Magnificent Mihirungs: The Colossal Flightless Birds of the Australian Dreamtime*. Bloomington: Indiana University Press.

Newell, N. D., 1963. "Crises in the History of Life." *Scientific American*, 208 (2): 76–92.

NGRIP Dating Group. 2008. *Greenland Ice Core Chronology 2005 (GICC05) 60,000 Year, 20 Year Resolution*. ftp://ftp.ncdc.noaa.gov/pub/data/paleo/icecore/greenland/summit/ngrip/gicc05-60ka-20yr.txt.

Nikolskiy, P. A., L. D. Sulerzhitsky, and V. V. Pitulko. 2011. "Last Straw versus Blitzkrieg Overkill: Climate-Driven Changes in the Arctic Siberian Mammoth Population and the Late Pleistocene Extinction Problem." *Quaternary Science Reviews* 30:2309–28.

O'Brien, H. D., J. T. Faith, K. E. Jenkins, D. J. Peppe, T. W. Plummer, Z. L. Jacobs, B. Li, et al. 2016. "Unexpected Convergent Evolution of Nasal Domes between Pleistocene Bovids and Cretaceous Hadrosaur Dinosaurs." *Current Biology* 26:1–6.

O'Connell, J. F., and J. Allen. 2015. "The Process, Biotic Impact, and Global Implications of the Human Colonization of Sahul about 47,000 Years Ago." *Journal of Archaeological Science* 56:73–84.

O'Connell, J. F., J. Allen, M. A. J. Williams, W. A. N. Williams, C. S. M. Turney, N. A. Spooner, J. Kamminga, G. Brown, and A. Cooper, A. 2018. "When Did *Homo sapiens* First Reach Southeast Asia and Sahul?" *Proceedings of the National Academy of Sciences USA*. 115 (34): 8482–90. https://doi.org/10.1073/pnas.1808385115.

Oliveira-Santos, L. G. R., and F. A. S. Fernandez. 2010. "Pleistocene Rewilding, Frankenstein Ecosystems, and an Alternative Conservation Agenda." *Conservation Biology* 24:4–6.

Orlando, L., J. A. Leonard, A. Thenot, V. Laudet, C. Guerin, and C. Hanni. 2003. "Ancient DNA Analysis Reveals Woolly Rhino Evolutionary Relationships." *Molecular Phylogenetics and Evolution* 28:485–99.

Oskam, C. L., M. E. Allentoft, R. Walter, R. P. Scofield, J. Haile, R. N. Holdaway, M. Bunce, and C. Jacomb. 2012. "Ancient DNA Analyses of Early Archaeological Sites in New Zealand Reveal Extreme Exploitation of Moa (Aves: Dinornithiformes) at All Life Stages." *Quaternary Science Reviews* 52:41–48.

Oskarsson, M. C. R., C. F. C. Klütsch, U. Boonyaprakob, A. Wilton, Y. Tanabe, and P. Savolainen. 2012. "Mitochondrial DNA Data Indicate an Introduction through Mainland Southeast Asia for Australian Dingoes and Polynesian Domestic Dogs." *Proceedings of the Royal Society, London B* 279:967–74.

Owen, R. 1840. "Part 1: Fossil Mammalia." In *The Zoology of the Voyage of* HMS Beagle, edited by C. R. Darwin (London: Smith, Elder).

———. 1857. "On the Scelidothere (*Scelidotherium leptocephalum*, Owen)." *Philosophical Transactions of the Royal Society of London* 147:101–10.

———. 1859a. "On the Fossil Mammals of Australia: Part I: Description of a Mutilated Skull of a Large Marsupial Carnivore (*Thylacoleo carnifex*, Owen), from a Calcareous Conglomerate Stratum, Eighty Miles S.W. of Melbourne, Victoria." *Philosophical Transactions of the Royal Society of London* 149:309–22.

———. 1859b. "Description of Some Remains of a Gigantic Land-Lizard (Megalania Prisca, Owen) from Australia." *Philosophical Transactions of the Royal Society of London* 149: 43–48.

———. 1870a. "On the Fossil Mammals of Australia: Part III: *Diprotodon australis*." *Philosophical Transactions of the Royal Society of London*. 160:519–78.

———. 1870b. "On the Molar Teeth, Lower Jaw, of *Macrauchenia patachonica*, Ow." *Philosophical Transactions of the Royal Society of London* 160:79–81.

———. 1871. "On the Fossil Mammals of Australia: Part IV: Dentition and Mandible of *Thylacoleo carnifex*, with Remarks on the Arguments for Its Herbivority." *Philosophical Transactions of the Royal Society of London* 161:213–66.

———. 1873. "*Procoptodon goliah*, Owen." *Proceedings of the Royal Society of London* 21:387.

———. 1874. "On the Fossil Mammals of Australia: Part IX: Family Macropodidae: Genera *Macropus*, *Pachysaigon*, *Leptosaigon*, *Procoptodon*, and *Palorchestes*." *Philosophical Transactions of the Royal Society of London* 164:783–803.

———. 1880. "Description of Some Remains of the Gigantic Land-Lizard (*Megalania prisca*; Owen), from Australia: Part II." *Philosophical Transactions of the Royal Society of London* 171:1037–50.

Pacher, M., and A. J. Stuart. 2009. "Extinction Chronology and Palaeobiology of the Cave Bear *Ursus spelaeus*." *Boreas* 38:189–206.

Paijmans, J. L. A., R. Barnett, M. T. P. Gilbert, M. L. Zepeda-Mendoza, J. W. F. Reumer, J. de Vos, G. Zazula, et al. 2017. "Evolutionary History of Saber-Toothed Cats Based on Ancient Mitogenomics." *Current Biology* 27:1–7.

Paunero, R.S., G. Rosales, J. L. Prado, and M. T. Alberdi. 2008. "Cerro Bombero: registro de *Hippidion saldiasi* Roth, 1899 (Equidae, Perissodactyla) en el Holoceno temprano de Patagonia (Santa Cruz, Argentina)." *Estudios Geológicos*, 64:89–98.

Pedersen, M. W., A. Ruter, and C. Schweger. 2016. "Postglacial Viability and Colonization in North America's Ice-Free Corridor." *Nature* 537:45–49.

Pedrono, M., O. W. Griffiths, A. Clausen, L. L. Smith, C. J. Griffiths, L. Wilmé, and D. A. Burney. 2013. "Using a Surviving Lineage of Madagascar's Vanished Megafauna for Ecological Restoration." *Biological Conservation* 159:501–6.

Penck, A., and E. Brückner. 1901/1909. *Die Alpen im Eiszeitalter*. 3 vols. Leipzig: Tauchnitz.

Penkman, K. E. M., D. S. Kaufman, D. Maddy, and M. J. Collins. 2008. "Closed-System Behaviour of the Intra-crystalline Fraction of Amino Acids in Mollusc Shells." *Quaternary Geochronology* 3:2–25.

Perez, V. R., L. R. Godfrey, M. Nowak-Kemp, D. A. Burney, J. Ratsimbazafy, and N. Vasey. 2005. "Evidence of Early Butchery of Giant Lemurs in Madagascar." *Journal of Human Evolution* 49:722–42.

Perry, G. L. W., A. B. Wheeler, J. R. Wood, and J. M. Wilmshurst. 2014. "A High-Precision Chronology for the Rapid Extinction of New Zealand Moa (Aves, Dinornithiformes)." *Quaternary Science Reviews* 105:126–35.

Pfizenmayer, E. W. 1939. *Siberian Man and Mammoth*. London: Blackie.

Piñero, J. M. L. 1988. "Juan Bautista Bru (1740–1799) and the Description of the Genus *Megatherium*." *Journal of the History of Biology* 21:147–63.

Pinhasi, R., T. F. G. Higham, L. V. Golovanova, and V. B. Doronichev. 2011. "Revised Age of Late Neanderthal Occupation and the End of the Middle Paleolithic in the Northern Caucasus." *Proceedings of the National Academy of Sciences USA* 108:8611–16.

Pinter, N., A. C. Scott, T. L. Daulton, A. Podoll, C. Koeberl, R. S. Anderson, and S. E. Ishman. 2011. "The Younger Dryas Impact Hypothesis: A Requiem." *Earth Science Reviews* 106:247–64.

Pitulko, V., E. Pavlova, and P. Nikolskiy. 2017. "Revising the Archaeological Record of the Upper Pleistocene Arctic Siberia: Human Dispersal and Adaptations in MIS 3 and 2." *Quaternary Science Reviews* 165:127–48.

Pledge, N. S. 1981. "The Giant Rat-Kangaroo *Propleopus oscillans* (De Vis), (Potoroidae: Marsupialia) in South Australia." *Transactions of the Royal Society of South Australia* 105:41–47.

Politis, G., and P. Messineo. 2007. "The Campo Laborde Site: New Evidence for the Holocene Survival of the Pleistocene Megafauna in the Argentine Pampa." *Quaternary International* 191:198–94.

Politis, G. G., M. A. Gutiérrez, D. J. Rafuse, and A. Blasi. 2016. "The Arrival of *Homo sapiens* into the Southern Cone at 14,000 Years Ago." *PLOS ONE* 11 (9): e0162870. https://doi.org/10.1371/journal.pone.0162870.

Potter, B. A. 2008. "Radiocarbon Chronology of Central Alaska: Technological Continuity and Economic Change." *Radiocarbon* 50:181–204.

Prado, J. L., and M. T. Alberdi. 2010. "Quaternary Mammalian Faunas of the Pampean Region." *Quaternary International* 212:176–86.

Prado, J. L., M. T. Alberdi, B. Azanza, B. Sánchez, and D. Frassinetti. 2005. "The Pleistocene Gomphotheriidae (Proboscidea) from South America." *Quaternary International* 126–128:21–30.

Prado, J. L., C. Martinez-Maza, and M. T. Alberdi. 2015. "Megafauna Extinction in South America: A New Chronology for the Argentine Pampas." *Palaeogeography, Palaeoclimatology, and Palaeoecology* 425:41–49.

Prevosti, F. J., and F. M. Martin. 2013. "Paleoecology of the Mammalian Predator Guild of Southern Patagonia during the Latest Pleistocene: Ecomorphology, Stable Isotopes, and Taphonomy." *Quaternary International* 305:74–84.

Prevosti, F. J., and S. F. Vizcaíno. 2006. "Paleoecology of the Large Carnivore Guild from the Late Pleistocene of Argentina." *Acta Palaeontologica Polonica* 51:407–22.

Price, G. J. 2008. "Taxonomy and Palaeobiology of the Largest-Ever Marsupial, *Diprotodon* Owen 1838 (Diprotodontidae, Marsupialia)." *Zoological Journal of the Linnean Society* 153:389–417.

Price, G. J. 2012a. "Diprotodon's Big Day Out." Diprotodon.com. http://www.diprotodon.com/diprotodons-big-day-out/

———. 2012b. "Digging Up *Diprotodon*." Diprotodon.com. http://www.diprotodon.com/digging-up-diprotodon/.

———. 2016. "The Ice Age Lizards of Oz." *Australasian Science*, April 2016, 20–23.

Price, G. J., K. J. Ferguson, G. E. Webb, Y.-x. Feng, P. Higgins, A. D. Nguyen, J.-x. Zhao, R. Joannes-Boyau, and J. Louys. 2017. "Seasonal Migration of Marsupial Megafauna in

Pleistocene Sahul (Australia–New Guinea)." *Proceedings of the Royal Society B* 284 (1863). https://doi.org/10.1098/rspb.2017.0785.

Price, G. J., J. Louys, J. Cramb, Y.-x. Feng, J.-x. Zhao, S. A. Hocknull, G. E. Webb, A. D. Nguyen, and R. Joannes-Boyau. 2015. "Temporal Overlap of Humans and Giant Lizards (Varanidae; Squamata) in Pleistocene Australia." *Quaternary Science Reviews* 125:98–105.

Price, G. J., and K. J. Piper. 2009. "Gigantism of the Australian *Diprotodon* Owen 1838 (Marsupialia, Diprotodontoidea) through the Pleistocene." *Journal of Quaternary Science* 24:1029–38.

Prideaux, G. J., L. K. Ayliffe, L. R. G. DeSantis, B. W. Schubert, P. F. Murray, M. K. Gagan, and T. E. Cerling. 2009. "Extinction Implications of a Chenopod Browse Diet for a Giant Pleistocene Kangaroo." *Proceedings of the National Academy of Sciences USA* 106:11646–650.

Prideaux, G. J., J. A. Long, L. A. Ayliffe, J. C. Hellstrom, B. Pillans, W. E. Boles, M. N. Hutchinson, et al. 2007. "An Arid-Adapted Middle Pleistocene Vertebrate Fauna from South-Central Australia." *Nature* 445:422–25. https://doi.org/10.1038/nature05471.

Prokopenko, A. A., L. A. Hinnov, D. F. Williams, and M. I. Kuzmin. 2006. "Orbital Forcing of Continental Climate during the Pleistocene: A Complete Astronomically Tuned Climatic Record from Lake Baikal, SE Siberia." *Quaternary Science Reviews* 25:3431–57.

Protopopov, A. V., O. Potapova, A. Kharlamova, G. G. Boeskorov, E. N. Maschenko, B. Shapiro, A. Soares, et al. 2016. "The Frozen Cave Lion (*Panthera spelaea*, Goldfuss, 1810): New-born Cubs from Eastern Siberia, Russia: The First Data on Early Ontogeny of the Extinct Species." *PA Journal of Vertebrate Paleontology, Program and Abstracts* 2016:209.

Rakotovao, M., Y. Lignereux, M. J. Orliac, F. Duranthon, and P.-O. Antoine. 2014. "*Hippopotamus lemerlei* Grandidier, 1868 et *Hippopotamus madagascariensis* Guldberg, 1883 (Mammalia, Hippopotamidae): anatomie crânio-dentaire et révision systématique." *Geodiversitas* 36 (1): 117–61. http://doi.org/10.5252/g2014n1a3.

Ramis, D., J. A. Alcover, J. Coll, and M. Trias. 2002. "The Chronology of the First Settlement of the Balearic Islands." *Journal of Mediterranean Archaeology* 15 (1): 3–24.

Rasmussen, M., S. L. Anzick, M. R. Waters, P. Skoglund, M. DeGiorgio, T. W. Stafford Jr., S. Rasmussen, et al. 2014. "The Genome of a Late Pleistocene Human from a Clovis Burial Site in Western Montana." *Nature* 506:225–29.

Raup, D. M., and J. J. Sepkoski Jr. 1982. "Mass Extinctions in the Marine Fossil Record." *Science* 215:1501–3. https://doi.org/10.1126/science.215.4539.1501.

Rawlence, N. J., and A. Cooper. 2013. "Youngest Reported Radiocarbon Age of a Moa (Aves: Dinornithiformes) Dated from a Natural Site in New Zealand." *Journal of the Royal Society of New Zealand* 43:100–107.

Reed, E. H. 2006. "*In situ* Taphonomic Investigation of Pleistocene Large Mammal Bone Deposits from the Ossuaries, Victoria Fossil Cave, Naracoorte, South Australia." *Helictite* 39:5–15.

Reich, D., R. E. Green, M. Kircher, J. Krause, N. Patterson, E. Y. Durand, B. Viola, et al. 2010. "Genetic History of an Archaic Hominin Group from Denisova Cave in Siberia." *Nature* 468:1053–60.

Reumer, J. W. F., L. Rook, K. van der Bourg, K. Post, D. Mol, and J. de Vos. 2003. "Late Pleistocene Survival of the Saber-Toothed Cat *Homotherium* in Northwestern Europe." *Journal of Vertebrate Paleontology* 23:260–62.

Ride, W. D. L., P. A. Pridmore, R. E. Barwick, R. T. Wells, and R. D. Heady. 1997. "Towards a Biology of *Propleopus oscillans* (Marsupialia: Propleopinae, Hypsiprymnodontidae)." *Proceedings- Linnean Society of New South Wales* 117:243–328.

Roberts, R. G., T. F. Flannery, L. K. Aycliffe, H. Yoshida, J. M. Olley, G. J. Prideaux, G. M. Laslett, et al. 2001. "New Ages for the Last Australian Megafauna: Continent-Wide Extinction about 46,000 Years Ago." *Science* 292:1888–92.

Rosenberger, A. L., L. R. Godfrey, K. M. Muldoon, G. F. Gunnell, H. Andriamialison, L. Ranivoharimanana, J. F. Ranaivoarisoa, et al. 2015. "Giant Subfossil Lemur Graveyard Discovered, Submerged, in Madagascar." *Journal of Human Evolution* 81:8–87.

Rosenmüller, J. C. 1794. "Quaedam de ossibus fossilibus animalis cuiusdam, historiam eius et cognitionem accuratiorem illustrantia, dissertatio, quam d. 22. Octob. 1794 ad disputandum proposuit Ioannes Christ. Rosenmüller Heßberga-Francus, LL.AA.M." In *Theatro anatomico Lipsiensi Prosector assumto socio Io. Chr. Aug. Heinroth Lips. Med. Stud. Cum tabula aenea—34 S.*, Leipzig, 1–34.

Rossetti, D. F., P. M. de Toledo, H. M. Moraes-Santos, and A. E. A. Santos Jr. 2004. "Reconstructing Habitats in Central Amazonia Using Megafauna, Sedimentology, Radiocarbon, and Isotope Analyses." *Quaternary Research* 61:289–300.

Rule, S., B. W. Brook, S. G. Haberle, C. S. M. Turney, A. P. Kershaw, and C. N. Johnson. 2012. "The Aftermath of Megafaunal Extinction: Ecosystem Transformation in Pleistocene Australia." *Science* 335:1483–86.

Runnegar, B. 1983. "A *Diprotodon* Ulna Chewed by the Marsupial Lion, *Thylacoleo carnifex*." *Alcheringa* 7:23–26.

Ruppel, C. D., and J. D. Kessler. 2017. "The Interaction of Climate Change and Methane Hydrates." *Reviews of Geophysics* 55:126–68. https://doi.org/10.1002/2016RG000534.

Saarinen, J., J. Eronen, M. Fortelius, H. Seppä, and A. M. Lister. 2016. "Patterns of Diet and Body Mass of Large Ungulates from the Pleistocene of Western Europe, and Their Relation to Vegetation." *Palaeontologia Electronica*, article 19.3.32A, 58pp. https://palaeo-electronica.org/content/2016/1567-pleistocene-mammal-ecometrics.

Saltré, F., B. W. Brook, M. Rodríguez-Rey, A. Cooper, C. N. Johnson, C. S. M. Turney, and C. J. A. Bradshaw. 2015. "Uncertainties in Dating Constrain Model Choice for Inferring Extinction Time from Fossil Records." *Quaternary Science Reviews* 112:128–37.

Saltré, F., M. Rodríguez-Rey, B. W. Brook, C. N. Johnson, C. S. M. Turney, J. Alroy, A. Cooper, et al. 2016. "Climate Change Not to Blame for Late Quaternary Megafauna Extinctions in Australia." *Nature Communications* 7:10511. https://doi.org/10.1038/ncomms10511.

Samonds, K. E., B. E. Crowley, T. R. N. Rasolofomanana, M. C. Andriambelomanana, H. T. Andrianavalona, T. N. Ramihangihajason, R. Rakotozandry, et al. 2019. "A New Late Pleistocene Subfossil Site (Tsaramody, Sambaina Basin, Central Madagascar) with Implications for the Chronology of Habitat and Megafaunal Community Change on Madagascar's Central Plateau." *Journal of Quaternary Science*: in press.

Sánchez, B., J. L. Prado, and M. T. Alberdi. 2004. "Feeding Ecology, Dispersal, and Extinction of South American Pleistocene Gomphotheres (Gomphotheriidae, Proboscidea)" *Paleobiology* 30:146–61.

Sanchez, G., V. T. Holliday, E. P. Gaines, J. Arroyo-Cabralese, N. Martínez-Tagüeña, A. Kowler, T. Lange, et al. 2014. "Human (Clovis)-Gomphothere (*Cuvieronius* sp.) Association ~13,390 Calibrated yBP in Sonora, Mexico." *Proceedings of the National Academy of Sciences USA* 111:10972–77.

Sandom, C., S. Faurby, B. Sandel, and J.-C. Svenning. 2014. "Global Late Quaternary Megafauna Extinctions Linked to Humans, Not Climate Change." *Proceedings of the Royal Society B.* 281:20133254, 2014. https://doi.org/10.1098/rspb.2013.3254.

Sankaraman, S., S. Mallick, M. Dannemann, K. Prüfer, J. Kelso, S. Pääbo, N. Patterson, and D. Reich. 2014. "The Genomic Landscape of Neanderthal Ancestry in Present-Day Humans." *Nature* 507:354–57.

Saunders, J. J. 1996. "North American Mammutidae." In *The Proboscidea*, edited by J. Shoshani and P. Tassy (Oxford: Oxford University Press), 271–79.

Scanlon, J. D., and M. S. Y. Lee. 2000. "The Pleistocene Serpent *Wonambi* and the Early Evolution of Snakes." *Nature* 403:416–20.

Schoch, W. H., G. Bigga, U. Böhner, P. Richter, and T. Terberger. 2015. "New Insights on the Wooden Weapons from the Paleolithic Site of Schöningen." *Journal of Human Evolution* 89:214–25.

Schubert, B. W. 2010. "Late Quaternary Chronology and Extinction of North American Giant Short-Faced Bears (*Arctodus simus*)." *Quaternary International* 217:188–94.

Schubert, B. W., R. W. Graham, H. G. McDonald, E. C. Grimm, and T. W. Stafford Jr. 2004. "Latest Pleistocene Paleoecology of Jefferson's Ground Sloth (*Megalonyx jeffersonii*) and Elk-Moose (*Cervalces scotti*) in Northern Illinois." *Quaternary Research* 61:231–40.

Schvyreva, A. K. 2015. "On the Importance of the Representatives of the Genus *Elasmotherium* (Rhinocerotidae, Mammalia) in the Biochronology of the Pleistocene of Eastern Europe." *Quaternary International* 379:128–34.

Scott, A. C., N. Pinter, M. E. Collinson, M. Hardiman, R. S. Anderson, A. P. R. Brain, S. Y. Smith, F. Marone, and M. Stampanone. 2010. "Fungus, Not Comet or Catastrophe, Accounts for Carbonaceous Spherules in the Younger Dryas 'Impact Layer.'" *Geophysical Research Letters* 37 (14). https://doi.org/10.1029/2010GL043345.

Scott, B., M. Bates, R. Bates, C. Coneller, M. Pope, A. Shaw, and G. Smith. 2014. "A New View from La Cotte de St Brelade, Jersey." *Antiquity* 88:13–29.

Scott, K. 1980. "Two Hunting Episodes of Middle Palaeolithic Age at La Cotte de Saint Brelade, Jersey." *World Archaeology* 12:137–52.

Semprebon, G. M., and F. Rivals. 2010. "Trends in the Paleodietary Habits of Fossil Camels from the Tertiary and Quaternary of North America." *Palaeogeography, Palaeoclimatology, Palaeoecology* 295:131–45.

Serangeli, J., U. Böhner, T. Van Kolfschoten, and N. J. Conard. 2015. "Overview and New Results from Large-Scale Excavations in Schöningen." *Journal of Human Evolution* 89:27–45.

Shapiro, B. 2015. *How to Clone a Mammoth: The Science of De-extinction*. Princeton, NJ: Princeton University Press.

Shapiro, B., A. J. Drummond, A. Rambaut, M. C. Wilson, P. E. Matheus, A. V. Sher, O. G. Pybus, et al. 2004. "Rise and Fall of the Beringian Steppe Bison." *Science* 306:1561–65.

Sharp, A. 2014. "Three-Dimensional Digital Reconstruction of the Jaw Adductor Musculature of the Extinct Marsupial Giant *Diprotodon optatum*." *PeerJ* 2:e514. https://doi.org/10.7717/peerj.514.

Signor, P. W. III, and J. H. Lipps. 1982. "Sampling Bias, Gradual Extinction Patterns, and Catastrophes in the Fossil Record." In *Geological Implications of Impacts of Large Asteroids and Comets on the Earth*, edited by L. T. Silver and P. H. Schultz, Geological Society of America Special Publication 190:291–96.

Simons, E. L., D. A. Burney, P. S. Chatrath, L. R. Godfrey, W. L. Jungers, and B. Rakotosamimanana. 1995. "AMS ^{14}C Dates for Extinct Lemurs from Caves in the Ankarana Massif, Northern Madagascar." *Quaternary Research* 43:249–54.

Simpson, G. G. 1940. "Mammals and Land Bridges." *Journal of the Washington Academy of Sciences* 30:137–63.

Smith, F. A., A. G. Boyer, J. H. Brown, D. P. Costa, T. Dayan, S. K. Morgan Ernest, A. R. Evans, et al. 2010. "The Evolution of Maximum Body Size of Terrestrial Mammals." *Science* 330:1216–19.

Smith, I. 2013. "Pre-European Maori Exploitation of Marine Resources in Two New Zealand Case Study Areas: Species Range and Temporal Change." *Journal of the Royal Society of New Zealand* 43:1–37. https://doi.org/10.1080/03036758.2011.574709.

Smith, M. 2013. "The Empty Desert: Inland Environments Prior to People." Chapter 3 of *The Archaeology of Australia's Deserts*, Cambridge World Archaeology (Cambridge: Cambridge University Press). https://doi.org/10.1017/CBO9781139023016.005.

Smith, M. J. 1985. "*Wonambi naracoortensis* Smith 1976—the Giant Australian Python." In *Kadimakara—Extinct Vertebrates of Australia*, edited by P. V. Rich, G. F. van Tets, and F. Knight (Lilydale, Victoria: Pioneer Design Studio), 156–59.

Smithsonian Miscellaneous Collections. 1915. *The Indiana Mastodon*. Vol. 66, 27–29.

Smyth, A. H. 1970. "The Writings of Benjamin Franklin." Vol. 5, 1767 to 1772. New York: Haskell House.

Soubrier, S., G. Gower, K. Chen, S. M. Richards, B. Llamas, K. J. Mitchell, S. Y. W. Ho, et al. 2016. "Early Cave Art and Ancient DNA Record the Origin of European Bison." *Nature Communications* 7:13158. https://doi.org/10.1038/ncomms13158.

Steadman, D. W., P. S. Martin, R. D. E. MacPhee, A. J. T. Jull, H. G. McDonald, C. A. Woods, M. Iturralde-Vinent, and G. W. L. Hodgkins. 2005. "Asynchronous Extinction of Late Quaternary Sloths on Continents and Islands." *Proceedings of the National Academy of Sciences USA* 102:11763–68.

Steadman, D. W., T. W. Stafford, D. J. Donahue, and A. J. T. Jull. 1991. "Chronology of Holocene Vertebrate Extinction in the Galapagos Islands." *Quaternary Research* 36:126–33.

Steele, T. E. 2007. "Vertebrate Records: Late Pleistocene of Africa." In *Encyclopedia of Quaternary Science*, edited by Scott Elias (Oxford: Elsevier), 3139–50.

Stiller, M., M. Molak, S. Prost, G. Rabeder, G. Baryshnikov, W. Rosendahl, S. Münzel, et al. 2014. "Mitochondrial DNA Diversity and Evolution of the Pleistocene Cave Bear Complex." *Quaternary International* 339–40:224–31.

Stirling, E. C., and A. H. C. Zeitz. 1900. "Fossil Remains of Lake Callabonna: Part I. *Genyornis newtoni*: A New Genus and Species of Fossil Struthious Bird." *Memoirs of the Royal Society of South Australia* 1:41–80.

Stock, C. S. 1920. "Origin of the Supposed Human Footprints of Carson City, Nevada." *Science* 51:514.

———. 1956. "Rancho La Brea: A Record of Pleistocene Life in California." Los Angeles County Museum of Natural History, Science Series No. 20.

Stock, C., and J. M. Harris. 1992. *Rancho La Brea: A Record of Pleistocene Life in California*. Science Series No. 37. Natural History Museum of Los Angeles County, Los Angeles, CA.

Stringer, C. B. 2012. "The Status of *Homo heidelbergensis* (Schoetensack 1908)." *Evolutionary Anthropology* 21:101–7.

———. 2016. "The Origin and Evolution of *Homo sapiens*." *Philosophical Transactions of the Royal Society B: Biological Sciences* 371 (1698): 20150237–20150237.

Stringer, C. B., and P. Andrews. 2012. *The Complete World of Human Evolution*. London: Thames & Hudson.

Stringer, C. B., and J. Galway-Witham. 2018. "When Did Modern Humans Leave Africa?" *Science* 359:389–90.

Stuart, A. J. 1982. *Pleistocene Vertebrates in the British Isles*. London: Longman.

———. 1987. "Pleistocene Occurrences of *Hippopotamus* in Britain." *Quartärpaläontologie, Berlin* 6:209–18.

———. 1991. "Mammalian Extinctions in the Late Pleistocene of Northern Eurasia and North America." *Biological Reviews* 66:453–562.

———. 2005. "The Extinction of Woolly Mammoth (*Mammuthus primigenius*) and Straight-Tusked Elephant (*Palaeoloxodon antiquus*) in Europe." *Quaternary International* 126–28: 171–77.

Stuart, A. J., P. A. Kosintsev, T. F. G. Higham, and A. M. Lister. 2004. "Pleistocene to Holocene Extinction Dynamics in Giant Deer and Woolly Mammoth." *Nature* 431:684–89.

Stuart, A. J., and A. M. Lister. 2007. "Patterns of Late Quaternary Megafaunal Extinctions in Europe and Northern Asia." *Courier Forschungsinstitut Senckenberg* 259:287–97.

———. 2011. "Extinction Chronology of the Cave Lion *Panthera spelaea*." *Quaternary Science Reviews* 30:2329–40.

———. 2012. "Extinction Chronology of the Woolly Rhinoceros *Coelodonta antiquitatis* in the Context of Late Quaternary Megafaunal Extinctions in Northern Eurasia." *Quaternary Science Reviews* 51:1–17.

———. 2014. "New Radiocarbon Evidence on the Extirpation of the Spotted Hyaena (*Crocuta crocuta* [Erxl.]) in Northern Eurasia." *Quaternary Science Reviews* 96:108–16.

Stuenes, S. 1989. "Taxonomy, Habits, and Relationships of the Subfossil Madagascan Hippopotami *Hippopotamus lemerlei* and *H. madagascariensis*." *Journal of Vertebrate Paleontology* 9:241–68.

Sulerzhitsky, L. D., and F. A. Romanenko. 1997. "Age and Dispersal of 'Mammoth' Fauna in Asian Polar Region (According to Radiocarbon Data)." *Kriosfera Zemli* [Earth cryosphere] 1 (4): 12–19. [In Russian.]

Surovell, T. A., V. T. Holliday, J. A. M. Gingerich, C. Ketron, C. V. Haynes Jr., I. Hilman, D. P. Wagner, E. Johnson, and P. Claeys. 2009. "An Independent Evaluation of the Younger Dryas Extraterrestrial Impact Hypothesis." *Proceedings of the National Academy of Sciences USA* 106:18155–58.

Sutikna, T., M. W. Tocheri, M. J. Morwood, E. W. Saptomo, Jatmiko, R. Due Awe, Sri Wasisto, et al. 2016. "Revised Stratigraphy and Chronology for *Homo floresiensis* at Liang Bua in Indonesia." *Nature* 532:366–69.

Sutton, A., M.-J. Mountain, K. Aplin, S. Bulman, and T. Denham. 2009. "Archaeozoological Records for the Highlands of New Guinea: A Review of Current Evidence." *Australian Archaeology* 69:41–58.

Tamm, E., T. Kivisild, M. Reidla, M. Metspalu, D. G. Smith, C. J. Mulligan, C. M. Bravi, et al. 2007. "Beringian Standstill and Spread of Native American Founders." *PLOS ONE* 2 (9): e829. https://doi.org/10.1371/journal.pone.0000829.

Tankersley, K. B. 2011. "Evaluating the Co-occurrence of *Platygonus compressus* and *Mylohyus nasutus* at Sheriden Cave, Wyandot County, Ohio." *Current Research in the Pleistocene* 28:173–75.

Tasmanian Museum and Art Gallery. 2013. "Skull of *Zygomaturus trilobus*—the Marsupial Hippopotamus: c. 45,000 Years Ago. Discovered 1920." http://www.abc.net.au/local/photos/2013/02/12/3688276.htm.

Tedford, R. H. 1973. "The Diprotodons of Lake Callabonna." *Australian Natural History* 17:349–54.

Titov, V. V. 2008. "Habitat Conditions for *Camelus knoblochi* and Factors in Its Extinction." *Quaternary International* 179:120–25.

Tonni, E. P., A. A. Carlini, G. J. Scillato-Yané, and A. J. Figini. 2003. "Cronología radiocarbónica y condiciones climáticas en la 'Cueva del Milodón' (sur de Chile) durante el Pleistoceno Tardío." *Ameghiniana* 40 (3): 7 pp.

Tryon, C. A., D. J. Peppe, J. T. Faith, A. Van Plantinga, S. Nightingale, J. Ogondo, and D. L. Fox, D.L. 2012. "Late Pleistocene Artefacts and Fauna from Rusinga and Mfangano Islands, Lake Victoria, Kenya." *Azania: Archaeological Research in Africa* 47:14–38.

Turney, C. S. M., T. F. Flannery, R. G. Roberts, C. Reid, L. K. Fifield, T. H. G. Higham, Z. Jacobs, N. Kemp, E. A. Colhoun, R. M. Kalin, and N. Ogle. 2008. "Late-Surviving Megafauna in Tasmania, Australia, Implicate Human Involvement in Their Extinction." *Proceedings of the National Academy of Sciences USA* 105:12150–53.

Turvey, S. T. 2009a. "In the Shadow of the Megafauna: Prehistoric Mammal and Bird Extinctions across the Holocene." In *Holocene Extinctions*, edited by S. T. Turvey (Oxford: Oxford University Press), 17–39.

———. 2009b. "Holocene Mammal Extinctions." In *Holocene Extinctions*, edited by S. T. Turvey (Oxford: Oxford University Press), 41–107.

Turvey, S. T., H. Tong, A. J. Stuart, and A. M. Lister. 2013. "Holocene Survival of Late Pleistocene Megafauna in China: A Critical Review of the Evidence." *Quaternary Science Reviews* 76:156–66.

Van Den Bergh, G. D, R. D. Awe, M. J. Morwood, T. Sutikna, T., Jatmiko, and E. W. Saptomo. 2008. "The Youngest *Stegodon* Remains in Southeast Asia from the Late Pleistocene Archaeological Site Liang Bua, Flores, Indonesia." *Quaternary International*. 182:16–48.

Van der Kaars, S., and P. De Deckker. 2002. "A Late Quaternary Pollen Record from Deep-Sea Core Fr10/95, GC17 Offshore Cape Range Peninsula, Northwestern Western Australia." *Review of Palaeobotany and Palynology* 120:17–39.

Van der Kaars, S., G. H. Miller, C. S. M. Turney, E. J. Cook, D. Nürnberg, J. Schönfeld, A. P. Kershaw, and S. J. Lehman. 2017. "Humans Rather Than Climate the Primary Cause of Pleistocene Megafaunal Extinction in Australia." *Nature Communications* 8, article 14142. https//doi.org/10.1038/ncomms14142.

Van der Made, J. , and H. W. Tong. 2008. "Phylogeny of the Giant Deer with Palmate Brow Tines: *Megaloceros* from West and *Sinomegaceros* from East Eurasia." *Quaternary International* 179:135–62.

Van der Plicht, J., and A. J. T. Jull. 2011. "Mammoth Extinction and Radiation Dose: A Comment." *Radiocarbon* 53 (4): 713–15.

Van der Plicht, J., V. I. van der Molodin, Y. V. Kuzmin, S. K. Vasiliev, A. V. Postnov, and V. S. Slavinsky. 2015. "New Holocene Refugia of Giant Deer (*Megaloceros giganteus* Blum.) in Siberia: Updated Extinction Patterns." *Quaternary Science Reviews* 114:182–88.

Vartanyan, S. L., K. A. Arslanov, J. A. Karhu, G. Possnert, and L. D. Sulerzhitsky. 2008. "Collection of Radiocarbon Dates on the Mammoths (*Mammuthus primigenius*) and Other Genera of Wrangel Island, Northeast Siberia, Russia." *Quaternary Research* 70:51–59.

Vartanyan, S. L., V. E. Garutt, and A. V. Sher. 1993. "Holocene Dwarf Mammoths from Wrangel Island in the Siberian Arctic." *Nature* 362:337–40.

Villavicencio, N. A., E. L. Lindsey, F. M. Martin, L. A. Borrero, P. I. Moreno, B. R. Marshall, and A. D. Barnosky. 2015. "Combination of Humans, Climate, and Vegetation Change Triggered Late Quaternary Megafauna Extinction in the Última Esperanza Region, Southern Patagonia, Chile." *Ecography* 38:125–40.

Vickers-Rich, P. 1987. "A Giant Bird of the Pleistocene." In *The Antipodean Ark*, edited by S. Hand and M. Archer (North Ryde: Angus and Robertson Publishers), 48–50.

———. 1991. "Mesozoic and Tertiary Birds on the Australian Plate." In *Vertebrate Palaeontology of Australasia*, edited by P. Vickers-Rich, J. M. Monaghan, R. F. Baird, and T. H. Rich (Melbourne: Monash University), 722–808.

Vickers-Rich, P., and N. W. Archbold. 1991. "Squatters, Priests and Professors: A brief history of Vertebrate Palaeontology in Terra Australis." In *Vertebrate Palaeontology of Australasia*, edited by P. Vickers-Rich, J. M. Monaghan, R. F. Baird, and T. H. Rich (Melbourne: Monash University), 20–29.

Virah-Sawmy, M., K. J. Willis, and L. Gillson. 2010. "Evidence for Drought and Forest Declines during the Recent Megafaunal Extinctions in Madagascar." *Journal of Biogeography* 37:506–19.

Vizcaíno, S. F., M. Zárate, M. S. Bargo, and A. Dondas. 2001. "Pleistocene Burrows in the Mar del Plata Area (Argentina) and Their Probable Builders." *Acta Palaeontologica Polonica* 46 (2): 289–301.

Walker, M. J. C., 2005. *Quaternary Dating Methods*. Chichester: John Wiley and Sons.

Walker, M. J. C., M. Berkelhammer, S. Björck, L. C. Cwynar, D. A. Fisher, A. J. Long, J. J. Lowe, et al. 2012. "Formal Subdivision of the Holocene Series/Epoch: A Discussion Paper by a Working Group of INTIMATE and the Subcommission on Quaternary Stratigraphy." *Journal of Quaternary Science*, 27:649–59. http://doi.org/10.1002/jqs.2565.

Wallace, A. R. 1876. *The Geographical Distribution of Animals: With a Study of the Relations of Living and Extinct Faunas as Elucidating the Past Changes of the Earth's Surface*. New York: Harper & Brothers. 2 vols.

Warren, J. C. 1852. *The Mastodon Giganteus of North America*. Boston: John Wilson and Son.

Waters, M. R., and T. W. Stafford. 2007. "Redefining the Age of Clovis: Implications for the Peopling of the Americas." *Science* 315:1122–26.

Waters, M. R., T. W. Stafford Jr., B. Kooyman, and L. V. Hills. 2015. "Late Pleistocene Horse and Camel Hunting at the Southern Margin of the Ice-Free Corridor: Reassessing the Age of Wally's Beach, Canada." *Proceedings of the National Academy of Sciences USA* 112:4263–67.

Webb, S. D. 1965. "The Osteology of *Camelops*." *Bulletin of the Los Angeles County Museum, Science Bulletin 1*.

Webb, S. 2013. *Corridors to Extinction and the Australian Megafauna*. London: Elsevier.

Weinstock, J., E. Willerslev, A. V. Sher, W. Tong, S. Y. W. Hol, D. Rubenstein, J. Storer, et al. 2005. "Evolution, Systematics, and Phylogeography of Pleistocene Horses in the New World: A Molecular Perspective." *PLOS Biology* 3:1373–1379:e241.

Welker, F., M. J. Collins, J. A. Thomas, M. Wadsley, S. Brace, E. Cappellini, S. T. Turvey, et al. 2015. "Ancient Proteins Resolve the Evolutionary History of Darwin's South American Ungulates." *Nature* 522:81–87.

Wells, R. T., and A. B. Camens. 2018. "New Skeletal Material Sheds Light on the Palaeobiology of the Pleistocene Marsupial Carnivore, *Thylacoleo carnifex*." *PLOS ONE* 13 (12): e0208020. https://doi. org/10.1371/journal.pone.0208020.

Wells, R. T., K. C. Moriarty, D. L. G. Williams. 1984. "The Fossil Vertebrate Deposits of Victoria Fossil Cave Naracoorte: An Introduction to the Geology and Fauna." *Australian Zoologist* 21:305–33.

Westaway, K. E., J. Louys, R. Due Awe, M. J. Morwood, G. J. Price, J.-x. Zhao, M. Aubert, et al. 2017. "An Early Modern Human Presence in Sumatra 73,000–63,000 Years Ago." *Nature* 548:322–25.

Westaway, M., J. Olley, and R. Grün. 2017. "At Least 17,000 Years of Coexistence: Modern Humans and Megafauna at the Willandra Lakes, South-Eastern Australia." *Quaternary Science Reviews* 157:206–11.

Weston, E. M., and A. M. Lister. 2009. "Insular dwarfism in hippos and a model for brain size reduction in *Homo floresiensis*." *Nature* 459:85–89.

White, A. W., T. H. Worthy, S. Hawkins, S. Bedford, and M. Spriggs. 2010. "Megafaunal Meiolaniid Horned Turtles Survived until Early Human Settlement in Vanuatu, Southwest Pacific." *Proceedings of the National Academy of Sciences USA* 107:15512–16.

White, M., P. Pettitt, and D. Schreve. 2016. "Shoot First, Ask Questions Later: Interpretative Narratives of Neanderthal Hunting." *Quaternary Science Reviews*. 140:1–20.

Whiteside, J. H., P. E. Olsen, T. Eglinton, M. E. Brookfield, and R. N. Sambrotto. 2010. "Compound-Specific Carbon Isotopes from Earth's Largest Flood Basalt Eruptions Directly Linked to the End-Triassic Mass Extinction." *Proceedings of the National Academy of Sciences USA* 107:6721–25.

Widga, C., T. L. Fulton, L. D. Martin, and B. Shapiro. 2012. "*Homotherium serum* and *Cervalces* from the Great Lakes Region, USA: Geochronology, Morphology and Ancient DNA." *Boreas* 41:546–56.

Wilkinson, C. S. 1885. "President's Address, Annual General Meeting." *Proceedings of the Linnean Society of New South Wales* 9:1207–41.

Williams, D. R. 2009. "Small Mammal Faunal Stasis in Natural Trap Cave (Pleistocene-Holocene), Bighorn Mountains, Wyoming." Master's thesis, University of Kansas, Lawrence.

Williams, J. W., B. N. Shuman, T. Webb III, P. J. Bartlein, and P. L. Leduc. 2004. "Late-Quaternary Vegetation Dynamics in North America: Scaling from Taxa to Biomes." *Ecological Monographs* 74:309–34.

Willis, P. M. A., and R. E. Molnar. 1997. "A Review of the Plio-Pleistocene Crocodilian Genus *Pallimnarchus*." *Proceedings of the Linnean Society of New South Wales* 117:224–42.

Wilmshurst, J. M., T. L. Hunt, C. P. Lipoc, and A. J. Anderson. 2011. "High-Precision Radiocarbon Dating Shows Recent and Rapid Initial Human Colonization of East Polynesia." *Proceedings of the National Academy of Sciences USA* 108:1815–20.

Wojtal, P., G. Haynes, J. Klimowicz, and K. Sobczyk. 2019. "The Earliest Direct Evidence of Mammoth Hunting in Central Europe—the Kraków Spadzista Site (Poland)." *Quaternary Science Reviews* 213:162–66.

Wood, J. R., and J. M. Wilmshurst. 2013. "Age of North Island Giant Moa (*Dinornis novaezealandiae*) Bones Found on the Forest Floor in the Ruahine Range." *Journal of the Royal Society of New Zealand* 43:250–55.

Woodburne, M. O. 2010. "The Great American Biotic Interchange: Dispersals, Tectonics, Climate, Sea Level and Holding Pens." *Journal of Mammalian Evolution* 17:245–64.

Woodman, N., and N. B. Athfield. 2009. "Post-Clovis Survival of American Mastodon in the Southern Great Lakes Region of North America." *Quaternary Research* 72:359–63.

Woodward, J. 2014. *The Ice Age: A Very Short Introduction*. Oxford: Oxford University Press.

Worthy, T. H. 2002. "The Youngest Giant: Discovery and Significance of the Remains of a Giant Moa (*Dinornis giganteus*) near Turangi, in Central North Island, New Zealand." *Journal of the Royal Society of New Zealand* 32:183–87.

———. 2015. "Moa." *Te Ara—the Encyclopedia of New Zealand*. http://www.TeAra.govt.nz/en/moa/print.

Worthy, T. H., and R. N. Holdaway. 2002. *The Lost World of the Moa*. Bloomington: Indiana University Press.

Worthy, T. H., A. J. D. Tennyson, M. Archer, A. M. Musser, S. J. Hand, C. Jones, B. J. Douglas, et al. 2006. "Miocene Mammal Reveals a Mesozoic Ghost Lineage on Insular New Zealand, Southwest Pacific." *Proceedings of the National Academy of Sciences USA* 103:19419–23.

Wroe, S. 2002. "A Review of Terrestrial Mammalian and Reptilian Carnivore Ecology in Australian Fossil Faunas, and Factors Influencing Their Diversity: The Myth of Reptilian Domination and Its Broader Ramifications." *Australian Journal of Zoology*. 50:1–24.

Wroe, S., J. H. Field, M. Archer, D. K. Grayson, G. J. Price, J. Louys, J. T. Faith, et al. 2013a. "Climate Change Frames Debate over the Extinction of Megafauna in Sahul (Pleistocene Australia-New Guinea)." *Proceedings of the National Academy of Sciences USA* 110:8777–81.

———. 2013b. "Reply to Brook et al: No Empirical Evidence for Human Overkill of Megafauna in Sahul." *Proceedings of the National Academy of Sciences USA* 110:E3369.
Wroe, S., J. Field, R. Fullagar, and L. S. Jermin. 2004. "Megafaunal Extinction in the Late Quaternary and the Global Overkill Hypothesis." *Alcheringa* 28:291–331.
Wroe, S., C. McHenry, and J. Thomason. 2005. "Bite Club: Comparative Bite Force in Big Biting Mammals and the Prediction of Predatory Behaviour in Fossil Taxa." *Proceedings of the Royal Society B* 272:619–25. https://doi.org/10.1098/rspb.2004.2986.
Wroe, S., T. Myers, F. Seebacher, B. Kear, A. Gillespie, M. Crowther, and S. Salisbury. 2003. "An Alternative Method for Predicting Body Mass: The Case of the Pleistocene Marsupial Lion." *Paleobiology* 29:403–11.
Zalasiewicz, J., C. N. Waters, C. P. Summerhayes, A. P. Wolfe, A. D. Barnosky, A. Cearreta, P. Crutzen, et al. 2017. "The Working Group on the Anthropocene: Summary of Evidence and Interim Recommendations." *Anthropocene* 19:55–60.
Zamorano, M., G. J. Scillato-Yané, and A. E. Zurita. 2014. "Revisión del género *Panochthus* (Xenarthra, Glyptodontidae)." *Revista del Museo de La Plata Sección Paleontología*, 14: 1–46.
Zazula, G. D., R. D. MacPhee, J. Z. Metcalfe, A. V. Reyes, F. Brock, P. S. Druckenmiller, P. Groves, et al. 2014. "American Mastodon Extirpation in the Arctic and Subarctic Predates Human Colonization and Terminal Pleistocene Climate Change." *Proceedings National Academy Sciences USA*. 111:18460–65.
Zazula, G. D., D. G. Turner, B. C. Ward, and J. Bond. 2011. "Last Interglacial Western Camel (*Camelops hesternus*) from Eastern Beringia." *Quaternary Science Reviews* 30:2355–60.
Zhegallo, V., N. Kalandadze, A. Shapavolov, Z. Bessudnova, N. Noskova, and E. Tesakova. 2005. "On the Fossil Rhinoceros *Elasmotherium* (Including the Collections of the Russian Academy of Sciences)." *Cranium* 22:17–40.
Zimov, S. A. 2005. "Pleistocene Park: Return of the Mammoth's Ecosystem." *Science* 308:(5723): 796–98.

INDEX

The letter *f* following a page number denotes a figure; *t* denotes a table.